T0338178

Geosynthetic Reinforced Soil (GRS) Walls

Geosynthetic Reinforced Soil (GRS) Walls

Jonathan T.H. Wu

Professor of Civil Engineering
University of Colorado Denver
Director of the Reinforced Soil Research Center
Editor-in-Chief of the Journal of Transportation Infrastructure Geotechnology

This edition first published 2019
© 2019 John Wiley & Sons Ltd

All rights reserved. No part of this publication may be reproduced, stored in a retrieval system, or transmitted, in any form or by any means, electronic, mechanical, photocopying, recording or otherwise, except as permitted by law. Advice on how to obtain permission to reuse material from this title is available at http://www.wiley.com/go/permissions.

The right of Jonathan T.H. Wu to be identified as the author of this work has been asserted in accordance with law.

Registered Offices
John Wiley & Sons, Inc., 111 River Street, Hoboken, NJ 07030, USA
John Wiley & Sons Ltd, The Atrium, Southern Gate, Chichester, West Sussex, PO19 8SQ, UK

Editorial Office
9600 Garsington Road, Oxford, OX4 2DQ, UK

For details of our global editorial offices, customer services, and more information about Wiley products visit us at www.wiley.com.

Wiley also publishes its books in a variety of electronic formats and by print-on-demand. Some content that appears in standard print versions of this book may not be available in other formats.

Limit of Liability/Disclaimer of Warranty
While the publisher and authors have used their best efforts in preparing this work, they make no representations or warranties with respect to the accuracy or completeness of the contents of this work and specifically disclaim all warranties, including without limitation any implied warranties of merchantability or fitness for a particular purpose. No warranty may be created or extended by sales representatives, written sales materials or promotional statements for this work. The fact that an organization, website, or product is referred to in this work as a citation and/or potential source of further information does not mean that the publisher and authors endorse the information or services the organization, website, or product may provide or recommendations it may make. This work is sold with the understanding that the publisher is not engaged in rendering professional services. The advice and strategies contained herein may not be suitable for your situation. You should consult with a specialist where appropriate. Further, readers should be aware that websites listed in this work may have changed or disappeared between when this work was written and when it is read. Neither the publisher nor authors shall be liable for any loss of profit or any other commercial damages, including but not limited to special, incidental, consequential, or other damages.

Library of Congress Cataloging-in-Publication Data

Names: Wu, Jonathan T. H., author.
Title: Geosynthetic reinforced soil (GRS) walls / by Jonathan T.H. Wu.
Description: Hoboken, NJ : John Wiley & Sons, 2019. | Includes bibliographical references and index. |
Identifiers: LCCN 2017044670 (print) | LCCN 2017056353 (ebook) | ISBN 9781119375852 (pdf) |
    ISBN 9781119375869 (epub) | ISBN 9781119375845 (cloth)
Subjects: LCSH: Reinforced soils. | Geosynthetics.
Classification: LCC TA760 (ebook) | LCC TA760 .W8 2018 (print) | DDC 624.1/62–dc23 LC record
    available at https://lccn.loc.gov/2017044670

Cover Design: Wiley
Cover Image: © Alena Hovorkova/Shutterstock
Cover Illustrations: Courtesy of Jonathan T.H. Wu

Set in 10/12pt Warnock by SPi Global, Pondicherry, India
Printed and bound in Singapore by Markono Print Media Pte Ltd

10 9 8 7 6 5 4 3 2 1

Contents

Preface

Since the reintroduction of reinforced soil in the early 1960s, many innovative reinforced soil wall systems have been developed to deal with earthwork construction where an abrupt change in grade is desired or needed. Reinforced soil wall systems deploy horizontal layers of tensile inclusion in the fill material to improve or achieve stability. These wall systems have demonstrated many distinct advantages over their conventional counterparts such as cantilever reinforced concrete earth retaining walls, gravity concrete walls, crib walls, etc. In addition to high load-carrying capacity, reinforced soil walls are typically more ductile (hence less susceptible to sudden collapse), more flexible (hence more tolerant to differential settlement), faster and easier to construct, more adaptable to low quality backfill, require less over-excavation, more economical to construct, and have lower life-cycle maintenance costs. To date, reinforced soil walls are being constructed at a rate of over 100,000 m^2 (in terms of total face area) annually in the U.S. alone.

Modern technologies of reinforced soil walls incorporate metallic strips/mats or synthetic polymeric sheets (termed geosynthetics) as tensile inclusion in the backfill during fill placement. Reinforced soil walls have commonly been designed by considering tensile inclusion as quasi-tieback elements to stabilize the fill material through soil–reinforcement interface bonding, and are collectively referred to as mechanically stabilized earth (MSE) walls. MSE walls with geosynthetics as reinforcement have been referred to as geosynthetic mechanically stabilized earth, or simply GMSE. To date, over 60,000 GMSE walls have been built along highways in the U.S.

Reinforcement spacing used in GMSE walls has been relatively large. This stems from a fundamental design concept that spacing of quasi-tieback elements hardly matters to performance, and that larger spacing would result in shorter construction time. The beneficial effect of deploying geosynthetic reinforcement on tight spacing, however, is gaining increased attention. The significant benefits of close reinforcement spacing were first realized through actual wall construction, and later verified by field-scale loading experiments. It has been shown that close reinforcement spacing will increase considerably the load-carrying capacity and, more importantly, improves stability of the reinforced soil mass. Studies have suggested that the behavior of reinforced soil mass with closely spaced reinforcement can be accurately characterized as soil–geosynthetic composites.

Geosynthetic reinforced soil (GRS) emerged as a viable alternative to GMSE in the early 2000s. GRS takes advantage of soil–geosynthetic interaction by which the soil mass is reinforced internally. To activate a significant beneficial effect of soil–geosynthetic interaction, reinforcement spacing in GRS is much smaller than in GMSE. Note that in the literature the term "GRS" has sometimes been used for all soil structures reinforced by geosynthetic inclusion without any regard to reinforcement spacing or the design concept.

GRS bears strong resemblance to GMSE, in that both systems are composed of three major components: facing, compacted backfill, and horizontal geosynthetic inclusion. The main difference between the two systems lies in the design concept. GRS considers closely spaced geosynthetic inclusion as a reinforcing element of a soil–geosynthetic composite (hence the term "reinforced" in GRS). GMSE, on the other hand, considers the geosynthetic inclusion as frictional tieback tension members to stabilize potential failure wedges (hence the term "stabilized" in GMSE). Because of this difference, the role of facing for the two systems is also very different. In GRS, the soil mass is *internally* reinforced to form a stable mass. The wall facing serves primarily as an aesthetic façade. It also serves to prevent soil sloughing and as a construction aid. In GMSE, however, facing is a major load-carrying component; if facing fails, failure of the GMSE wall will usually be imminent.

In today's practice, GMSE is enjoying a much wider popularity than GRS. This is in part because there is a lack of understanding of GRS, and in part because GMSE is similar to conventional earth retaining walls in design concept. Most designers are not entirely comfortable with a soil wall that achieves stability through internal reinforcing of the soil behind the wall rather than through the resistance offered *externally* by the facing.

Lately, a number of renowned designers and wall builders have estimated about 5–10% failure rate for GMSE, with a majority being associated with serviceability (i.e., excessive deformation). The National Concrete Masonry Association has also estimated a 2–8% failure rate of various types for GMSE walls. Whether it is structural failure or serviceability failure, the failure rate is much too high compared to other types of earth structures. Studies into the causes of failure have not lead to conclusive solutions to the problem. By employing tight reinforcement spacing to form soil–geosynthetic composites of higher stiffness and ductility, GRS has slowly but gradually affirmed itself as a viable alternative wall system to GMSE. GRS has promised some advantages as a sound wall system of the future, including (i) closely spaced reinforcement of GRS improves fill compaction efficiency and relaxes requirement of stiffness/strength of geosynthetics, (ii) GRS tends to be much less susceptible to long-term creep when well-compacted granular fill is employed, (iii) GRS provides much better seismic stability, and (iv) GRS mass exerts less earth pressure against facing and improves facing stability. Failure of GRS is practically nonexistent as long as well compacted granular fill is used. This is likely because GRS does not rely on the stability of any single structural component (e.g., facing or tension anchors) to maintain overall stability.

This book addresses both GRS and GMSE, with a much stronger emphasis on the former. Details of GMSE have been given by several design guides, such as the AASHTO bridge design specifications, the Federal Highway Administration NHI MSE walls and steepened slopes manual, and the National Concrete Masonry Association design manual. For completeness, this book begins with a review of shear strength of soils (Chapter 1) and classical earth pressure theories (Chapter 2). Chapter 3 addresses the observed behavior of soil–geosynthetic composites, reinforcing mechanisms of GRS, and GRS walls of different types of facing. Chapter 4 addresses geosynthetics as reinforcement, with emphasis on mechanical properties of geosynthetics, including load–deformation properties, creep properties, stress–relaxation properties, and soil–geosynthetic interface properties. Chapter 5 discusses design concepts of GRS walls and describes a number of prevalent design methods for GRS walls. In addition, recent advances on design of GRS and a new design method incorporating the recent advances is delineated. Design examples for each of the design method are given to help illustrate the design methods. Chapter 6 addresses construction of GRS walls, including construction procedure of GRS walls and general construction guidelines. It is my hope that the civil engineering community will become more familiar with GRS through this book, and makes better use of this novel technology in earthwork construction.

This book would not have been possible without the contribution of many of my colleagues and friends. Foremost is Professor Gerald A. Leonards, who was an inspiration for my lifelong interest in the theories and practice of geotechnical engineering. I must acknowledge Bob Barrett, a true innovator of reinforced soil technology, with whom I have had the privilege to work on many fact-exploring projects over the past three decades. From Bob I leaned many key issues of GRS. I also wish to thank Mike Adams and Jennifer Nicks, two relentless FHWA researchers whose field-scale experiments allowed me to learn the behavior of GRS. I also wish to acknowledge an outstanding wall builder, Calvin VanBuskirk, who had the vision to suggest separation of GRS from GMSE. I am especially in debt to Fumio Tatsuoka, who kindly shared many valuable experimental techniques and his unique experiences during my two sabbatical leaves at the University of Tokyo. Fumio was extremely instrumental for many field-scale experiments of GRS that I was involved in.

I was fortunate to have worked with many outstanding research associates on GRS and related subjects, including (in alphabetical order) Noom Aksharadananda, Daniel Alzamora, Vasken Arabian, John Ballegeer, Bill Barreire, Michael Batuna, Melissa Beauregard, Richard Beck, John Billiard, David Bixler, Harold Blair, Eric Y. Chen, Nick S.-K. Cheng, Nelson N. Chou, Alan Claybourn, Phil Crouse, David Curran, Mark Davis, Gary Dieward, Gene Dodd, Robert Duncanson, Nicolas El-Hahad, Chris Ellis, Egbal Elmagre, Zeynep Erdogen, Barbara Evans, Tony Z-Y. Feng, Seth Flutcher, Brian Francis, Chris Gemperline, Dave Gilbert, Justin Hall, Khamis Haramy, Mark Hauschild, Matthew Hayes, Sam Helwany, Dennis Henneman, Jason Hilgers, Zhenshun Hong, Kanop Ketchart, Elaheh Kheirkhahi, Cassie Klump, Ilyess Ksouri, W.T. Hsu, Kevin Lee, S.L. Lee, K.H. Lee, J.C. Lin, C.W. Ma, Paul Macklin, David Manka, Mike May, Rick McCain, Araya Messa, John Meyers, Greg Monley, Larry Moore, Mike Nelson, James Olson, Jean-Baptiste Payeur, Breden Peters, Thang Pham, John Pierce, Michele Pollman, Xiaopei Qi, Mark Reiner, Zac Robinson, Jerzy Salamon, Brett Schneider, Pauline Serre, Rennie Seymour, Daming Shi, Barry Siel, Saeed Sobhi, C.K. Su, Omar Takriti, Damon Thomas, Sheldon C.-Y.Tung, Alan Tygesen, Mark Vessely, Diane TeAn Wang, Roy Wittenburg, Derek Wittwer, Temel Yetimoglu, and Sam S.-H. Yu.

Finally, I wish to express my gratitude to my mother Umeko to whom I owe everything.

Jonathan T.H. Wu
Greenwood Village, Colorado

1

Stresses and Shear Strength of Soils

In the design of earth structures, good knowledge of shear stiffness/strength of soil is of importance because excessive deformation or failure of earth structures may occur as a result of insufficient resistance to shear stress. This chapter presents a review of shear behavior and shear strength of soils that are relevant to the design of earth structures. We begin the chapter with an explanation of stress at a point, followed by a brief explanation of effective stress and Mohr–Coulomb failure criterion. We then discuss commonly used laboratory and field tests for evaluation of shear behavior and determination of the shear strength of soils. We conclude the chapter with a discussion of the design consideration of the shear strength for soils under different loading conditions.

1.1   Stress at a Point

In engineering analysis and design of structures, *stress* has been proven to be an extremely useful parameter to quantify the effects of internal and external influences on a structure. For earth structures, common influences include external loads, self-weight of soil and water, seepage force, and temperature change. Stress in a body is commonly referred to a plane. Stress on a plane with cross-sectional area A when subjected to a force system denoted F can be evaluated by a simple equation: stress $= F/A$. This equation, however, is useful only if the force on the given plane is known and if the stress can be approximated as being uniform on that plane. This is generally not the case for a soil mass where the stresses of interest typically occur on a plane other than the plane of load applications, and the stresses may be far from being uniform. To this end, we shall begin the discussion with a review of the *stress at a point*, a subject we were first exposed to when studying "mechanics of materials" as undergraduate engineering students. We shall discuss three topics that will help us gain a better understanding of *stresses at a point*: (a) the concept of stress at a point in terms of stress vector, (b) the computation of stress vector on any given plane by the Cauchy formula, and (c) graphical representation of stress at a point by a Mohr circle and the pole of a Mohr circle. A good understanding of these topics will allow us to gain insights into Rankine analysis, a prevailing lateral earth pressure theory, which we will discuss in detail in Chapter 2.

Geosynthetic Reinforced Soil (GRS) Walls, First Edition. Jonathan T.H. Wu.
© 2019 John Wiley & Sons Ltd. Published 2019 by John Wiley & Sons Ltd.

1.1.1 Stress Vector

Let us begin by considering an earth retaining wall subjected to concentrated and distributed loads over the "crest" (top surface of a wall), as shown in Figure 1.1(a). Figure 1.1(b) shows a plane cutting through the soil mass behind the wall at point P. The force acting over a very small area ΔA on the plane of cut surrounding point P is denoted ΔF. If the plan of cut is referred to as the "$\vec{n}$ plane" (i.e., a plane with its outward normal being a unit vector $\vec{n}$; the arrow above the notation "n" is merely a symbol indicating that it is a vector, i.e., it has a magnitude, an orientation, and a sense), the intensity of force at point P on the $\vec{n}$ plane can be expressed by a *stress vector* $\overrightarrow{T_n}$ as:

$$\text{stress vector } \overrightarrow{T_n} = \lim_{\Delta A \to 0} \frac{\Delta F}{\Delta A} \tag{1-1}$$

Note that the stress vector can be viewed as the *resultant* of stresses on the $\vec{n}$ plane.

We shall find it convenient to identify two major components of $\overrightarrow{T_n}$ which are perpendicular and tangent to the $\vec{n}$ plane:

normal stress (σ) = the component of $\overrightarrow{T_n}$ that is normal to the $\vec{n}$ plane

shear stress (τ) = the component of $\overrightarrow{T_n}$ that is tangent to the $\vec{n}$ plane

The stress vector $\overrightarrow{T_n}$ is therefore a resultant of normal stress σ and shear stress τ on the given plane.

If we had chosen a different plane of cut through point P, the stress vector on that plane would generally be different from what we obtained previously. In fact, for every plane passing through point P, there will generally be a different stress vector associated with that plane. The number of planes passing through point P is infinite; hence, there

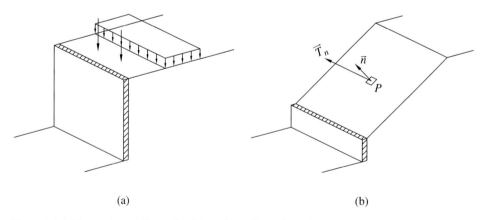

(a)   (b)

Figure 1.1 (a) An earth retaining wall subjected to self-weight and external loads on the crest and (b) stress resultant $\overrightarrow{T_n}$ on a plane $\vec{n}$ through point P

are infinite stress vectors at point P. Stress at point P is *the totality of all the stress vectors at point P*. To state that we know the stress at a given point, we must know all the stress vectors at that point.

1.1.2 Cauchy Formula

Since there is an infinite number of stress vectors at a point, it would appear that it is not possible to know the stress at a point. This problem was resolved thanks to the *Cauchy formula*. Cauchy (circa 1820) showed that the stress vector on any plane at a point could be determined provided that the stress vectors on three orthogonal planes at that point are known, i.e.,

$$\overrightarrow{T_n} = \overrightarrow{T_n}\, n_x + \overrightarrow{T_n}\, n_y + \overrightarrow{T_n}\, n_z \qquad (1\text{-}2)$$

where

$\overrightarrow{T_n}$ = the stress vector on a plane with its outward normal unit vector $\bar{n}$

$\overrightarrow{T_x}, \overrightarrow{T_y}, \overrightarrow{T_z}$ = the stress vectors on the x-, y-, and z-planes, respectively

n_x, n_y, n_z = directional cosines of $\bar{n}$ in the x-, y-, and z-directions, respectively, where $n_x = \cos\theta_x$, $n_y = \cos\theta_y$, $n_z = \cos\theta_z$ (see Figure 1.2).

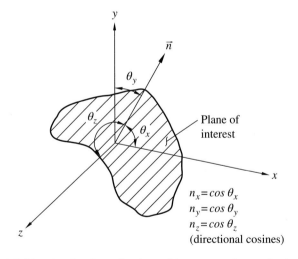

Figure 1.2 Directional cosines of a plane with an outward normal unit vector $\bar{n}$

The Cauchy formula, Eqn. (1-2), allows the stress vectors on any plane to be determined by knowing only the stress vectors on three orthogonal planes; therefore, stress at a point can now be fully defined by knowing only the stress vectors on three perpendicular planes $\overrightarrow{T_x}, \overrightarrow{T_y}$, and $\overrightarrow{T_z}$. As shown in Figure 1.3, $\overrightarrow{T_x}$ can be considered as the resultant of σ_x, τ_{xz}, and τ_{xy}. Similarly, $\overrightarrow{T_y}$ can be considered as the resultant of σ_y, τ_{yz}, and τ_{yx}; and $\overrightarrow{T_z}$, the resultant of σ_z, τ_{zx}, and τ_{zy}.

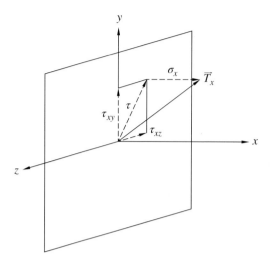

Figure 1.3 Stress vector $\overrightarrow{T_x}$ as the resultant of stress components σ_x, τ_{xz}, and τ_{xy}

Since the stress at a point can be defined completely by $\overrightarrow{T_x}$, $\overrightarrow{T_y}$, and $\overrightarrow{T_z}$, it follows that the stress at a point can be described by nine stress components, known as the *stress tensor*:

$$
\begin{bmatrix}
\sigma_x & \tau_{xy} & \tau_{xz} \\
\tau_{yx} & \sigma_y & \tau_{yz} \\
\tau_{zx} & \tau_{zy} & \sigma_z
\end{bmatrix}
$$

All of us may have seen a graph like Figure 1.4 in engineering mechanics books whenever the subject of *stress* comes up. The graph shows the nine components of the

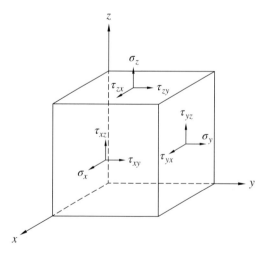

Figure 1.4 Representation of stress at a point by a stress tensor

stress tensor, hence stress at a point. Even though the graph shows the nine components on a cube, we should picture it in our mind as a point. The cube is just a convenient way to show stresses on different planes. With an assumption that the body-moment and couple-stress do not exist in the system, the stress tensor must be symmetrical, i.e., $\tau_{xy} = \tau_{yx}, \tau_{xz} = \tau_{zx}, \tau_{yz} = \tau_{zy}$ (Fung, 1977). The stress tensor therefore has only six independent components.

In a plane-strain condition (see Section 1.4.3), a three-dimensional structure can be analyzed as being two-dimensional. Stress at a point in a plane-strain condition is therefore fully defined if the stress vectors on two orthogonal planes at the point of interest are known (an extension of this statement can be seen in Examples 1.2 and 1.3). The stress tensor in a plane-strain condition can be expressed as:

$$\begin{bmatrix} \sigma_x & \tau_{xz} \\ \tau_{xz} & \sigma_z \end{bmatrix}$$

This stress tensor has three independent stress components ($\sigma_x, \sigma_z, \sigma_{xz}$). If these three stress components are known, stresses on any plane at the point can be determined by the Cauchy formula. For example, the normal stress (σ) and shear stress (τ) on a plane with an outward normal unit vector $\bar{n}$ making an angle θ with the x-axis (as shown in Figure 1.5) can be calculated by:

$$\sigma = \frac{\sigma_x + \sigma_y}{2} + \frac{\sigma_x - \sigma_y}{2}\cos 2\theta + \tau_{xy}\sin 2\theta$$
$$\tau = -\frac{\sigma_x - \sigma_y}{2}\sin 2\theta + \tau_{xy}\cos 2\theta \tag{1-3}$$

Eqn. (1-3) is a simplified form of the Cauchy formula (Eqn. (1-2)) in a two-dimensional condition.

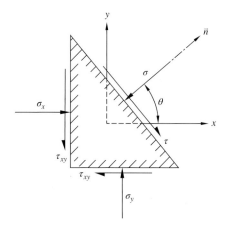

Figure 1.5 Two-dimensional representation of a plane with its outward normal unit vector $\bar{n}$

It should be noted that the use of the concept of stress in the analysis of earth structures is based on an assumption that soil is a continuum. Since soil is in fact a particulate material, in that the particles are typically not bonded and voids are always present, the assumption of a continuum maybe questionable (see Figure 1.6). However, we shall continue to make use of "stress" for the design and analysis of earth structures because it has proven to be an extremely useful tool. Keep in mind, however, that stress in soil is merely a "defined" parameter. When referring to stresses in a soil, it should be viewed on a macro-scale, and the soil is considered a continuum for the purposes of engineering analysis.

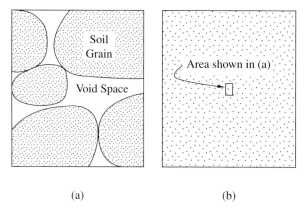

(a)   (b)

Figure 1.6 Stress at a point in a soil mass: (a) reality (micro-scale) and (b) idealized as being a uniform continuum (macro-scale)

1.1.3   Mohr Circle of Stress

The *Mohr circle of stress*, as shown in Figure 1.7(a), is a plot of normal stress vs. shear stress of all permissible stresses at a point under two-dimensional conditions. Every point on a Mohr circle represents the normal and shear stresses on a particular plane at that point. A Mohr circle of stress therefore can be regarded as a graphical representation of stresses at a point under two-dimensional conditions. There are an infinite number of points on a Mohr circle, and each point corresponds to a plane passing through the point of interest. Two distinct planes exist on a Mohr circle where shear stress $\tau = 0$. These planes are called *principal planes*. The stresses on the principal planes are known as *principal stresses*, denoted by σ_1 and σ_3, as shown in Figure 1.7(a). The principal stress σ_1 is the largest normal stress (or major principal stress), whereas the principal stress σ_3 is the smallest normal stress (or minor principal stress).

Figure 1.7(b) shows the sign conventions for plotting a Mohr circle of stress. In soil mechanics, we denote compressive normal stress as positive and tensile normal stress as negative, which is opposite to the sign convention commonly used in structural mechanics. We do so because soil has little tensile resistance, and most normal stresses in geotechnical engineering analysis are compressive. The sign convention of considering compressive stress as positive avoids having to show nearly every normal stress in geotechnical engineering analysis with a negative sign. A shear stress that makes a clockwise rotation about any point *outside* of the plane is considered a positive shear

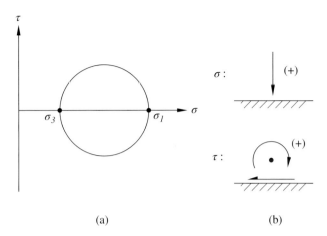

(a)   (b)

Figure 1.7 (a) Two-dimensional representation of stress at a point by a Mohr circle of stress and (b) sign conventions of normal and shear stresses for plotting Mohr circles

stress (see Figure 1.7(b)). Conversely, a shear stress making counterclockwise rotation about any point outside of the plane is considered a negative shear stress.

1.1.4 Pole of Mohr Circle

We recall that every point on a Mohr circle (with coordinates σ and τ) corresponds to a given plane at the point of interest. It would be very useful to associate the stresses σ and τ with the plane. To that end, the *pole of planes* (or simply *pole*) of a Mohr circle is the most useful. The pole, also referred to as the *origin or center of planes*, is a unique point on a Mohr circle that allows us to determine the orientation of the plane for a given set of permissible σ and τ; it also allows us to determine σ and τ on any given plane.

Once the pole of a Mohr circle is located, we can connect the pole with any point on the circle by a straight line; the orientation of the straight line is then the orientation of the plane. Take a point of coordinates (σ_a, τ_a) on a Mohr circle shown in Figure 1.8(a), for example. The orientation of the plane on which stresses σ_a and τ_a act is indicated by a straight line connecting the pole with point (σ_a, τ_a). Figure 1.8(b) shows the orientation of the two planes of maximum shear stress, each determined by a straight line connecting the pole with the point on the Mohr circle having the largest magnitude of τ. There is one unique plane denoted by a line tangent to the Mohr circle at the pole. The orientation of that tangent line is the orientation of the plane on which the stresses σ_{pole} and τ_{pole} act.

The only question remaining is: how do we locate the pole of a Mohr circle? The answer is simple. There is one and only one pole for each Mohr circle. Therefore, if one permissible set of stress (σ_a and τ_a) and the orientation of the plane of (σ_a, τ_a) are known, the pole can be located by back-tracking the step described above. In other words, all we have to do is draw a straight line through point (σ_a, τ_a) parallel to the given orientation to the plane; the pole of the Mohr circle is the point where the line intersects the Mohr circle. Example 1.1 illustrates how the pole of a Mohr circle of stress is determined. Since $\sigma = 5$ psi and $\tau = 2$ psi (denoted by point A on the Mohr circle) are known to act on the *vertical* plane, a *vertical* line can be drawn through point A to locate the pole of the Mohr circle.

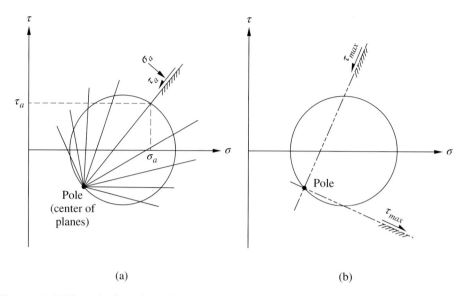

(a)   (b)

Figure 1.8 (a) The pole of a Mohr circle of stress and (b) orientation of planes of maximum shear stress

Note that the pole can also be located by the stresses on the horizontal plane, $\sigma = 1.5\,\text{psi}$ and $\tau = -2\,\text{psi}$. Again, there is only one pole for a Mohr circle, so the location of the pole as determined by any sets of σ and τ will be the same. Note that the orientation of a plane deduced from a pole reflects the *actual* orientation of the plane. The orientation can be described by an angle that it makes with the horizontal, the vertical, or any other reference plane.

Example 1.1   The stress at a point under a plane-strain condition is defined by the stresses on the vertical and horizontal planes, as shown in Figure Ex. 1.1(a). Determine, by Mohr circle of stress, (a) the principal stresses and (b) the orientation of the planes of maximum shear stress.

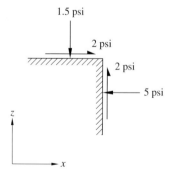

Figure Ex. 1.1(a)

Solution:

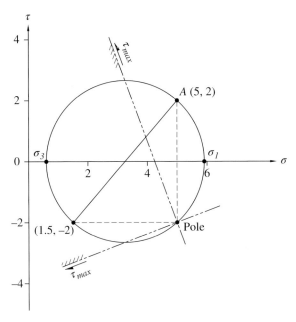

Figure Ex. 1.1(b)

a) The Mohr circle of stress at the point is constructed by the stresses on the two given planes, ($\sigma = 5$, $\tau = 2$) and ($\sigma = 1.5$, $\tau = -2$). From the Mohr circle, the principal stress can be determined graphically as: $\sigma_1 \approx 5.9$ psi and $\sigma_3 \approx 0.6$ psi (note: more precise values can be obtained by sketching the Mohr circle using engineering software, such as AutoCAD).

b) As shown in Figure Ex. 1.1(b), the pole of the Mohr circle can be determined by drawing a vertical line through ($\sigma = 5$, $\tau = 2$) or drawing a horizontal line through ($\sigma = 1.5$, $\tau = -2$). The orientation and respective sense of each of the two τ_{max} planes are as indicated on the Mohr circle, see Figure Ex. 1.1(b).

Examples 1.2 and 1.3 provide additional exercises on how the concept of pole can be used to determine stresses on a plane.

Example 1.2 The stresses on plane A–A and the horizontal plane at the same point are as shown in Figure Ex. 1.2(a). Note that the shear stress on plane A–A is missing. Determine (a) the shear stress on plane A–A, τ_{A-A}, and (b) the magnitude of τ_{max} and the orientation of the planes of τ_{max}, at the point.

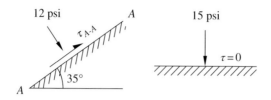

Figure Ex. 1.2(a)

Solution:

A point on the Mohr circle ($\sigma = 15$, $\tau = 0$) corresponding to the horizontal plane is known to be on the Mohr circle, point A. Another point on the Mohr circle corresponding to stresses on plane A–A is along a vertical line through ($\sigma = 12$, $\tau = 0$).

The steps in the solution to (a) and (b) (see Figure Ex. 1.2(b)) are as follows:

i) Sketch a line through ($\sigma = 15$, $\tau = 0$) making an angle ($90° - 35°$) = $55°$ from the horizontal (clockwise, toward the general direction of the vertical line through ($\sigma = 12$, $\tau = 0$)); this line will intersect the vertical line (at $\sigma = 12$ psi) at point B, which is also a point on the Mohr circle. The magnitude of τ_{A-A} is the ordinate of point B, hence $\tau_{A-A} = 4.28$ psi.

ii) Construct the Mohr circle with two known points on the circle (points A and B) by first finding the center (determined by the intersection of the horizontal σ-axis and the perpendicular bisector of line $\overline{AB}$). The ordinate of the peak points on the Mohr circle is $\tau_{max} = 4.56$ psi (this can also be determined graphically or from geometric relations in Figure Ex. 1.2(b)).

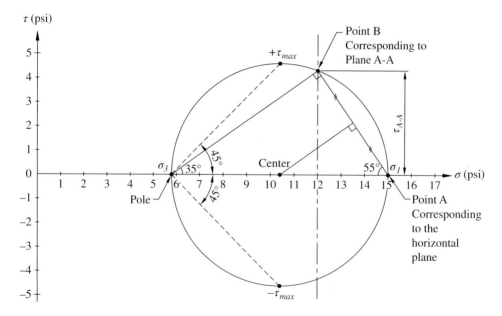

Figure Ex. 1.2(b)

iii) Locate the pole of the Mohr circle by drawing a horizontal line through point A.

iv) Determine the orientation of the two planes of τ_{max}, as shown in Figures Ex. 1.2(b) and (c).

Figure Ex. 1.2(c)

Example 1.3 At a point of interest in a soil mass, τ_{max} is known to be 15 psi, and the stresses on a plane inclined 30° with the horizontal are measured as shown in Figure Ex. 1.3(a). Determine the maximum and minimum normal stresses at point P and the stresses on the 45° plane.

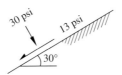

Figure Ex. 1.3(a)

Solution:

The solution can be found by the following procedure (see Figure Ex. 1.3(b)):

i) Locate point P with coordinates ($\sigma = 30$ psi, $\tau = 13$ psi).

ii) Draw a line through point P making an angle of 30° to the horizontal. The line will intersect the horizontal axis at point Q.

iii) Determine the center of the Mohr circle, point A (by connecting points P and Q with a straight line, then sketch a normal line through the midpoint of the line between P and Q; point A will be the intersection between the normal line and the horizontal axis).

iv) Construct the Mohr circle using point A as the center and distance AP or AQ as the radius (this circle is referred to as *Circle A*). The maximum and minimum normal stresses can readily be determined from the Mohr circle; they are the coordinates of the two points where the circle intersects the horizontal axis, σ_1 and σ_3 in Figure Ex. 1.3(b).

v) Locate the pole of the Mohr circle, which turns out to coincide with point Q.

vi) Draw a straight line through point Q making an angle of 45° to the horizontal. The coordinates of the point where the line intersects Circle A are σ and τ on the 45° plane: $\sigma = 22\,\text{psi}$ and $\tau = 15\,\text{psi}$ (as shown in Figure Ex. 1.3(b)).

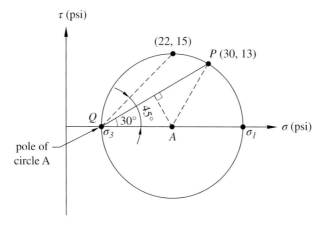

Figure Ex. 1.3(b)

There is another possible scenario where the pole may be located at $\sigma = 30\,\text{psi}$ and $\tau = 13\,\text{psi}$ (i.e., point P). In that case, the Mohr circle will be a circle tangent to the 30° line at point P (as shown in Figure Ex. 1.3(c)). Such a circle will have its center at point B,

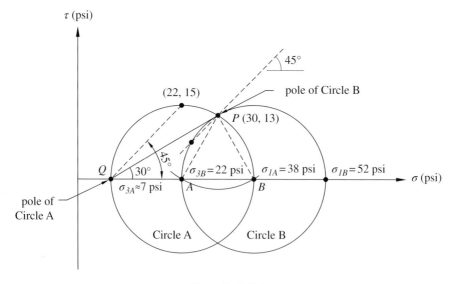

Figure Ex. 1.3(c)

and is referred to as *Circle B*. The maximum and minimum normal stresses are readily determined from Circle B. The stresses on the 45° plane can be determined by sketching a straight line from point P (i.e., from the pole of Circle B) making a 45° angle to the horizontal plane.

From the two Mohr circles seen in Figure Ex. 1.3(c), σ_1 and σ_3 for the two scenarios, as well as respective stresses on the 45° planes, can readily be determined. Their values are shown in Figure Ex. 1.3(d).

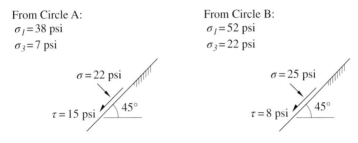

From Circle A:
$\sigma_1 = 38$ psi
$\sigma_3 = 7$ psi

From Circle B:
$\sigma_1 = 52$ psi
$\sigma_3 = 22$ psi

$\sigma = 22$ psi

$\tau = 15$ psi   45°

$\sigma = 25$ psi

$\tau = 8$ psi   45°

Figure Ex. 1.3(d)

1.2 Concept of Effective Stress

Natural soil is generally a three-phase material, comprising solid particles, pore water, and pore air. Since the pore fluids (pore water and pore air) offer no resistance to static shear stress, the conventional definition of stress has to be revised when dealing with the shear strength of soil. This brings about the concept of effective stress. The effective normal stress, commonly denoted as σ', is the part of applied normal stress that *controls the shear resistance and volume change of a soil* when water is present in the pores. In 1923, Terzaghi presented the principle of effective stress, an intuitive relationship based on experimental data of fully saturated soils. The relationship is:

$$\sigma' = \sigma - u \tag{1-4}$$

where σ is the total stress and u is the porewater pressure. The total stress is the stress calculated by picturing the soil as a single-phase continuum, i.e., the definition commonly used for normal stress.

Although the shear resistance of soil is controlled by effective stress, it is sometimes simpler to perform stability analysis of earth structures in terms of total stress because it does not require us to know the porewater pressure. This, however, is only warranted when a valid relationship can be established between shear strength and total stress. Such a relationship is only available in a limited number of cases where variations of in situ porewater pressure or drainage conditions do not deviate significantly from those in the laboratory tests. The short-term stability of saturated clays is a distinct example where total stress analysis with undrained shear strength can be carried out for stability analysis. This point will be addressed further in Section 1.5.2.

1.3 Mohr–Coulomb Failure Criterion

If shear stress on a plane reaches a limiting maximum value on that plane, shear failure is said to occur. A number of failure criteria for soils have been proposed to define the limiting condition, such as Mohr–Coulomb criterion, triangular conical criteria, extended von Mises criteria, curved extended von Mises criteria (Lade, 2005). Among these failure criteria, the first two have frequently been used for analysis of earth structures conducted by the finite element methods of analysis. The Mohr–Coulomb failure criterion is by far the most commonly used criterion for limiting equilibrium analysis and routine design of earth structures.

In the Mohr–Coulomb failure criterion, the shear strength (τ_f) of soil can be described by the following equation:

$$\tau_f = \sigma_f \tan\phi + c \tag{1-5}$$

in which σ_f is normal stress at failure, ϕ is the *angle of internal friction* (or simply *friction angle*), and c is *cohesion*. Even though Eqn. (1-5) is a common expression of the Mohr–Coulomb failure criterion, it falls short of describing the stresses involved in a rigorous manner. We know from Section 1.1 that stresses are generally different on different planes. However, Eqn. (1-5) only describes the relationship between the normal and shear stresses *at failure*, and does not address the associated plane of the stresses. Yet it has been found to be a useful expression for performing total stress analysis in which the strength of a soil is defined by total stress strength parameters c and ϕ. This point is discussed further in Section 1.5.

A more rigorous expression of Mohr–Coulomb failure criterion is given in terms of effective stress as:

$$\tau_{ff} = \sigma'_{ff} \tan\phi' + c' \tag{1-6}$$

In Eqn. (1-6), the double subscript ff denotes *on failure plane at failure*, τ_{ff} is the shear stress *on failure plane at failure* (commonly referred to as *shear strength*), σ'_{ff} is the effective normal stress *on failure plane at failure*, and c' and ϕ' are effective shear strength parameters. Recall Terzaghi's effective stress principle that the shear resistance of soil is dictated by effective stress, as seen in Eqn. (1-6).

As shown in Figure 1.9, the straight line given by Eqn. (1-6) is a line tangent to all effective stress Mohr circles at failure. Failure is said to occur when the Mohr circle "touches" the straight line, known as the *Mohr–Coulomb failure envelope*. The point of tangency (such as point F in Figure 1.9) represents the effective normal stress and shear stress on failure plane at failure. Note that this is true only when the normal stress is expressed in terms of effective stress. Also note that the failure envelopes for many soils may not necessarily be straight lines, especially for dense soils (dense sands and stiff clays), but a straight-line approximation can be taken over the range of stress of interest. This is discussed further in Section 1.5.1.

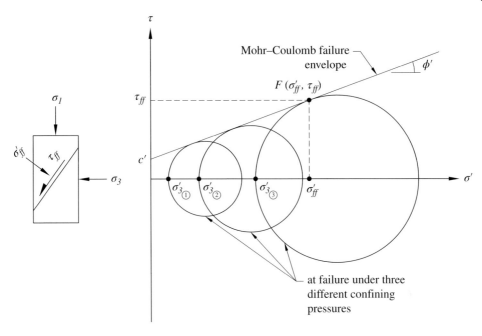

Figure 1.9 The Mohr–Coulomb failure criterion in terms of effective stress, where stresses on the failure plane at failure (σ_{ff}', τ_{ff}) are represented by the point of tangency of the Mohr circle

In terms of principal stresses, the Mohr–Coulomb failure criterion can be expressed as

$$\sigma_{1f}' = \sigma_3' \tan^2\left(45° + \frac{\phi'}{2}\right) + 2c' \tan\left(45° + \frac{\phi'}{2}\right)$$
(1-7)

or

$$\left(\sigma_1' - \sigma_3'\right)_f = \frac{2c\cos\phi' + 2\sigma_3 \sin\phi'}{1 - \sin\phi'}$$
(1-8)

The subscript *f* in Eqns. (1-7) and (1-8) means "at failure." These equations are useful in some applications of the Mohr–Coulomb failure criterion.

The use of the Mohr–Coulomb failure criterion requires determination of the *Mohr–Coulomb strength parameters c′ and φ′ (or c and φ)*. It is important to keep in mind that these strength parameters are not "inherent" soil properties. Their values have been found to vary with the drainage condition, stress path, and stress history of a soil.

1.4   Shear Strength Tests

We will now look at some common tests used for determining the shear strength of soils, with emphasis on determination of the Mohr–Coulomb shear strength parameters *c′* and *φ′* (referred to as *Mohr–Coulomb drained strength parameters*) or *c* and *φ* (referred to as *Mohr–Coulomb undrained strength parameters*). The shear strength parameters can be determined by conducting laboratory tests or in situ (a Latin phrase meaning "on site") tests. The most commonly used laboratory tests are the direct shear test, the triaxial compression test, and the unconfined compression test. The most common field tests in North America are the standard penetration test and the cone penetration test. For cohesive soils, vane shear tests have sometimes been used in the laboratory as well as in the field for determination of undrained shear strength.

There is an abundance of literature on shear strength tests (e.g., Bishop and Henkel, 1962; Head and Epps, 1982; Holtz et al., 2011). Only a brief description of commonly used shear strength tests is given here.

1.4.1 Direct Shear Test

Figure 1.10 shows a schematic diagram and a photo of direct shear test setup and apparatus. In direct shear test, a specimen of soil prepared at prescribed density and moisture content is confined laterally in a metal box of a square or circular cross-section. The box is split horizontally in half with a small clearance between the upper and lower boxes. The test is typically carried out by fixing the position of the lower box and applying horizontal forces to move the upper box relative to the lower box. A constant normal load (N) is first applied to the top of the specimen by a dead weight or an air bladder, then the shear force (T) exerted on the upper box is gradually increased until failure occurs. The test is usually repeated with a few different normal loads to determine the Mohr–Coulomb strength parameters c and ϕ.

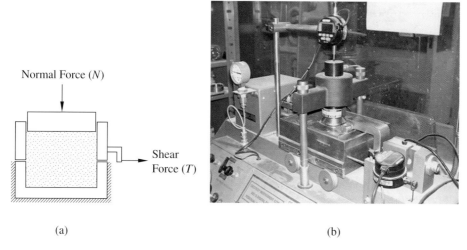

(a)   (b)

Figure 1.10 Direct shear test: (a) schematic test setup and (b) a photo of test apparatus

A typical set of direct shear test results are shown in Figure 1.11(a). In addition to the relationships between the relative displacements of the upper and lower boxes (δ) and the shear stress (*shear stress = T/A*), vertical displacement of the soil specimen

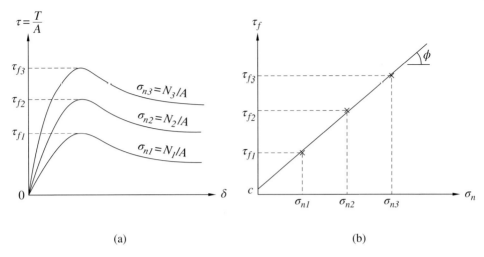

Figure 1.11 Results of direct shear tests under three normal loads: (a) relationship between relative displacement (δ) and shear stress (τ), and (b) relationship between normal stress (σ_n) and shear stress at failure (τ_f); all stresses are on the failure plane (the horizontal plane)

during shear is often measured as well. By plotting the relationship between the shear stress at failure ($\tau_f = T_f/A$) vs. the corresponding normal stress ($\sigma_n = N/A$), as shown in Figure 1.11(b), the Mohr–Coulomb failure envelope of the soil can be obtained. The failure envelope is formed by simply drawing a *best-fit straight line* through the data points. Unlike in triaxial tests (see Section 1.4.2), there is no need to sketch Mohr circles at failure and find the common tangent of the circles in this case. This is because the data points, such as those shown in Figure 1.11(b), are stresses on the failure plane (the horizontal plane in this case) at failure. Clean cohesionless soils have $c = 0$. For moist cohesionless soil or soil containing some fines, however, the best-fit straight line failure envelope will likely have a non-zero intercept (i.e., cohesion, $c \neq 0$). This cohesion is called *apparent cohesion* and should be ignored in design as it cannot be counted on during the service life of an earth structure.

For sands in a dense condition, the relationship between relative displacement (δ) and shear stress (τ) typically shows a "drop" after a peak value of τ is reached, and reduces to a *residual* value (see Figure 1.12(a)). This behavior is termed *softening*. In such a case, friction angles corresponding to peak strength and residual strength can be obtained. Note that the initial area (A), rather than corrected shear areas occurred during testing, is usually used in the calculations of σ_n and τ_f. This is because the ϕ-value is little affected by the use of the corrected area. The former is usually larger than the latter. In some cases, the reverse can occur; i.e., $\phi_{residual} > \phi_{peak}$, as seen in Figure 1.12(b). This is due to the difference in apparent cohesion, which can be larger for the peak strength than for the residual strength. For design purposes, ϕ_{peak} should usually be used. If large deformation due to progressive failure is expected, $\phi_{residual}$ should be used instead.

1.4.2 Triaxial Test

Most triaxial tests are performed by applying vertical axial loads to a cylindrical soil specimen while it is subjected to pressure confinement. The setup and a photo of triaxial

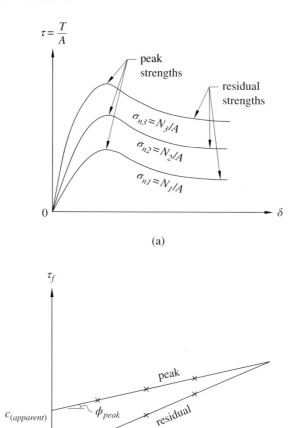

Figure 1.12 Direct shear test results: (a) peak and residual strengths, and (b) failure envelopes of peak and residual strengths (for some soils, $\phi_{residual}$ may be larger than ϕ_{peak}, as shown)

test are shown in Figure 1.13. In a triaxial test, a cylindrical soil specimen is fitted between two rigid end caps, with its vertical surface covered by a flexible latex membrane, placed inside an enclosed cell/chamber that is filled with water (although air pressure, positive or negative, can also be used), and loaded vertically until failure occurs.

A triaxial test is conducted in two stages. In stage 1, cell pressure (or confining pressure) is applied to the cylindrical soil specimen to provide pressure confinement (and consolidate the specimen if a consolidated triaxial test is performed). In stage 2, increasingly larger vertical axial compressive loads are applied to the specimen through a piston connected to the top of specimen until failure occurs. Triaxial tests with this mode of loading are referred to as *axial compression* triaxial tests.

Figure 1.14 shows Mohr circles corresponding to the two stages of a triaxial test. In stage 1, the soil specimen is subjected to an equal cell pressure (σ_c) in all directions, the

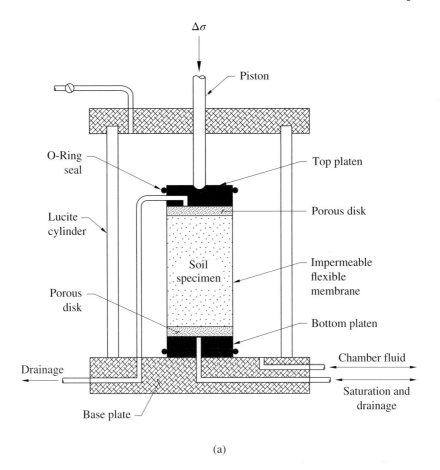

$\Delta\sigma$

Piston

O-Ring seal

Top platen

Lucite cylinder

Porous disk

Soil specimen

Impermeable flexible membrane

Porous disk

Bottom platen

Chamber fluid

Drainage

Saturation and drainage

Base plate

(a)

(b)

Figure 1.13 Triaxial test: (a) schematic test setup and (b) a photo of test specimen and test apparatus

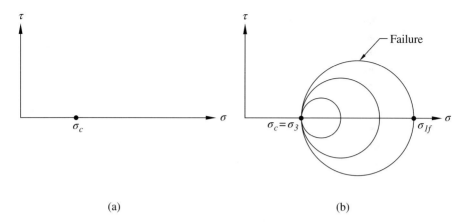

Figure 1.14 Mohr circles of an axial compression (AC) triaxial test: (a) stage 1 (hydrostatic loading) and (b) stage 2 (shear)

corresponding Mohr circle is simply a *point* on the σ-axis, and the soil is not subject to any shear stress on any plane; the soil specimen is said to be subject to *hydrostatic loading* (in stage 1). In stage 2, however, the Mohr circle becomes a circle, and the circle will grow larger in size as increasing vertical load is applied. Other than the principal planes, all other planes are subjected to shear stress of different magnitudes; the soil specimen is said to be subject to *shear* (in stage 2).

Figure 1.15 depicts schematically the results of triaxial tests conducted by using different confining pressures, high, medium, and low. The upper part of the figure shows stress–strain relationships in stage 2 (shear) loading, plotted in terms of change in axial stress versus change in axial strain. The change in axial stress is commonly

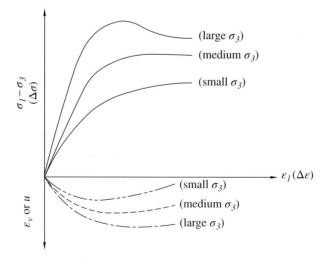

Figure 1.15 Results of axial compression triaxial tests under different confining pressures

expressed in terms of *deviator stress* which is the difference in principal stresses $(\sigma_1 - \sigma_3)$ versus the change in axial strain which is the change in vertical strain (ε_1 or $\varepsilon_{vertical}$). As seen in Figure 1.15, the stiffness (i.e., the slopes of the stress–strain curves) and strength of the soil are usually higher under larger consolidation/confining pressure. The lower part of Figure 1.15 shows the change in volumetric strain ε_v (in a *drained* test) or the change in porewater pressure u (in an *undrained* test). In a drained test, most soils would exhibit compression (reduction in volume) at small axial strains, follow by dilation (increase in volume) as strain increases. In an undrained test, volume change is prohibited, hence the "tendency" for volume change is reflected by change in porewater pressure. A tendency of dilation is reflected by a decrease in porewater pressure, and a tendency of compression by an increase in porewater pressure.

One major advantage of the triaxial test over the direct shear test is that the drainage condition of the soil specimen during testing can be controlled to simulate the drainage condition anticipated in the field; also, porewater pressure during testing can be measured in an undrained test. Three common types of triaxial tests for different drainage conditions are the unconsolidated undrained (UU) test, the consolidated undrained (CU) test, and the consolidated drained (CD) test. The first word of these terms (either U or C) refers to stage 1 testing; namely, whether drainage (consolidation) is allowed in stage 1 of triaxial testing while the soil specimen is being subjected to a prescribed confining pressure. The second word (either U or D) refers to stage 2 of testing; namely, whether drainage is permitted during application of deviator stress.

Assuming that vertical stress and confining pressure are principal stresses (denoted by σ_1 and σ_3), soil strength parameters c' and ϕ' (or c and ϕ) can readily be determined from the principal stresses at failure in triaxial tests. Example 1.4 illustrates how the effective strength parameters c' and ϕ' are determined from the results of three consolidated drained (CD) triaxial tests. The example also shows how the orientation of failure planes is determined with the aid of Mohr circles.

Example 1.4 Three triaxial compression tests were performed on a dry sand under different confining pressures. The confining pressures and corresponding deviator stresses at failure were:

σ_3 (psi)	$(\sigma_1 - \sigma_3)_f$ (psi)
5	16
10	31
15	47

Determine the strength parameters c' and ϕ', and the orientation of the failure plane at failure.

Solution:

Three Mohr circles at failure cooresponding to the three tests are shown in Figure Ex. 1.4.

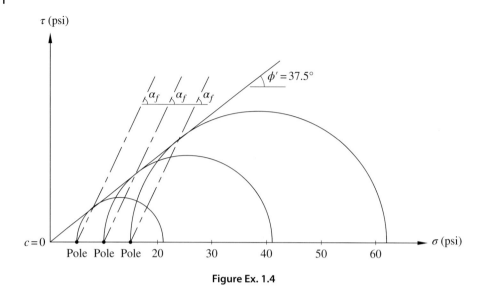

Figure Ex. 1.4

Since the sand is dry, the strength parameters obtained from results of the triaxial tests are effective strength parameters (i.e., $u = 0$). From Figure Ex. 1.4,

$$c' = 0, \ \phi' = 37.5°$$

and the failure plane would occur on a plane inclined at

$$\alpha_f = 45° + \left(\frac{\phi'}{2}\right) = 64° \text{ from the horizontal.}$$

The shear strength of a soil in a drained condition is usually quite different from that in an undrained condition. A drained condition is usually assumed for soils of high permeability in that the excess porewater pressure induced by applied loads will dissipate quickly without accumulation. On the other hand, an undrained condition is usually assumed for soils of low permeability immediately after load applications before any significant consolidation can take place. (Note: the term *consolidation* refers to "deformation accompanied by dissipation of porewater pressure".) Low-permeability soils in the field typically change gradually from an undrained condition to a drained condition after load applications. When consolidation is eventually completed, shear strength of low-permeability soils will become drained strength. The drained strength of a soil can be expressed in terms of effective stress with shear strength parameters c' and ϕ'. The undrained strength, on the other hand, can be expressed in terms of total stress, with shear strength parameters c_u and ϕ_u (or simply denoted as c and ϕ).

The drained strength of a soil may be obtained by performing either drained tests or undrained tests. If an undrained test is used, the effective stress can be computed by simply subtracting measured porewater pressure from the applied (total) stress if the soil is fully saturated. Example 1.5 shows how drained strength parameters (c' and ϕ') and undrained strength parameters (c and ϕ) of a saturated clay can be determined

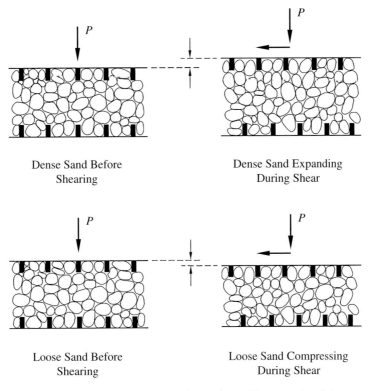

Figure 1.16 Conceptual volume change behavior during shear of loose sand and dense sand (Leonards, 1962)

from results of CU triaxial tests with measured porewater pressure. Loose sands and normally consolidated clays typically have positive porewater pressure at failure, whereas dense sands and overconsolidated clays typically have negative porewater pressure at failure. The behavior for a loose sand to contract and a dense sand to dilate during shear is explained by Leonards (1962) using the sketches shown in Figure 1.16.

Example 1.5   Three consolidated undrained (CU) triaxial tests were performed on a fully saturated clay. The principal stress and porewater pressure at failure for each test were measured. The stresses and porewater pressures at failure are shown in the following table:

Confining Pressure (psi)	$(\sigma_1 - \sigma_3)_f$ (psi)	Pore Pressure at Failure (psi)
15.0	10.3	8.2
30.0	20.2	16.0
45.0	30.5	25.2

Determine the Mohr–Coulomb drained and undreained strength parameters (c', ϕ' and c, ϕ).

Solution:

Three total and effective Mohr circles at failure for the three CU tests are shown in Figure Ex. 1.5. Also shown in the figure are the total and effective Mohr–Coulomb failure envelopes.

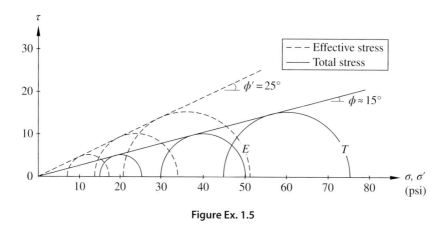

Figure Ex. 1.5

From the failure envelopes,

$c' = 0$, $\phi' = 25°$

$c = 0$, $\phi = 15°$

Figure 1.17 shows idealized failure envelopes of typical loose sand, dense sand, normally consolidated clay, and overconsolidated clay, in terms of both total and effective stresses. The strength parameters $c = c' = 0$ as determined in Example 1.5 suggest that the clay in the example is normally consolidated. It is important to point out that failure envelopes are usually curved (concave downward) for dense sands and overconsolidated clays. This point is addressed further in Section 1.5.

The stress–strain behavior and volume change (or porewater pressure change) behavior of a soil is significantly affected by its initial state. Figure 1.18 shows comparative triaxial test results of loose sand/normally consolidated clay, medium sand/lightly overconsolidated clay, and dense sand/heavily overconsolidated clay. Dense sands and heavily overconsolidated clays usually show strain softening, i.e., a peak deviator stress is reached then reduces to a residual value. For loose sands and normally consolidated clays, on the other hand, a peak deviator stress is usually reached at failure with little or no strain softening. In terms of volume change (in a drained test) and porewater pressure change (in an undrained test), dense sands and heavily overconsolidated clays usually show dilation (in a drained test) or negative porewater pressure (in an undrained test) at failure, whereas loose sands and normally consolidated

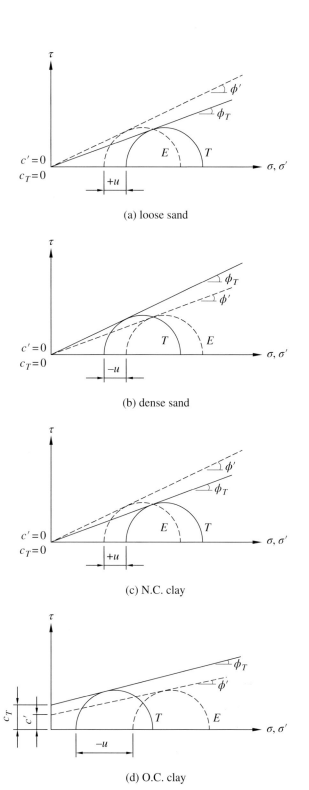

(a) loose sand

(b) dense sand

(c) N.C. clay

(d) O.C. clay

Figure 1.17 Idealized failure envelopes of (a) loose sand, (b) dense sand, (c) normally consolidated clay, and (d) overconsolidated clay (compare with Figures 1.29 and 1.35 in Section 1.5 for nonidealized dense sand and overconsolidated clay, respectively)

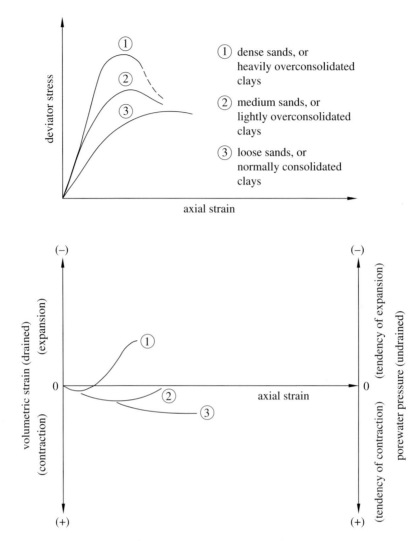

Figure 1.18 Comparative results of axial compression triaxial test for loose to dense sands, and normally consolidated to overconsolidated clays (modified after Japanese Foundation Engineering Society, 1995)

clays usually show contraction (in a drained test) or positive porewater pressure (in an undrained test) at failure.

In addition to being able to control drainage conditions during testing, a triaxial test can also control the *loading path* to simulate the anticipated loading conditions of an earth structure. Figure 1.19 shows four loading paths that can be simulated by a triaxial test. Also shown is an example application of each loading path. The loading paths are axial compression (AC), lateral extension (LE), axial extension (AE), and

Example Application	Laboratory Loading Path

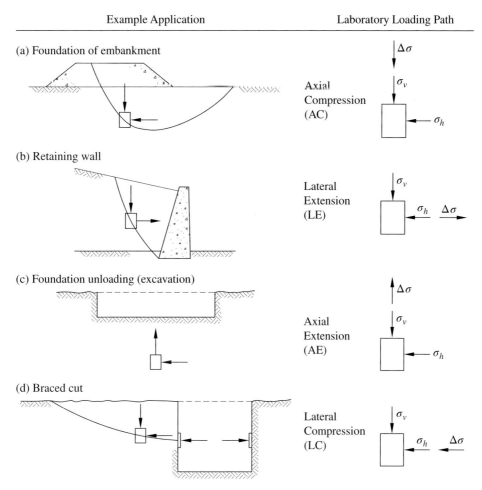

Figure 1.19 Loading paths of triaxial tests and example applications (modified after Holtz et al., 2011)

lateral compression (LC). Our discussions thus far have been focused on AC, even though other loading paths may be more applicable to a certain loading condition. Note that AC and LE tests make a specimen shorter and wider, and AE and LC will have the opposite effect, making a specimen longer and narrower. The results of AC and LE tests for most soils have been found to be nearly identical, and so have the results of AE and LC tests. The ϕ'-value measured from triaxial AE and LC tests is generally slightly higher than from AC and LE tests (typically about 10% higher for cohesionless soils, and about 20% higher for normally consolidated cohesive soils) (Kulhawy and Mayne, 1990). In some instances, however, the strength obtained from AE and LC tests of normally consolidated clays have been found to be much lower than those from AC and LE tests because of soil anisotropy (Hunt, 1986).

1.4.3 Plane-Strain Test

A *plane-strain condition* prevails if the geometry, material properties, and loads do not vary along the longitudinal direction of a very long structure. When the longitudinal dimension of a structure is about 10 times or greater than other dimensions, it is generally considered *very long*. Figure 1.20 shows examples of earth structures that can be approximated as being in a plane-strain condition. For a structure in a plane-strain condition, the longitudinal strain is equal to zero and there is no variation of stress or displacement in the longitudinal direction. Therefore, analysis of a structure in a plane-strain condition can be performed by considering only a *slice* of the structure of unit thickness; i.e., a two-dimensional analysis is applicable.

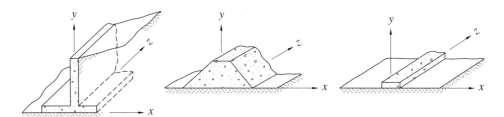

Figure 1.20 Examples of plane-strain condition in which the longitudinal dimension of the structure (in the *z*-direction) is much greater than other dimensions; also, there is no variation in geometry, material properties, or loading in the longitudinal direction

A plane-strain test gives better simulation of many "long" earth retaining structures than a triaxial test. Figure 1.21(a) shows the setup of a test specimen in a plane-strain test, in which a cuboidal soil specimen is confined between two lubricated rigid (acrylic/plexiglass) plates maintained a fixed distance apart. The soil-plate assembly is then placed inside a pressure cell similar to that used in a triaxial test. The rigidity of the plates restrains the deformation of the soil specimen in the longitudinal direction to ensure zero longitudinal strain ($\varepsilon_z = 0$), while the lubricated surface allows the soil specimen to deform without artificial constraint on deformation in the other two directions. Tatsuoka et al. (1984) have developed a lubrication technique that is shown to result in an interface friction angle on the order of 0.05° at the contact surfaces of soil and a smooth plate (plexiglass). The technique, shown in Figure 1.21(b), involves introducing a latex membrane between the soil and the plexiglass, and applying a thin layer of silicon lubricant over the surface of the plexiglass. As seen in Figure 1.21(c), internal movement of the soil specimen can be monitored with very good accuracy by tracking the movement of a grid system printed on the latex membrane before testing.

Laboratory tests performed on medium and dense sands have indicated that the internal friction angle (ϕ') determined by plane-strain tests is approximately 10% higher than those determined by triaxial compression tests. For design of earth retaining walls, friction angles determined from triaxial tests are often used even when a wall is approaching a plane-strain condition. The use of such friction angles would provide an added safety margin in design.

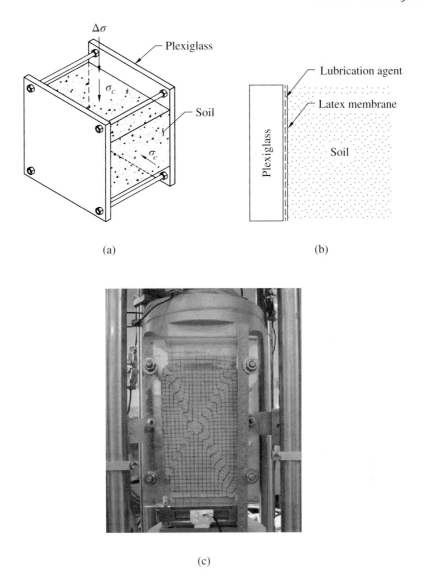

(a)

(b)

(c)

Figure 1.21 Plane-strain test: (a) schematic setup of test specimen, (b) a lubrication technique at the contact surface between soil and plexiglass developed by Tatsuoka, and (c) use of the lubrication technique to monitor internal movement of a soil specimen (photo courtesy of Hoe Ling)

1.4.4 Vane Shear Test

The vane shear test is a testing method for determination of the undrained shear strength of non-fissured, fully saturated clays. The vane shear test involves pushing a four-bladed vane (see Figure 1.22) on the end of a rigid rod into the soil, then rotating the rod at 6–12° per minute until a cylinder of soil (diameter d, length h) contained

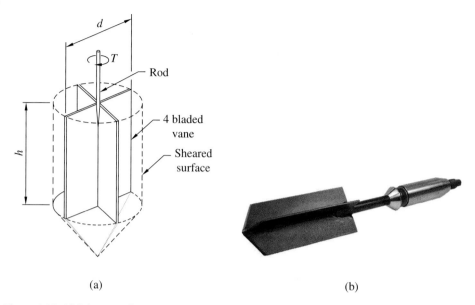

(a)   (b)

Figure 1.22 (a) Schematic diagram of a vane shear test and (b) photo of an electric vane (photo courtesy of Gouda Geo-Equipment B.V., the Netherlands)

within the blades is sheared off. By measuring the torque T required to induce failure of a soil, the undrained shear strength of the soil S_u can be evaluated as:

$$S_u = \frac{T}{\pi d^2 \left(\dfrac{h}{2} + \dfrac{d}{6} \right)}$$

(1-9)

The vane shear test has been performed the in the field and in the laboratory. The former is typically performed in conjunction with drill-hole explorations in soft to medium clays. The test can also be carried out in soft soil without a bore hole, by direct penetration of the vane from the ground level.

1.4.5 Standard Penetration Test

The standard penetration test (SPT) is the most widely used field test in North America. It is usually conducted as part of soil boring operations. The test involves lowering a standard size *split-spoon sampler* (Figure 1.23(a)) to the bottom of a bore hole, and using a hammer weighing 140 lb (0.62 kN) free-falling 30 inches to advance the sampler deeper into the ground (see Figure 1.23(b)). The number of blows of the hammer required to advance the split-spoon sampler three successive 6-inch (152-mm) intervals into the bottom of the bore hole is recorded. The number of blows for the last 12 inches (305 mm) is termed the *standard penetration resistance* or *blow count*, commonly denoted by N.

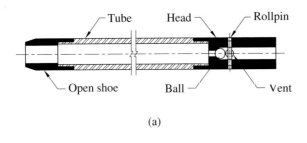

(a)

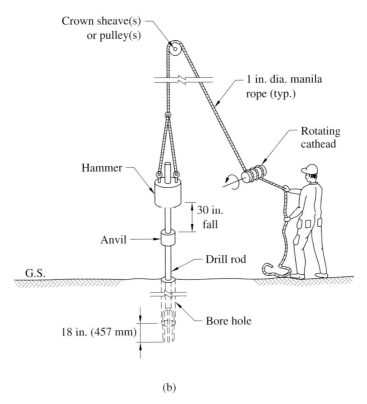

(b)

Figure 1.23 The standard penetration test (SPT): (a) schematic diagram of a split-spoon sampler and (b) typical SPT setup (Kovacs et al., 1981)

A number of relationships between N and shear strength have been suggested. The relationships for large gravels and clays are generally unreliable and can serve only as a reference. A simple yet rather reliable empirical correlation between N and ϕ' for sands is shown Figure 1.24(a). This relationship has been used widely in routine geotechnical engineering design. A commonly used correlation between N and unconfined compressive strength, U_c ($U_c = 2S_u$ for saturated clays, S_u = undrained shear strength) for cohesive soils of low to high plasticity is shown in Figure 1.24(b).

Studies have suggested that hammer efficiency (η) can vary over a wide range (30–90%), depending on the hammer type. Seed et al. (1985) and Skempton (1986)

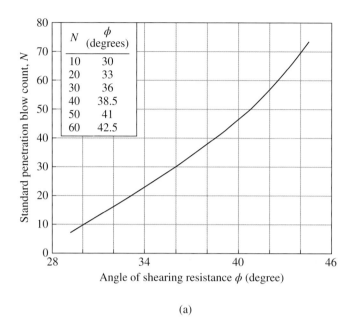

(a)

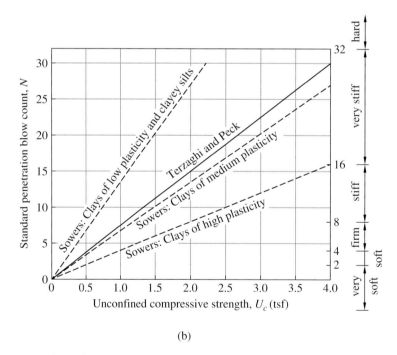

(b)

Figure 1.24 Correlations between SPT blow count *N* and (a) angle of internal friction ϕ' of granular soils and (b) unconfined compressive strength U_c of cohesive soils (NAVFAC, 1986)

proposed that raw field N-values be standardized to N_{60} to reflect $\eta = 60\%$. This has become common practice. When not stated specifically, the N-values in most correlations may be taken as being N_{60}.

1.4.6   Cone Penetration Test

Estimation of the shear strength of a soil under static loading by means of a *pseudo dynamic* test, such as the standard penetration test, is inherently difficult, especially when free water level is encountered. More reliable estimation may be obtained from static penetration tests, of which the most commonly used in North America is the cone penetration test (CPT).

In the cone penetration test, a penetrometer with a cone-shaped tip attached to a string of solid rods running inside hollow outer rods is pushed into the ground at a controlled rate. Figure 1.25 shows an overview of CPT testing with a truck-mounted rig. Using the weight of the truck as reaction, a hydraulic ram located inside the truck pushes the cone into the ground. The end resistance of the cone at a given depth is recorded as the cone resistance, q_c, and $q_c = \dfrac{\text{force required to advance the cone}}{\text{end area of the cone}}$. After q_c is determined, the outer rods are pushed downward. The cone and sleeve thus advance together after the range of motion inside the cone has been taken up. Such a sequence allows end bearing resistance and side friction to be determined separately. In some cone penetrometers (e.g., electric cones), a load cell is mounted near the conical tip and allows continuous recording of cone resistance. Other cone penetrometers, referred to as *piezocones*, can also measure porewater pressure at the conical tip.

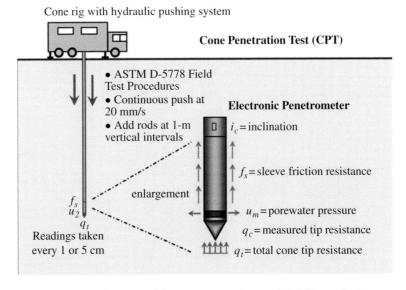

Figure 1.25 Overview of the cone penetration test (CPT) (Mayne, 2007)

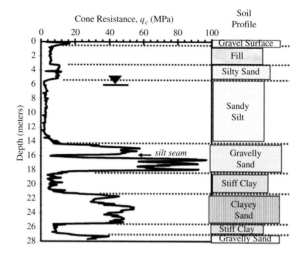

Figure 1.26 A set of cone penetration test (CPT) data: cone resistance vs. depth with soil boring log (Mayne, 2007)

A set of CPT test results is shown in Figure 1.26. The variation of cone resistance with soils of different types is evident from the cone resistance data. Various theoretical and empirical correlations have been developed to relate q_c to ϕ' for sands. One such correlation is given in Figure 1.27. For cohesive soils, correlations between q_c and undrained shear strength (S_u) have also been proposed. Sanglerat (1972) suggests $S_u = q_c/15$

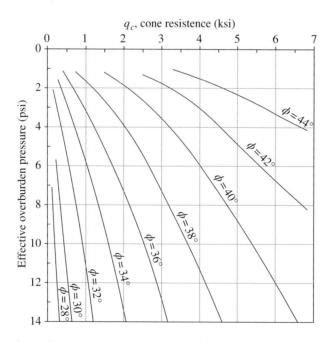

Figure 1.27 Correlations between cone resistance, overburden pressure, and internal friction angle (Robertson and Campanella, 1983)

for soft to stiff clays, and $S_u = q_c/30$ for stiff fissured clays. Schmertmann (1978) proposes the following correlation:

$$S_u = \frac{q_c - \Sigma \gamma z}{N_c} \tag{1-10}$$

where $\Sigma \gamma z =$ total overburden pressure at depth z and $N_c =$ a correction factor (10 for stiff clays and 16 for soft clays with a *Fugro friction cone*).

A major disadvantage of CPT as compared with SPT is that soil samples are not obtained as part of the CPT test operations.

1.4.7 Plate Load Test

The behavior of a clay is known to be significantly affected by the stress history of the clay. The degree of prestress in a clayey deposit can be determined by performing a laboratory consolidation (oedometer) test on undisturbed field samples. The engineering behavior of granular soils is also affected strongly by the stress history. A granular soil that has been *prestressed* is known to be much stronger than the same soil in a non-prestressed state. Since undisturbed field samples of granular soils are extremely difficult to obtain, the degree of prestress in a granular soil deposit needs to be determined by in situ tests. Among the many in situ shear tests, the plate load test is among the few methods that can capture directly the extent of prestress in a granular deposit.

The setup of a plate load test is shown in Figure 1.28(a), which involves applying increasing vertical loads to a 1-ft diameter circular plate (or 1 ft × 1 ft square plate) on the bottom of a test pit and recording the accompanied settlement. The load–settlement

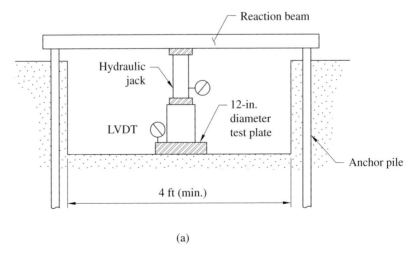

(a)

Figure 1.28 The plate load test: (a) schematic test setup, (b) "the" load–settlement curves for non-prestressed sands in a drained condition (Leonards, circa 1978), and (c) a measured load–settlement curve from an actual plate load test vs. the load–settlement curve of non-prestressed sand for $N = 10$ (Leonards, circa 1978)

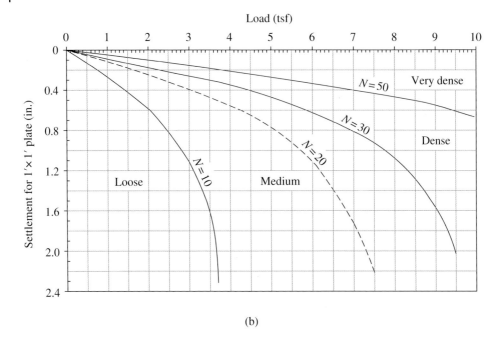

(b)

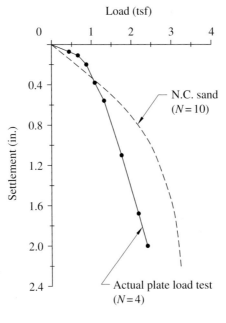

(c)

Figure 1.28 (Continued)

curves for non-prestressed sands (under drained conditions) of different N-values have been established by Leonards (personal communication, circa 1978), as shown in Figure 1.28(b). By sketching the load–settlement curve obtained from an actual plate load test conducted on a granular deposit and comparing it with the curves in Figure 1.28(b), the level of prestress of the granular deposit can be estimated. For example, Figure 1.28(c) shows the load–settlement curve obtained from an actual plate load test performed on a sand with an SPT blow count of 4. A comparison of the load–settlement curve for non-prestressed sand of $N = 10$ reveals that the soil in this case has been prestressed to approximately 0.8 tons/ft^2. The level of prestress will affect the settlement significantly in response to applied loads. Prevailing methods for predicting settlement in sands (e.g., methods proposed by Terzaghi and Peck, 1967; Peck et al., 1974; Schmertmann, 1978) are all for non-prestressed sands. In this example, for loads ranging between 0 and 0.8 tons/ft^2, the corresponding settlement can be conservatively estimated as about 1/5 of that of the non-prestressed sand. On the other hand, for the part of the load exceeding 0.8 tons/ft^2, the corresponding settlement is that of the non-prestressed sand.

1.5 Design Considerations

1.5.1 Shear Strength of Granular Soils

When a granular soil is subjected to static loads, pore water would typically drain rapidly with little build up of excess porewater pressure due to high permeability; therefore, drained strength generally prevails in design. As granular soil has little or no cohesion ($c' = 0$), the shear strength of sands or gravels is typically characterized only by the Mohr–Coulomb strength parameter ϕ', the effective angle of internal friction. In practice, it is often referred to as ϕ (without a prime) for simplicity, but the friction angle of granular soils always means the *effective* friction angle for granular soils.

The value of ϕ' of a granular soil can be determined by performing laboratory tests, such as the direct shear test, triaxial test, plane-strain test, etc. For triaxial tests, CU tests (with porewater pressure measurement) or CD tests may be performed. For a given granular soil, the value of ϕ' determined by direct shear tests is generally slightly lower than that by triaxial tests, which in turn is generally slightly lower than that by plane-strain tests. When performing these laboratory tests, large particles will need to be removed prior to testing. Large particles are those larger than about 1/10 of the specimen diameter for triaxial tests, and about 1/6 of the specimen diameter (or the side of a square specimen) for direct shear tests. Note that when preparing laboratory test specimens with large particles removed, it is the *relative density* of the in situ soil that should be replicated, not the void ratio.

If a granular soil is indeed cohesionless (i.e., without appreciable amount of fines, say more than 5–10%, which are "sticky when wet"), proper data interpretation for determination of ϕ' is to find the *best-fit* straight line through the data points or tangent to Mohr circles at failure and ignore the cohesion intercept, rather than forcing a straight line through the origin, even though the cohesion is known to be zero. The Mohr–Coulomb failure envelope of compacted granular soils is generally curved over a range of lower normal stress, as seen in Figure 1.29; the denser the soil, the more "curved" the failure envelope in that range. For design purposes, a *secant* value over the range of relevant normal stress could be used. Example 1.6 illustrates this procedure.

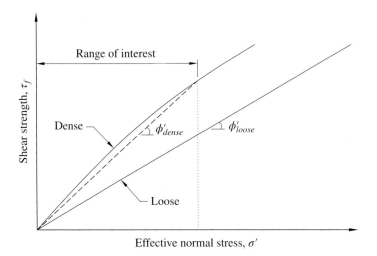

Figure 1.29 Mohr–Coulomb strength envelopes of compacted granular soils; the denser the soil, the more *curved* the strength envelope

An alternative approach to plotting Mohr circles at failure for determination of the strength parameters c' and ϕ' is to make use of a *p–q diagram*. In this approach, the values of $p = (\sigma_1' + \sigma_3')_f/2$ and $q = (\sigma_1' - \sigma_3')_f/2$ at failure are calculated and plotted as shown in Figure 1.30. After a best-fit straight line is drawn through the data points, the slope (β) and intercept (α) of the straight line can be determined. The Mohr–Coulomb strength parameters can then be calculated as $\phi' = \sin^{-1}(\tan \beta)$ and $c' = (\alpha/\cos\phi')$. The *p–q* approach usually yields slightly more consistent values of strength parameters c'

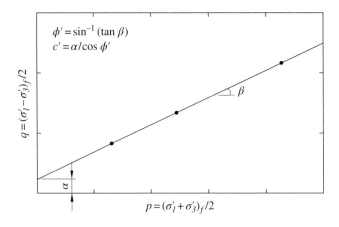

Figure 1.30 Determination of strength parameters c' and ϕ' by a *p–q* diagram

and ϕ' than drawing a common tangent line to Mohr circles at failure because the "common tangent" of multiple circles often involves a higher degree of personal judgment.

Example 1.6 The results of drained (CD) triaxial compression tests performed on a compacted crushed diabase at confining pressures ranging from 5 psi to 110 psi are shown in Figure Ex. 1.6(a). The soil used in the tests was prepared at an as-compacted target unit weight of 140 lb/ft³. The test at confining pressure of 70 psi was repeated as a check. This diabase is to be used as the fill material of a 17-ft high geosynthetic reinforced soil (GRS) bridge abutment of which the design load on the bridge sill is 30 psi. Determine the Mohr–Coulomb strength parameters c' and ϕ' of the fill material to be used in the design.

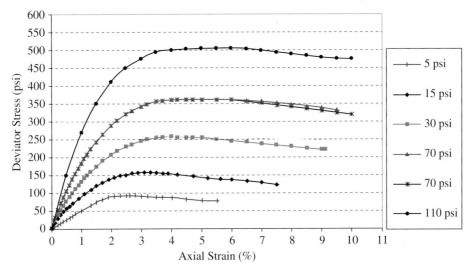

Figure Ex. 1.6(a)

Solution:

The Mohr circles at failure (as defined by the peak strengths) are shown in Figure Ex. 1.6(b). It is seen that the failure envelope is highly curved. A curved failure envelope implies that ϕ varies with the normal stress. For a curved failure envelope, a *secant* strength parameter could be employed in the design because the secant slope is the average value of the "tangent" slope over the range of normal stress relevant to the problem. To that end, the range of normal stress relevant to the abutment needs to be estimated first. At the mid-depth of the GRS abutment under a sill pressure of 30 psi, vertical stress is calculated:

$$\sigma_{v\,(mid\text{-}depth)} \approx \frac{(140\ \text{lb/ft}^3)(17\ \text{ft/2})}{144\ (\text{in}^2/\text{ft}^2)} + 30\ (\text{lb/in}^2) = 38\ (\text{lb/in}^2)$$

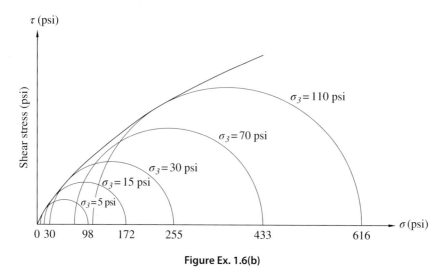

Figure Ex. 1.6(b)

Assuming σ_3 at the mid-depth is about equal to the vertical stress (for a conservative estimate of secant ϕ' of a "compacted" granular fill), the range of σ_3 of interest is therefore between 0 and 38 lb/in^2.

To determine a "secant" friction angle for the curved failure envelope, we should first determine if the fines content is "appreciable" (say, more than 5–10%) and, if so, are the fines "cohesive" (i.e., sticky when wet). For the purposes of illustration, we will consider two conditions: (a) the fines content is small (≤ 5–10%) *or* the fines are cohesionless ($c = 0$), and (b) the fines content is high (> 5–10%) *and* the fines are cohesive ($c > 0$). When uncertain, it would be conservative to assume that the soil is cohesionless (and ignore the cohesion). However, remember this rule of thumb: *for a soil with a high ϕ-value, a little bit of c will go a long way.* Thus, an assumption of $c = 0$ may be overly conservative. Physically checking the soil condition (are the fines sticky when wet?) is very important to avoid over-conservatism.

a)   If the fines content is low or the fines are cohesionless, i.e., $c = 0$

For a $c = 0$ soil, the secant ϕ-value can be determined by a simple four-step procedure:

1)   Draw a smooth curved failure envelope tangent to all at-failure Mohr circles that are considered reliable (see Figure Ex. 1.6(b)).

2)   Estimate the largest relevant value of σ_3 (or σ_1) for the problem at hand.

3)   Sketch a failure circle tangent to the failure envelope constructed in step 1 at the estimated largest value of σ_3 (or σ_1) (see Figure Ex. 1.6(c)) and determine the corresponding point of tangency of the circle on the curved failure envelope.

4)   Determine the *secant ϕ-value*, which is the slope of a straight line connecting the origin and the point of tangency determined in step 3.

For the fill material in this example, the relevant Mohr circle is a circle tangent to the curved failure envelope at $\sigma_3 = 38$ psi, and step 4 would involve drawing a straight line connecting the origin and the point of

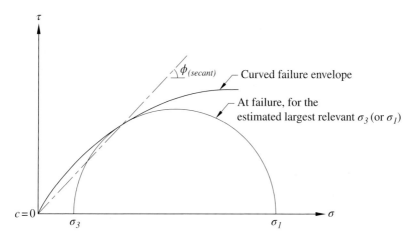

Figure Ex. 1.6(c)

tangency determined in step 3. The secant friction angle for σ_3 in the range of normal stress between 0 and 38 psi is determined to be approximately 53°. For the purposes of analysis, $c' = 0$ and $\phi' = 53°$ should be used. If an upper limit of $\phi' = 40°$ is selected for conservative design (as stipulated by AASHTO (2014) for the design of MSE walls), the design friction angle would be 40° (i.e., use $c' = 0$ and $\phi' = 40°$ in the design).

b)   If the fines content is high and the fines are cohesive, i.e., $c > 0$

For a $c > 0$ soil, the first order of business is to determine the value of cohesion, c. This can be accomplished by conducting at least two triaxial tests at lower confining pressures and using the corresponding Mohr circles at failure to determine the c-value. Alternatively, the c-value can be determined by a *p–q diagram* (see Figure 1.30). The *p–q diagram* for $\sigma_3 = 5$ psi and 15 psi is shown in Figure Ex. 1.6(d). The c-value is determined to be 10.2 psi (or 70 kPa). As a general rule, when selecting confining pressures

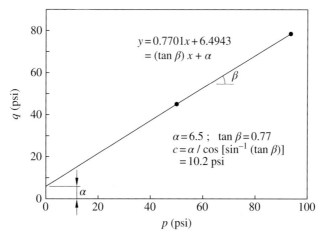

Figure Ex. 1.6(d)

to conduct a series of triaxial tests, it is recommended that at least two "low" values of confining pressure (say, 5 psi and 10 psi) be selected if the fines of the soil may be cohesive.

The remaining task is to determine the ϕ-value associated with $c = 10.2$ psi. The secant ϕ-value can be determined by a procedure similar to the four-step procedure for $c = 0$:

1) Construct a smooth curved failure envelope tangent to all failure Mohr circles that are considered reliable.

2) Estimate the largest relevant value of σ_3 (or σ_1) for the problem at hand.

3) Sketch a failure circle tangent to the failure envelope established in step 1 at the estimated largest value of σ_1 or σ_3 (see Figure Ex. 1.6(e)) and determine the corresponding point of tangency of the circle on the failure envelope.

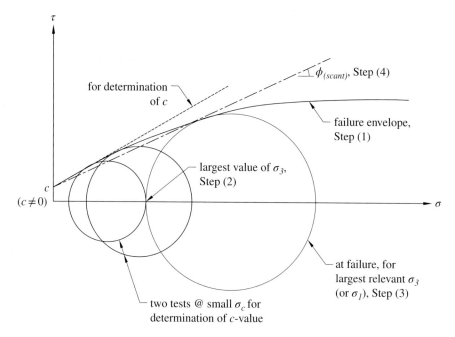

Figure Ex. 1.6(e)

4) Determine the *secant ϕ-value*, which is the slope angle of a straight line connecting the *cohesion intercept* and the point of tangency determined in step 3 (see Figure Ex. 1.6(e)).

Using the procedure described above, the secant ϕ-value for the fill material is determined as 48°. In summary, $c = 10.2$ psi and $\phi = 48°$ should be used in the analysis and design.

Table 1.1 Typical values of internal friction angle (ϕ') for granular soils (Leonards, 1962)

Size of Grain	State of Compaction	Values of ϕ' (Degrees)	
		Rounded Grains Uniform Gradation	Angular Grains Well Graded
Medium sand	Very loose	28–30	32–34
	Moderately dense	32–34	36–40
	Very dense	35–38	44–46
Gravel–sand ratio (%)			
65–35	Loose	–	39
65–35	Moderately dense	37	41
80–20	Dense	–	45
80–20	Loose	34	–
Blasted rock fragments	–	45–55	

Typical values of ϕ' for granular soils are given in Table 1.1. It is seen that ϕ' is affected by the gradation, state of compaction, and particle shapes. Since it is very difficult to obtain undisturbed samples of sands, correlations with results of in situ tests (such as the standard penetration test and the cone penetration test) are often used to determine the ϕ' of granular foundation soils. Example correlations are seen in Figures 1.24(a) and 1.27.

1.5.2 Shear Strength of Clays

Unlike granular soil, the shear strength of clays usually vary significantly before, during, and long after construction. Therefore, the shear strength value of a clay used in design must correspond to the most critical condition, which is often corresponding to the condition immediately after applications of external loads (there are exceptions to this; the exceptions are discussed later in this section). At this time, the shear strength derives nearly entirely from cohesion of the clay (which is unaffected by normal stress, i.e., $\phi = 0$), and can be expressed by a single strength parameter called *undrained shear strength*, commonly denoted by S_u or c_u.

Undrained shear strength is the strength of a clay before application of external loads. Drainage of clays is very limited during application of external loads, therefore the shear strength of the clay will remain the undrained shear strength. As time progresses, however, porewater will slowly drain out of the clay and result in an increase in intergranular stress (i.e., an increase in effective stress), and lead to increasing shear strength and a higher safety margin with time.

The undrained shear strength of a saturated clay can be determined by conducting unconsolidated-undrained (UU) triaxial tests on undisturbed samples representative of the in situ condition. The failure envelope of a *saturated* clay from UU triaxial tests, as

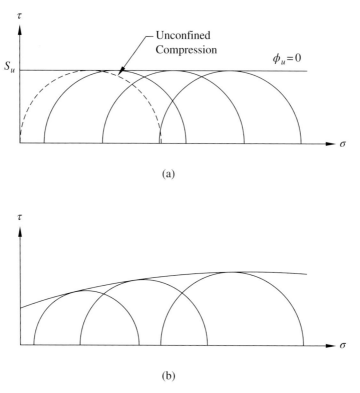

Figure 1.31 Failure envelopes from UU triaxial tests for (a) fully saturated clay and (b) partially saturated clay and fissured clay

seen in Figure 1.31(a), is a horizontal line that can be described as $\tau_f = S_u$ and $\phi_u = 0$. Figure 1.31(b) shows the Mohr–Coulomb failure envelope of a partially saturated clay obtained from UU triaxial tests. Examples of partially saturated clays are compacted clays and clays situated above a steady free water level. Since the failure envelope of a partially saturated clay is usually curved at lower confining pressure, it would be overly conservative to obtain the undrained shear strength by an unconfined compression (UC) test, a special case of the UU triaxial test with zero confining pressure. To determine the strength parameters of a partially saturated clay, a series of UU tests at different confining pressures covering the range of interest of the normal stress should be conducted, and the undrained shear strength should be expressed in terms of c_u and ϕ_u. Just as for dense sands (see Section 1.5.1), a straight secant line over the applicable range of σ_c may be used if the failure envelope is curved. Fissured clays have failure envelopes similar to that of a partially saturated clay (see Figure 1.31(b)). The shear strength parameters c_u and ϕ_u of fissured clays can be obtained by using the same procedure as for partially saturated clays.

When feasible, UU tests should be performed on the vertical, horizontal, and inclined directions of a clay deposit to determine the undrained strengths of the clay in different orientations. This is because most clays are anisotropic (i.e., shear strength is direction dependent). If a test specimen is *manufactured* in the laboratory to simulate the anticipated conditions of a compacted clay, the clay specimen should replicate the void ratio and degree of saturation anticipated in the field.

The undrained shear strength of a clay can also be obtained from other laboratory tests (other than the UU triaxial test) or in situ tests, such as the vane shear test, the Swedish fall-cone test, and the pocket penetrometer test. The undrained shear strength of a clay determined by laboratory tests has been found to be lower than that determined by in situ tests. This is likely due to sample disturbance during sampling operation, sample transport, and laboratory test specimen preparation.

The values of undrained shear strength listed in Table 1.2 have been used to describe the consistency of a clay. Simple field identification methods are also given in the table for on-site estimate of undrained shear strength.

Table 1.2 Qualitative and quantitative expressions for consistency of clays (Peck et al., 1974)

Consistency	Field Identification	Undrained Shear Strength, S_u (lb/ft^2)
Very soft	Easily penetrated several inches by fist	< 250
Soft	Easily penetrated several inches by thumb	250–500
Medium	Can be penetrated several inches by thumb with moderate effort	500–1,000
Stiff	Readily indented by thumb but penetrated only with great effort	1,000–2,000
Very stiff	Readily indented by thumbnail	2,000–4,000
Hard	Indented with difficulty by thumbnail	> 4,000

The consolidated-undrained (CU) triaxial test enables the undrained shear strength of a clay to be determined after the clay specimen has been consolidated under a prescribed consolidation pressure (p) before shear. The results of CU tests, in terms of total stress, can be expressed by an equation relating S_u ($\phi_u = 0$) with the corresponding consolidation pressure p. For normally consolidated clays, the relationship between S_u and p is linear and passes through the origin, as seen in Figure 1.32. For overconsolidated clays, however, the relationship is nonlinear before p exceeds the previous maximum pressure (i.e., preconsolidation pressure, σ_p') of the clay (see Figure 1.32).

The S_u/σ_v' ratio (σ_v' = vertical effective overburden pressure), commonly referred to as the *c–p ratio* or *undrained strength ratio*, for normally consolidated sedimentary clays typically falls between 0.20 and 0.35. Several empirical formulae have been proposed for the S_u/σ_v' ratio of clays; some formulae have the S_u/σ_v' ratio equal to a constant while others have it as a function of plasticity index, liquidity index, or effective friction

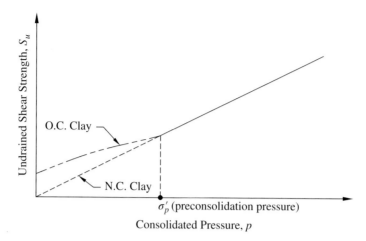

Figure 1.32 Relationship between undrained shear strength and consolidation pressure for normally consolidated and overconsolidated clays

angle. A well-accepted empirical equation of the S_u/σ_v' ratio for normally consolidated clays is given by Skempton (1957) as:

$$\frac{S_{u(VST)}}{\sigma_v'} = 0.11 + 0.0037\,PI \tag{1-11}$$

where $S_{u(VST)}$ is the undrained shear strength obtained from the vane shear test and *PI* is the plasticity index of the clay (expressed as a percentage).

For overconsolidated clays, the S_u/σ_v' ratio has been found to increase with the overconsolidation ratio, OCR, where OCR = (past maximum vertical effective stress)/ (in situ vertical effective stress), and the ratio is typically greater than 0.6. The following two equations have been proposed for overconsolidated clays (Ladd et al., 1977):

$$\left(\frac{S_u}{\sigma_v'}\right)_{O.C.} = \left(\frac{S_u}{\sigma_v'}\right)_{N.C.} (OCR)^{0.8} \tag{1-12}$$

$$\left(\frac{S_u}{\sigma_v'}\right)_{O.C.} = \frac{\sin\phi'}{2}(OCR)^{0.8} \tag{1-13}$$

Ladd (1991) suggests using $(S_u/\sigma_v')_{N.C.} = 0.22$. Jamiolkowski et al. (1985) suggests the S_u/σ_v' ratio be taken as 0.23 ± 0.04 for lightly overconsolidated clays of OCR ≤ 2. Mesri (1989) and Terzaghi et al. (1996) recommend that the S_u/σ_v' ratio be set as 0.22 for stability analysis involving inorganic soft clays and silts. For organic soils, excluding peats, an S_u/σ_v' ratio of 0.26 has been recommended by Terzaghi et al. (1996).

The empirical correlations for the S_u/σ_v' ratio as noted above can be used for preliminary estimate of undrained shear strength S_u as a function of depth for a clay deposit.

Example 1.7 illustrates how to estimate the undrained shear strength as a function of depth for a clay stratum where the OCR can be estimated. It is important to bear in mind that the value of the undrained shear strength of a clay obtained by different test methods may vary significantly (Mayne, 2006). For example, S_u/σ_v' ratio of 0.14 as determined by unconfined compression tests, 0.275 by UU triaxial tests, 0.21 by field vane shear tests, and 0.34 by plane-strain tests have been reported for Boston Blue clay. Use of the empirical correlations must be made judiciously.

Example 1.7   As shown in Figure Ex. 1.7(a), the lowest free water level in the history of a clay deposit is 10 ft below the ground surface (G.S.). The current free water level is at the ground surface.

   a)   Determine your best-estimated variation of the overconsolidation ratio with depth.

   b)   Obtain a conservative profile of the undrained shear strength as a function of depth for the clay deposit.

(modified after Jewell, 1996)

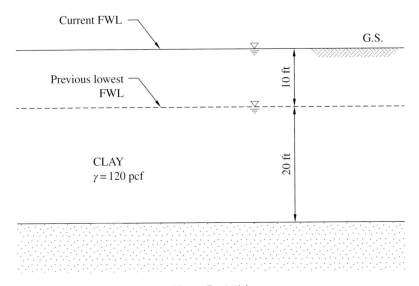

Figure Ex. 1.7(a)

Solution:

   The effective vertical stress profiles for the current free water level (σ_o') and the previous lowest free water level (σ_p') are shown in Figure Ex. 1.7(b). It is assumed that the effective vertical stress for the previous lowest free water level is the highest vertical stress ever experienced by the clay deposit.

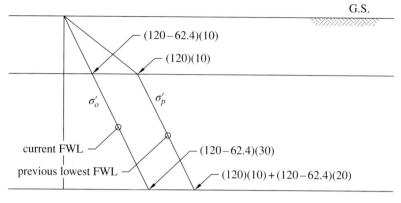

Figure Ex. 1.7(b)

With $OCR = \sigma'_p / \sigma'_o$, the profile of OCR as a function of depth can be determined as shown in Figure Ex. 1.7(c).

Adopting $(S_u / \sigma'_o)_{N.C.} = 0.22$ (a conservative correlation suggested by Ladd, 1991) and Eqn. (1-12), the undrained shear strength at selected depths can be determined as follows.

At a depth of 10 ft,

$$(S_u / \sigma'_o)_{O.C.} = 0.22(2.1)^{0.8} = 0.40; \text{ hence, } S_u = 0.40\left[(120 - 62.4)(10)\right] = 230 \left(\text{lb/ft}^2\right)$$

At a depth of 20 ft,

$$(S_u / \sigma'_o)_{O.C.} = 0.22(1.54)^{0.8} = 0.31; \text{ hence, } S_u = 0.31\left[(120 - 62.4)(20)\right] = 357 \left(\text{lb/ft}^2\right)$$

At a depth of 30 ft,

$$(S_u / \sigma'_o)_{O.C.} = 0.22(1.36)^{0.8} = 0.28; \text{ hence, } S_u = 0.28\left[(120 - 62.4)(30)\right] = 484 \left(\text{lb/ft}^2\right)$$

Variation of S_u with depth is shown in Figure Ex. 1.7(d).

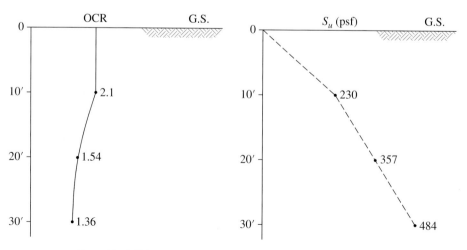

Figure Ex. 1.7(c)

Figure Ex. 1.7(d)

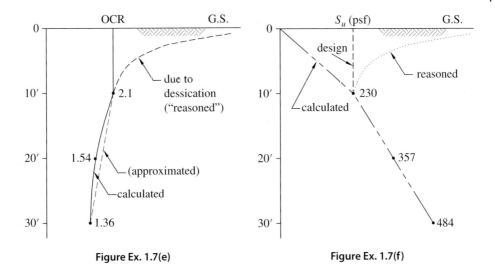

Figure Ex. 1.7(e)   **Figure Ex. 1.7(f)**

If dessication near the ground surface is taken into account, the top 10 ft of the clay deposit will have much higher values of OCR, labelled as the "reasoned" OCR profile in Figure Ex. 1.7(e). The "reasoned" profile of undrained shear stregth would then be much higher than the calculated profile in the top 10 ft (see Figure Ex. 1.7(f)). For design purposes, a constant undrained shear strength in the upper 10 ft (as shown in Figure Ex. 1.7(f)) may be used.

Some clays are quite sensitive to remolding and can lose considerable strength due to disturbance. The ratio of undrained shear strengths in an undisturbed state and in a remolded state (at the same water content) is called *sensitivity*. Sensitivity for most clays is between 2 and 4, and could be as high as 16 or even higher for highly sensitive clays. Possible loss of strength of sensitive clays due to disturbance (e.g., pile driving during construction) should be accounted for in design.

As noted at the beginning of Section 1.5.2, the undrained condition is not always the most critical condition for designs involving clays. There are situations in which shear strength tends to *decrease* with time. Obviously, short-term undrained shear strength will no longer correspond to the most critical condition in these cases. Duncan and Wright (2005) have given a comprehensive list of causes where resisting shear strength increases with time and where shear stress increases with time. Three cases related closely to earth walls where shear strength of clay would reduce with time are as follows:

- Excavation in clay: Upon excavation in clay, the overburden above the new ground surface is removed. The clay underneath the new surface will have a tendency to swell and loses its shear strength with time.

- Rise of free water level in a clay stratum: As a result of a rise of free water level (usually as a result of snow melt or prolong heavy rainfall), the effective stress in the zone where rise of ground water occurred will reduce from moist unit weight to submerged unit weight, hence there will be a gradual decrease in shear strength with time. Rise of free water level usually occurs very slowly in clay. However, this is not the case with a stiff clay stratum where fissures are often present. Water can percolate into the fissures in stiff clays and gradually soften the clay and reduce shear strength with time.

- Non-durable geo-materials: Clay shale and clay stones have sometimes been broken up and used as fill material or foundation material for earth walls. The material can be compacted into seemingly stable fill. Over time, however, if wetted by ground water, the fill material will soften and the shear strength will reduce considerably.

We shall examine more closely the case of excavation in clay which is involved occasionally in construction of earth walls. Let us consider an excavation in a moderately overconsolidated clay where the excavation is being carried out fairly rapidly at an approximately uniform rate, and that no drainage of the clay occurs during excavation. Figure 1.33(a) shows the configuration of excavation and the free water levels (FWL) before and after excavation. Figure 1.33(b) shows the change in height of the soil above point P and the change in shear stress on a potential slip plane passing through point P. The shear stress is seen to increase linearly with time during excavation and becomes a constant thereafter. Figure 1.33(c) shows the porewater pressure at point P. It is seen that the initial and final values of porewater pressure are dictated by the original and final free water levels. Between the two, as a result of reduced load, the soil will swell and cause a reduction in porewater pressure during excavation, then try to reacquire equilibrium. Since there is little drainage during excavation, the shear strength will remain essentially constant. After excavation is completed, however, the strength will decrease with time because of swelling of the clay due to load reduction, as shown in Figure 1.33(d). Two observations can be made: (i) during excavation, the shear stress increases with time while the shear strength remains constant, and (ii) after excavation is completed, even though the shear stress is constant, the shear strength reduces with time, hence there is a continued decrease in the factor of safety, as seen in Figure 1.33(e). For a cut in clay, the factor of safety will be a minimum after the porewater pressure has increased to the hydrostatic value, which will occur long after excavation has been completed. Hence, use of the undrained shearing resistance before excavation to evaluate the stability of a cut slope would be unsafe. It is therefore necessary to use long-term strength, i.e., *effective* strength parameters, for stability analysis.

For the purposes of comparison, similar plots for an embankment constructed over a saturated moderately overconsolidated clay foundation are shown in Figure 1.34. In this case, the soil will begin consolidation right after the embankment construction is completed. The porewater pressure will dissipate with time and lead to increasing shear strength with time. As a result, the most critical condition is at the end of embankment construction. Because the shearing resistance of the soil is approximately constant during construction, we can assume the shear strength at the end of construction is the same as that before construction. Therefore, the undrained shear strength of the clay foundation should be used for stability analysis.

In the cases where drained shear strength should be used in design (such as the three cases described above), the drained strength of saturated clays can be determined either by conducting CU triaxial tests with porewater pressure measurement or by conducting CD triaxial tests. The value of c' is zero for normally consolidated clays

(a) Slope Considered

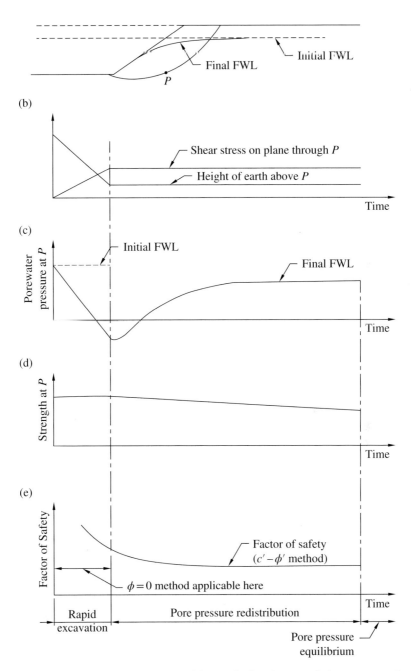

Figure 1.33 Changes in porewater pressure and factor of safety during and after a cut is made in a moderately overconsolidated clay (modified after Bishop and Bjerrum, 1960)

(a) Slope Considered

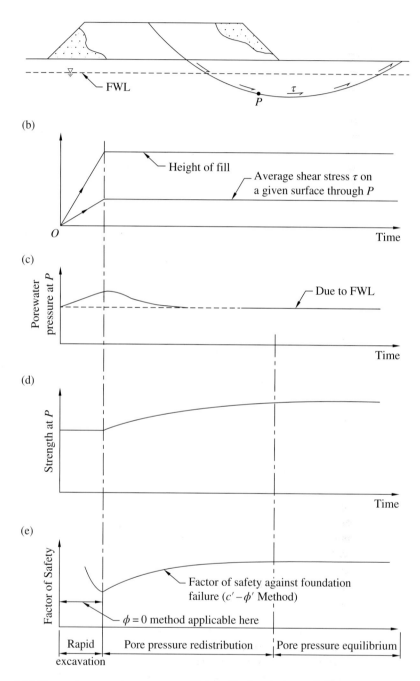

Figure 1.34 Change in porewater pressure and factor of safety during and after construction of an embankment over a saturated moderately overconsolidated clay foundation (modified after Bishop and Bjerrum, 1960)

and greater than zero for overconsolidated clays (see Figure 1.35). The value of c' of overconsolidated clays increases with the degree of overconsolidation and is usually less than 600 lb/ft² (29 kPa), whereas the value of ϕ' generally falls between 20° and 35°, with higher values being associated with lower plasticity index. Example 1.8 illustrates how to determine the strength parameters for the stability analysis of a cut slope.

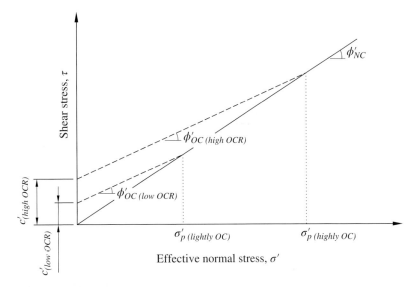

Figure 1.35 Effective strength envelopes for normally consolidated and overconsolidated clays (modified after Duncan and Wright, 2005)

Example 1.8 A highway cut is to be made in a medium stiff overconsolidated clay deposit. The results of two consolidated-undrained (CU) triaxial tests conducted on representative undisturbed samples are as shown in the table below:

Test No.	$\sigma'_c = \sigma_{3f}$ (psf)	$(\sigma'_1 - \sigma_{3f})$ (psf)	u_f (psf)
1	250	744	−58
2	1,000	984	+454

Determine the Mohr–Coulomb strength parameters that should be used for stability analysis of the cut.

(modified after Perloff and Baron, 1976.)

Solution:

The stability of the cut will decrease with time as expansion of the clay at the bottom of the cut will occur in response to the operation of excavation. Hence, we should evaluate the long-term stability of the cut with effective strength parameters. From CU triaxial compression test results, total and effective Mohr circles at failure and the corresponding Mohr–Coulomb failure envelopes can be determined, as shown in Figure Ex. 1.8.

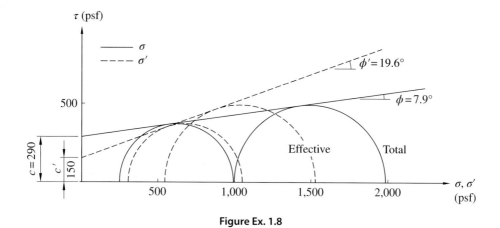

Figure Ex. 1.8

The effective strength parameters $c' = 150\,\text{lb/ft}^2$ and $\phi' = 20°$ should be used for stability analysis of the cut.

As noted earlier in this section, stiff clays are generally fissured. The spacing of fissures may vary from 5 mm near the ground surface to more than 50 mm at depth 30 m below the ground surface (Terzaghi et al., 1996). The orientation of the fissures is generally random. These fissures form localized weak planes in a stiff clay. When a small-size test specimen is used to determine the shear strength of a stiff clay, very different strength values are often obtained from different test specimens. For a test specimen that contains no fissure (such as sample A in Figure 1.36), the undrained shear strength will likely be quite high (say, $3{,}000\,\text{lb/ft}^2$ or 140 kPa); on the other hand, for a test specimen containing fissures (such as sample B in Figure 1.36), the undrained strength will likely be very low (say, $150\,\text{lb/ft}^2$ or 7 kPa). When determining the undrained shear

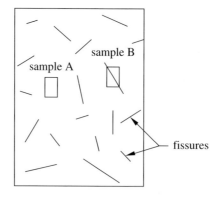

Figure 1.36 Schematics of two samples taken from an idealized large mass of fissured stiff clay

strength of a stiff clay, large-size specimens should be tested so that the strength values will represent the global strength of the clay. When it is difficult to test large-size specimens, an average value of shear strengths determined from a large number of tests with small-size specimens may be used as an alternative (remember not to throw away test data that appear odd!).

1.5.3   Shear Strength of Silts

Silts are typically non-plastic and behave in similar way to very fine sands. However, some silts are plastic and behave like clays. An example of plastic silt is San Francisco Bay mud, which has a liquid limit near 90, a plasticity index of about 45, and is classified as MH per the Unified Soil Classification System. Determination of the shear strength of a silt therefore may follow that of granular soils or that of clays, depending on whether it is non-plastic or plastic. Also, in terms of particle size and engineering behavior, silts fall between sands and clays. It is often difficult to anticipate whether a silt is better analyzed in a drained condition (as for sands) or an undrained condition (as for clays). When there is doubt over whether a silt should be analyzed drained or undrained, it is best to analyze both conditions to cover the range of possibilities.

For compacted silts, the same laboratory tests used for clays may be used to determine the shear strength of silts. Note that silts are generally moisture sensitive and their compaction characteristics are similar to those of clays. The undrained shear strength of silts has been found to be influenced strongly by water content. Unlike clays, however, reliable correlations are not available for the estimation of the undrained shear strengths of plastic silts. Data for S_u/σ_v' for different silts have been found to vary widely (Duncan and Wright, 2005).

References

AASHTO (2014). *LRFD Bridge Design Specifications*, 7th edition with 2016 interims. American Association of State Highway and Transportation Officials, Washington, D.C.

Bishop, A.W. and Bjerrum, L. (1960). The Relevance of the Triaxial Test to the Solution of Stability Problems. *Proceedings, ASCE Research Conference on the Shear Strength of Cohesive Soils, Boulder, Colorado.*

Bishop, A.W. and Henkel, D.J. (1962). *The Measurement of Soil Properties in the Triaxial Test*, 2nd edition. Edward Arnold Ltd., London, 228 pp.

Duncan, J.M. and Wright, S.G. (2005). *Soil Strength and Slope Stability*. John Wiley and Sons, Hoboken, New Jersey, 297 pp.

Fung, Y.C. (1977). *A First Course in Continuum Mechanics*, 2nd edition. Prentice Hall, New Jersey.

Head, K.H. and Epps, R. (editors) (1982). *Manual of Soil Laboratory Testing, Volume II*, 3rd edition. Whittles Publishing, Dunbeath, Caithness.

Holtz, R.D., Kovacs, W.D., and Sheahan, T.C. (2011). *An Introduction to Geotechnical Engineering*, 2nd edition. Prentice Hall, New Jersey.

Hunt, R.E. (1986). *Geotechnical Engineering Analysis and Evaluation*. McGraw-Hill Book Company, 729 pp.

Jamiolkowski, M., Ladd, C.C., Germaine, J.T., and Lancellotta, R. (1985). New Developments in Field and Laboratory Testing of Soils. *Proceedings, 11th International Conference on Soil Mechanics and Foundation Engineering, San Francisco, Volume 1*, pp. 57–153.

Japanese Foundation Engineering Society (1995). *Introduction to Soils and Foundations*, Series 21, 330 pp.

Jewell, R.A. (1996). *Soil Reinforcement with Geotextiles. Construction Industry Research and Information Association*. CIRIA Special Publication 123, London, 332 pp.

Kovacs, W.D., Salomone, L.A., and Yokel, F.Y. (1981). *Energy Measurements in the Standard Penetration Test*. Building Science Series 135, National Bureau of Standards, Washington, D.C.

Kulhawy, F.H. and Mayne, P.W. (1990). *Manual on Estimating Soil Properties for Foundation Design*. Electric Power Research Institute, Palo Alto, California.

Ladd, C.C. (1991). Stability Evaluation during Staged Construction. Terzaghi Lecture. *Journal of Geotechnical Engineering, ASCE*, 117(4), 540–615.

Ladd, C.C., Foott, R., Ishihara, K., Schlosser, F., and Poulos, H.G. (1977). Stress-deformation and strength characteristics. *Proceedings, 9th International Conference on Soil Mechanics and Foundation Engineering, Tokyo*, pp. 421–494.

Lade, P.V. (2005). *Overview of Constitutive Models for Soils*. ASCE Geotechnical Special Publication No. 139, Calibration of Constitutive Models, GeoFrontiers, Austin, Texas, pp. 1–34.

Leonards, G.A. (1962). Engineering Properties of Soils. In *Foundation Engineering*, McGraw-Hill, New York.

Mayne, P.W. (2006). *In-Situ Test Calibrations for Evaluating Soil Parameters*. Characterization and Engineering Properties of Natural Soils, Singapore Workshop.

Mayne, P.W. (2007). *Cone Penetration Testing*. NCHRP Synthesis 368, Transportation Research Board, Washington, D.C.

Mesri, G. (1989). A Re-evaluation of $S_{u(mob)} = 0.22\,\sigma_p'$ Using Laboratory Shear Tests. *Canadian Geotechnical Journal*, 26(1), 162–164.

NAVFAC (1986). *Design Manual 7.02 Foundations and Earth Structures*. Bureau of Yards and Docks, U.S. Navy.

Peck, R.B., Hanson, W.E., Thornburn, T.H. (1974). *Foundation Engineering*. John Wiley and Sons, New York, 514 pp.

Perloff, W.H. and Baron, W. (1976). *Soil Mechanics Principles and Applications*. John Wiley & Sons, New York, 745 pp.

Robertson, P.K. and Campanella, R.E. (1983). Interpretation of Cone Penetration Tests, Part II: Clay. *Canadian Geotechnical Journal*, 20(4), 734–745.

Sanglerat, G. (1972). *The Penetrometer and Soil Exploration: Interpretation of Penetration Diagrams – Theory and Practice*. Transportation Research Board, National Academy of Sciences, Washington, D.C., 488 pp.

Schmertmann, J.H. (1978). *Guidelines for Cone Penetration Test Performance and Design*. Report No. FHWA-TS-78-209, Federal Highway Administration, Washington, D.C.

Seed, H.B., Tokimatsu, K. Harder, L.F., and Chung, R.M. (1985). Influence of SPT Procedures in Soil Liquefaction Resistance Evaluations. *Journal of Geotechnical Engineering, ASCE*, 111(12), 1425–1445.

Skempton, A.W. (1957). Discussion: Further Data on the c/p Ratio in Normally Consolidated Clays. *Proceedings, Institution of Civil Engineers*, No. 7, pp. 305–307.

Skempton, A.W. (1986). Standard Penetration Test Procedures and the Effects in Sands of Overburden Pressure, Relative Density, Particle Size, Aging and Overconsolidation. *Geotechnique*, 36(3), 425–447.

Tatsuoka, F., Molenkamp, F., Torii, T., and Hino, T. (1984). Behaviour of Lubrication Layers of Platens in Element Tests. *Soils and Foundations*, 24(1), 113–128.

Terzaghi, K. (1923) *Erdbaumechanik auf Bodenphysikalischer Grundlage*. Franz Deuticke, Liepzig/Vienna.

Terzaghi, K. and Peck, R.B. (1967). *Soil Mechanics in Engineering Practice*. John Wiley & Sons, New York, 729 pp.

Terzaghi, K., Peck, R.B., and Mesri, G. (1996). *Soil Mechanics in Engineering Practice*, 3rd edition. John Wiley & Sons, New York, 549 pp.

2

Lateral Earth Pressure and Rigid Earth Retaining Walls

When a significant change in grade is needed or desired in earthwork construction, one of four methods can be taken: (i) provide a stable slope, about 2:1 (horizontal:vertical) for granular soil without any concern of seepage (and about 4:1 with concern of seepage), to allow for gradual transition between the grades, (ii) construct a stable steepened slope with geosynthetic inclusion near the face of the slope to allow tighter transition between the grades, (iii) construct an earth retaining wall to allow an abrupt change in grade, and (iv) a combination of the above measures. The choice of method depends largely on the availability of space between the grades. An earth retaining structure is the method of choice when there is severe space constraint.

Earth retaining walls can be grouped into two categories: rigid retaining walls (gravity walls, cantilever concrete walls, crib walls, etc.) and flexible retaining walls (cantilever sheet pile walls, anchored sheet pile walls, etc.). Design and analysis for the former is typically based on classical earth pressure theories, while the latter is based on empirical pressure diagrams and charts. In this chapter, we shall focus our attention on rigid retaining walls. Compared to "reinforced soil walls" where the soil mass is *internally stabilized* by embedding tensile inclusion within the soil mass, the soil mass in "rigid retaining walls" (and "earth retaining walls" in general) is *externally stabilized* by structural members constructed in front of the soil mass to retain the soil.

We shall begin the chapter with an explanation of the *at-rest earth* condition, to wit, a condition where there is no lateral deformation or lateral displacement in a soil mass. We shall then discuss two classical earth pressure theories (namely, Rankine analysis and Coulomb analysis) and applications of the theories. We shall conclude the chapter with a discussion of a few additional topics related to the design of rigid retaining walls.

For both Rankine and Coulomb analyses, the influence of submergence, seepage, relative wall movement, and seismic forces will be discussed in some detail. In-depth understanding of Rankine and Coulomb analysis methods will help us gain a better understanding of lateral earth pressure theories and the design of rigid earth retaining walls.

Geosynthetic Reinforced Soil (GRS) Walls, First Edition. Jonathan T.H. Wu.
© 2019 John Wiley & Sons Ltd. Published 2019 by John Wiley & Sons Ltd.

2.1   At-Rest Earth Pressure

Certain earth retaining structures do not allow the soil behind to deform or displace laterally. Examples of such retaining structures include rigid basement walls of buildings, rigid bridge abutment walls, and rigid concrete box culverts. For these structures, the soil behind the structure typically experience neither lateral deformation nor lateral displacement. Such a stress/deformation state is referred to as the *at-rest condition* or K_0-*condition*, and the lateral pressure in such a state is termed the *earth pressure at-rest*. In other words, the at-rest earth pressure refers to the lateral pressure against an unyielding and non-displaced structure.

The ratio between horizontal and vertical effective normal stresses in an at-rest condition is called the *coefficient of earth pressure at rest*, commonly denoted by K_0, i.e.,

$$K_0 = \left. \frac{\sigma'_x}{\sigma'_z} \right]_{@\,zero\,lateral\,deformation\,and\,zero\,displacement} \tag{2-1}$$

where σ'_x and σ'_z are the effective lateral stress and the effective vertical stress, respectively.

Figure 2.1 shows the Mohr circle of stresses in a granular soil ($c = 0$) at depth z below a level crest under a K_0-condition. Being under a K_0-condition, the soil experiences no lateral deformation or lateral displacement, hence cannot mobilize its shearing strength (i.e., cannot reach a failure condition). The Mohr circle, therefore, cannot *touch* the Mohr–Coulomb failure envelope. Note that in a closely spaced geosynthetic reinforced soil mass (which is the subject of interest in Chapters 3 to 6), if the frictional resistance at the soil–geosynthetic interface can provide perfect restraint to lateral deformation and if the geosynthetic reinforcement is infinitely stiff, the reinforced soil mass will be in a K_0-condition, hence will never reach a failure condition. With limited interface bonding and finite reinforcement stiffness, however, a closely spaced reinforced soil mass can *approach* a K_0-condition.

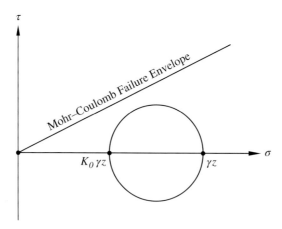

Figure 2.1 Mohr circle of stress at a depth z in a granular soil mass with a horizontal crest and under the K_0-condition (for $K_0 < 1$)

Since a failure condition cannot be developed in a K_0-condition, the value of K_0 cannot be determined by equations that describe stresses at failure, i.e., a limiting equilibrium analysis is of no help for determination of K_0. In the literature, the value of K_0 has been determined by laboratory or in situ tests. Laboratory measurement has been made by a modified consolidometer or a modified triaxial test apparatus. In situ measurements have also been made with self-boring pressuremeters, dilatometers, and Glötzl-type thin wall earth pressure cells. In design of earth retaining walls, the value of K_0 is usually obtained by empirical correlations.

The magnitude of K_0 for non-prestressed sands and normally consolidated clays has been reported to be in the range of 0.35 to 0.5. Available test data have indicated that K_0 for these soils can be related to the effective Mohr–Coulomb strength parameter ϕ' (Jaky, 1948) as:

$$K_{0(NC)} = 1 - \sin\phi' \tag{2-2}$$

The values of K_0 for in situ granular and cohesive soil deposits have been correlated with standard penetration blow count (N) and a normalized N-value (N_{pa}/σ'_{vo}), as shown in Figure 2.2, and can be expressed as:

$$K_0 = 0.52 - 0.01(N) \tag{2-3}$$

$$K_0 = 0.43 + 0.059\left(\frac{N_{pa}}{\sigma'_{vo}}\right) \tag{2-4}$$

The scatter of data seen in Figure 2.2 (coefficients of determination, $R^2 = 0.465$ for Eqn. (2-3); $R^2 = 0.841$ for Eqn. (2-4)) suggests that these empirical equations are largely simplified correlations. Other equations for correlations of K_0 with in situ tests have been proposed, but they have been found to give rather different values of K_0. Since all the equations are empirically based, they are applicable to the database from which the correlations were derived. It is important to bear in mind that the large differences of K_0-values determined by different correlations suggest that none of the correlations are accurate in general. They have, however, been used in routine designs for convenience.

For prestressed soils, including compacted granular soils and overconsolidated clays, the K_0-value has been found to range from 0.5 to 4.0 (and even higher for heavy machine compacted clays). The value of K_0 is influenced significantly by stress history, commonly expressed in terms of overconsolidation ratio (OCR). The value of K_0 is larger for soils of higher OCR. Mayne and Kulhawy (1982) proposed a correlation for clays of low to medium sensitivity and uncemented sands as:

$$K_0 = (1 - \sin\phi')(OCR)^{\sin\phi'} \tag{2-5}$$

In Eurocode 7, use of the equation below is suggested:

$$K_0 = (1 - \sin\phi')\sqrt{OCR} \tag{2-6}$$

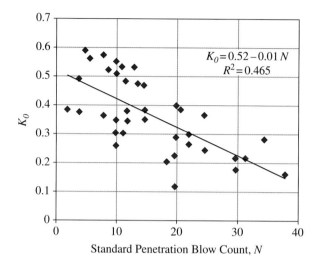

(a)

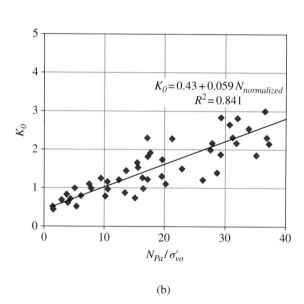

(b)

Figure 2.2 Correlations between K_0 and (a) standard penetration blow count, N, for sands (Eqn. (2-3)) (modified after Yasukawa et al., 2012) and (b) normalized N-value (Eqn. (2-4)) (modified after Kulhawy et al., 1989)

Brooker and Ireland (1965) provided an example of how K_0-value may vary with OCR and the plasticity index (*PI*) of clays, as shown in Figure 2.3. Eqns. (2-5) and (2-6), and Figure 2.3 have been used in routine designs involving prestressed soils.

Typical values of K_0 are given in Table 2.1. Values of K_0 suggested by a number of researchers and agencies are given in Table 2.2.

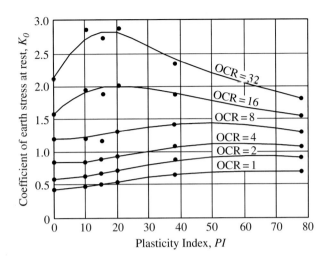

Figure 2.3 Correlations between K_0 and plasticity index (*PI*) for clay of different overconsolidation ratios (OCRs) (Brooker and Ireland, 1965)

Table 2.1 Typical range of K_0-values for different soil types and conditions

Soil	Condition	K_0
In situ granular deposit	Loose	0.5–0.6
	Dense	0.3–0.4
In situ clay deposit	Normally consolidated	0.4–0.6
	Overconsolidated	1.0–3.0
Granular backfill	End dumped	0.5–0.6
	Machine compacted	1.0–1.5
Clay backfill	Hand tempted	1.0–2.0
	Machine compacted	2.0–6.0

Table 2.2 K_0-values proposed by some researchers/agencies (modified after Yasukawa et al., 2012)

	Dense Sand	Loose Sand	Normally Consolidated Clay	Overconsolidated Clay
Bishop (1958)	0.36	0.46	0.43–0.57	
Sowers and Sowers (1979)	0.4	0.6	0.8–1.0	
Teng (1962)	0.35	0.6	0.45–0.75	1.0
Henry (1986)	0.4	0.6	0.5–0.75	1.0
Hunt (1986)	1.0–1.5	0.5	$1-\sin \phi'$	1.0–4.0
Bowles (1996)	0.4	0.6	0.4	0.8
Swedish Road Administration (2004)	0.35	0.55	0.4–0.7	

2.2 Rankine Analysis

Two classical earth pressure theories have commonly been used for the analysis and design of earth retaining structures: Rankine theory and Coulomb theory. In Rankine analysis, *every point* in the soil mass behind a retaining structure is assumed to be in a state of *limiting equilibrium*, i.e., the state of stress reaches Mohr–Coulomb shear failure criterion everywhere in the soil mass behind a retaining structure. Coulomb analysis, on the other hand, assumes that shear failure occurs only along a single planar failure surface in the soil mass behind a retaining structure. In Rankine analysis, the earth pressure acting on a retaining structure in a limiting equilibrium condition is determined mathematically or graphically. In Coulomb analysis, the *resultant force* acting on a retaining structure is determined by the equilibrium of resultant forces on soil wedges formed by (i) an assumed failure surface, (ii) the wall crest, and (iii) the backface of the retaining structure, under a limiting equilibrium condition. We shall discuss Rankine analysis in this section. Coulomb analysis is discussed next in Section 2.3.

2.2.1 Active and Passive Conditions and Graphical Solution

Rankine analysis can be performed graphically or mathematically. We shall begin with graphical analysis in this section, followed by mathematical analysis in Section 2.2.2. In order to understand how *limiting equilibrium* develops in a soil mass, let us consider the stress condition of an idealized scenario: a large mass of soil with a horizontal crest behind a rigid, perfectly smooth vertical wall of infinite depth (see Figure 2.4); the soil is assumed to be isotropic (i.e., its properties are *orientation-independent*) and homogeneous (i.e., its properties are *location-independent*). The initial state of stress at any depth z in the soil mass is assumed to be in an at-rest condition, as shown by circle ① in Figure 2.5 (assuming $K_0 < 1$). Note that the vertical stress (σ_z) and horizontal stress (σ_x), both due to the self-weight of the soil, are principal stresses because the crest is horizontal and there is no interface friction between the wall and the soil, thus there is no shear stress on the horizontal or vertical planes.

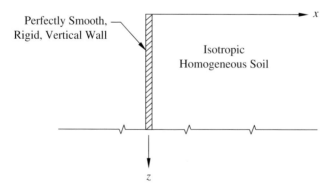

Figure 2.4 An isotropic homogeneous soil mass with a horizontal crest behind a frictionless rigid vertical wall of infinite depth

Now, let's visualize the rigid, perfectly smooth wall is moving horizontally to the left (away from the soil mass) due to the self-weight of the soil, and examine how the stress at the depth z will change as a result. As the wall is moving away from the soil mass, the lateral stress σ_x at a depth z in the soil will decrease in response to the movement, while the vertical stress σ_z (due to soil weight at that depth) will remain practically unchanged. In terms of Mohr circles, the decrease in σ_x will lead to a larger Mohr circle, as shown by circle ② in Figure 2.5. When the movement reaches a certain magnitude (the movement needed is addressed in Item 8 of Section 2.2.7), the soil will reach a failure state (a limiting equilibrium state), and the Mohr circle will touch the Mohr–Coulomb failure envelope, as shown by circle ③ in Figure 2.5, known as the *Rankine active failure* condition. It is called "active" because this mode of failure typically occurs due to the self-weight of the soil plus (if present) external loads on the crest in which the soil assumes an active role (i.e., soil is *pushing* the wall to a failure state).

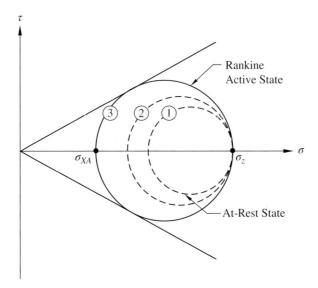

Figure 2.5 Progressive development of a Rankine active condition in terms of Mohr circles (numbers on Mohr circles indicate sequence of development)

If there is an *external* horizontal force applied to the wall to move it toward the soil mass (i.e., toward the right of the diagram), the lateral stress σ_x in the soil mass will increase in response to the movement, while the vertical stress σ_z will again remain practically unchanged. The change in Mohr circle in response to increasing σ_x is shown in Figure 2.6. Note that the size of the Mohr circle due to the movement will become smaller at first, becomes a point (at which time $\sigma_x = \sigma_z$), then grows larger (as σ_x becomes increasingly larger than σ_z) and eventually become large enough to touch the Mohr–Coulomb failure envelope. Such a limiting condition is known as the *Rankine passive failure* condition. In this mode of failure, soil is being *pushed* by the wall to failure (i.e., soil assumes a passive role), thus the term *passive*. Note that in order to develop a passive condition, there must be an external force applied to the wall.

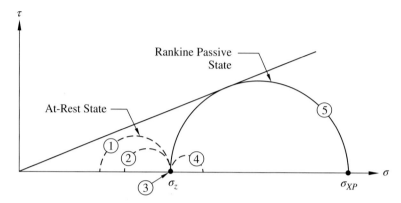

Figure 2.6 Progressive development of a Rankine passive condition in terms of Mohr circles (numbers on Mohr circles indicate sequence of development)

2.2.2 Mathematical Solution

We shall now examine the mathematical solution for Rankine active and passive conditions to help us gain a better understanding of Rankine analysis. The state of stresses of a continuum in a static condition must satisfy the *equations of static equilibrium*. Consider a two-dimensional condition and assume that the only force distributed within the soil mass (called the "body force") is the self-weight, the equations of static equilibrium can be written as:

$$\frac{\partial \sigma_x}{\partial x} + \frac{\partial \tau_{xz}}{\partial z} = 0$$

$$\frac{\partial \tau_{xz}}{\partial x} + \frac{\partial \sigma_z}{\partial z} = \gamma$$

(2-7)

where γ is the unit weight of the soil.

Eqn. (2-7) has three unknowns (σ_x, σ_z and τ_{xz}) with only two equations. A complete closed-form solution to Eqn. (2-7) does not exist. However, solutions to Eqn. (2-7) can be obtained for a limited set of problems with specific conditions. A set of the specific conditions could be as follows: the wall under consideration is vertical, rigid, perfectly smooth, and of infinite depth; the soil is homogeneous and isotropic; and the crest is horizontal and subjected to uniformly distributed inclined surcharge over the entire crest, as shown in Figure 2.7. Under these conditions, there would be no variation of stresses in the x-direction. Eqn. (2-7) then reduces to

$$\frac{d\sigma_z}{dz} = \gamma$$

$$\frac{d\tau_{xz}}{dz} = 0$$

(2-8)

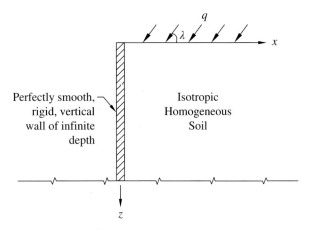

Figure 2.7 Conditions for mathematical solution of Rankine analysis: a perfectly smooth, rigid, vertical, and infinitely deep wall, homogeneous and isotropic soil, and subjected to a uniform inclined surcharge over the crest

Upon integration of Eqn. (2-8) and substitution of boundary conditions that $\sigma_z = q \sin \lambda$ and $\tau_{xz} = q \cos \lambda$ at $z = 0$, we obtain

$$\sigma_z = \gamma z + q \sin \lambda$$
$$\tau_{xz} = q \cos \lambda$$

$$(2\text{-}9)$$

Eqn. (2-9) provides a "partial" solution to Eqn. (2-7); however, our primary interest is to determine the lateral stress σ_x, which is not part of the solution given by Eqn. (2-9). Since this cannot be accomplished by the equations of equilibrium alone, we need to introduce another equation — the Mohr–Coulomb failure criterion which correlates σ_x with σ_z and τ_{xz} at failure. For a granular soil ($c = 0$), the Mohr–Coulomb failure criterion can be expressed as:

$$\left(\frac{\sigma_z - \sigma_x}{2} \right)^2 + \tau_{xz}^2 = \left(\frac{\sigma_x + \sigma_z}{2} \right)^2 \sin^2 \phi$$

$$(2\text{-}10)$$

Because of the quadratic nature of Eqn. (2-10), we can expect to have two solutions for σ_x. The solutions can be obtained by substituting Eqn. (2-9) into Eqn. (2-10) and solving for σ_x as:

$$\sigma_x' = \sigma_z' \left(\frac{1 + \sin^2 \phi'}{1 - \sin^2 \phi'} \right) - \frac{2 \sin \phi'}{1 - \sin^2 \phi'} \sqrt{ \sigma_z'^2 - \tau_{xz}^2 \left(\frac{1 - \sin^2 \phi'}{\sin^2 \phi'} \right) }$$

or

$$(2\text{-}11)$$

$$\sigma_x' = \sigma_z' \left(\frac{1 + \sin^2 \phi'}{1 - \sin^2 \phi'} \right) + \frac{2 \sin \phi'}{1 - \sin^2 \phi'} \sqrt{ \sigma_z'^2 - \tau_{xz}^2 \left(\frac{1 - \sin^2 \phi'}{\sin^2 \phi'} \right) }$$

As it turns out, the smaller of the two σ_x-values in Eqn. (2-11) is the Rankine active lateral pressure, and the larger is the Rankine passive earth pressure. Note that the two solutions of σ_x in Eqn. (2-11) are obtained by combining the *equilibrium* equations with an equation that describes the at-failure condition (a *limiting* condition). Such a solution method is referred to as the *limiting equilibrium method*.

For the condition of a vertical wall with a horizontal crest and subjected to vertical surcharge (i.e., $\lambda = 90°$, thus $\sigma_z = \gamma z + q$, $\tau_{xz} = 0$), Eqn. (2-11) reduces to

$$\sigma_{x\,(active)} = (\gamma z + q) \tan^2\left(45° - \frac{\phi}{2}\right)$$

$$\sigma_{x\,(passive)} = (\gamma z + q) \tan^2\left(45° + \frac{\phi}{2}\right)$$

(2-12)

or simply,

$$\sigma_{x\,(active)} = (\gamma z + q) K_A$$

$$\sigma_{x\,(passive)} = (\gamma z + q) K_P$$

(2-13)

where $K_A = \tan^2(45 - \phi/2) = (1 - \sin\phi)/(1 + \sin\phi)$ is termed the *coefficient of active earth pressure*; and $K_P = \tan^2(45 + \phi/2) = (1 + \sin\phi)/(1 - \sin\phi)$ is termed the *coefficient of passive earth pressure*.

Note that the same values of σ_x as those shown in Eqn. (2-13) can be obtained from the limiting Mohr circles seen in Figures 2.5 and 2.6, based entirely on geometric consideration. This confirms that the two solutions of σ_x given by Eqn. (2-13) are indeed the Rankine active and passive earth pressures.

For a cohesive soil ($c \neq 0$), the limiting lateral earth pressures in Eqn. (2-13) become

$$\sigma_{x\,(active)} = (\gamma z + q) K_A - 2c\sqrt{K_A}$$

$$\sigma_{x\,(passive)} = (\gamma z + q) K_P + 2c\sqrt{K_P}$$

(2-14)

Eqn. (2-14) reveals that tension cracks (which occur when $\sigma_x < 0$) would form for $c > 0$ soils in the active condition. The depth of tension cracks can be determined by setting the $\sigma_{x\,(active)}$ of Eqn. (2-14) equal to zero. For $q = 0$, the depth of tension crack $z_t = 2c/(\gamma\sqrt{K_A})$. Tension cracks, when present, will reduce the effective contact length between soil and wall, and should be taken into consideration in analysis and design.

Eqns. (2-13) and (2-14) have often been used in situations where they should not. It is important to keep in mind the conditions to which Eqns. (2-13) and (2-14) apply are: (i) the backfill is uniform, (ii) the backface of the wall is vertical, (iii) the crest is horizontal, (iv) the surcharge q is vertical and uniformly applied over the crest, and (v) the wall has a tendency to rotate about its base. The last point about wall movement is addressed in Item 7 of Section 2.2.7.

Example 2.1   A bulldozer operator found out that the maximum slope he could negotiate was 45° (see Figure Ex. 2.1(a)). The bulldozer weighs 4 tons and the blade width is 6 ft. If a gauge registering forces is placed behind the bulldozer blade (see Figure Ex. 2.1(b)):

a)   Estimate the maximum force (F_{max}) that would be registered if the dozer continues into a cut of height H = 5 ft.

b)   Estimate the maximum height of cut (H_{max}) the dozer could move.

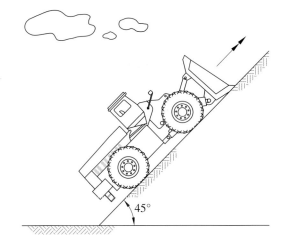

Figure Ex. 2.1(a)

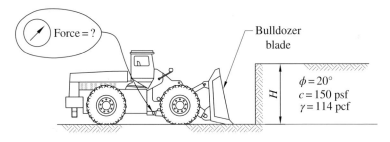

Figure Ex. 2.1(b)

Assume in (a) and (b) that the blade is high enough to accommodate any H and all metal parts are smooth enough that the adhesion between soil and blade can be ignored.

(source unknown)

Solution:

The maximum slope that can be negotiated by the bulldozer is 45°. This implies that when the bulldozer is moving up a 45° slope, the driving force of the bulldozer is approximately equal to the component of its weight parallel to the slope.

$$F_{max} = W\left(\sin 45°\right) = 8,000\left(\sin 45°\right) = 5,600\,(\text{lb})$$

Assuming the resistance of soil can be evaluated by Rankine passive resistance, from Eqn. (2-14), the force F_{max} is

$$F_{max} = \left(\frac{1}{2}K_p \gamma H^2 + 2c\sqrt{K_p}\,H\right)(6') = 5,660\,(\text{lb})$$

For γ = 114 pcf, c = 150 pcf, and ϕ = 20°,

$$H_{max} = 1.55 \text{ ft or } 18.5 \text{ in.}$$

Note: The above calculations ignore the side friction on the failure wedge moved by the bulldozer.

Example 2.2 A 10-ft high retaining wall supporting a dry cohesionless backfill is inclined at 10° from the vertical. Two cases, as shown in Figure Ex. 2.2(a), are considered. The backfill has a unit weight of 110 lb/ft³ and strength parameter $\phi = 35°$.

  a)   Use Rankine analysis to determine for both cases the magnitude, orientation, and location of the resultant force acting on the wall.

  b)   Although Rankine analysis considers neither strain nor displacements explicitly, what can you say about the relative displacements of the wall and backfill for the two cases?

(modified after Perloff and Baron, 1976)

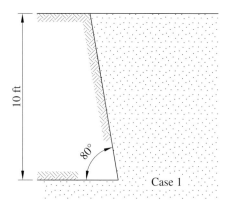
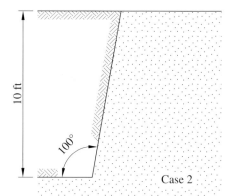

Figure Ex. 2.2(a)

Solution:

a) At z = 10 ft,

$$\sigma_z = \sigma_1 = \gamma z = 100(10) = 1{,}100 \text{ (psf)}$$

$$\sigma_x = \sigma_3 = \gamma z K_A = 1{,}100 \tan^2(27.5°) = 300 \text{ (psf)}$$

Mohr circle of stresses @ z = 10 ft is shown in Figure Ex. 2.2(b).

From the pole of the Mohr circle, stresses on the two planes inclined at 10° off the vertical can be determined graphically from the Mohr circle, see Figure Ex. 2.2(b).

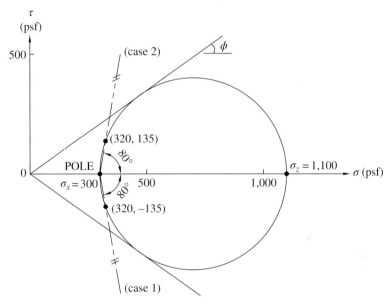

Figure Ex. 2.2(b)

Case 1:

For Case 1 (see Figure Ex. 2.2(c)),

$$F_s = \frac{135}{2}\left(\frac{10}{\cos 10°}\right) = 686 \left(\frac{\text{lb}}{\text{ft}}\right)$$

$$F_n = \frac{320}{2}\left(\frac{10}{\cos 10°}\right) = 1{,}625 \left(\frac{\text{lb}}{\text{ft}}\right)$$

$$\alpha = \tan^{-1}\frac{F_s}{F_n} = \tan^{-1}\left(\frac{686}{1{,}625}\right) = 22.9°$$

$$P_A = \frac{1{,}625}{\cos 22.9°} = 1{,}765 \left(\frac{\text{lb}}{\text{ft}}\right) \quad \left(\text{or } P_A = \sqrt{F_s^2 + F_n^2}\right)$$

Force P_A is acting at $\dfrac{H}{3}$ from the base, orientation: $22.9° + 10° = 32.9°$ from the horizontal.

Case 2:

For case 2 (see Figure Ex. 2.2(d)), F_s, F_n, P_A, and α are the same magnitude as for Case 1, except the orientation of α is on *the other side* of the normal.

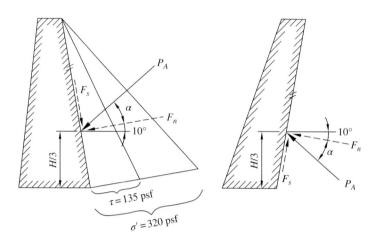

| Figure Ex. 2.2(c) | Figure Ex. 2.2(d) |

b)   The relative displacements of the wall and backfill can be deduced from the sense of shear force acting on the wall. For Case 1, the sand will move down relative to the wall; for Case 2, the sand will move` up relative to the wall.

2.2.3   Failure Surface

Figure 2.8 shows two Mohr circles corresponding to the active and passive conditions of a wall with a horizontal crest and subjected to a uniform vertical surcharge q. Since vertical stress, σ_z ($\sigma_z = \gamma z + q$), is acting on the horizontal plane, the poles for the active and passive Mohr circles can readily be determined by drawing a horizontal line through σ_z (see Figure 2.8). The orientation of failure surfaces for the active and passive conditions can thus be determined by connecting the respective pole with the respective points of tangency by a straight line. The failure surfaces of a wall for the two limiting equilibrium conditions are shown in Figure 2.9. Note that the orientation of the family of failure surfaces is the same for both granular and cohesive soils of the same effective friction angle. Since the orientation of the failure surfaces at any depth are the same, the failure surfaces are flat (not curved). In fact, the failure surfaces are a family of parallel planes. Note that the family of failure surfaces shown in Figure 2.9 are "interrupted" by the presence of a wall; they do not extend beyond the heel of the wall.

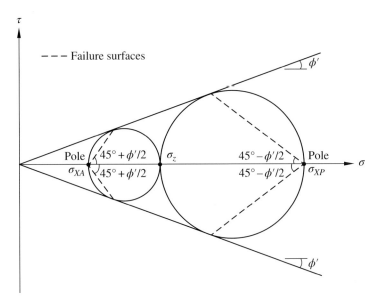

Figure 2.8 Limiting Mohr circles and the corresponding failure planes for Rankine active and passive conditions of a vertical wall with a horizontal crest and subject to uniform vertical surcharge. Note: The orientation of failure surfaces is the same for a granular backfill ($c = 0$) and a cohesive soil ($c > 0$)

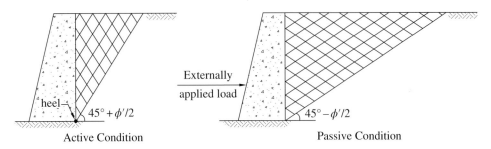

Figure 2.9 Family of failure surfaces for Rankine active and passive conditions of a vertical wall with a horizontal crest and subject to uniform vertical surcharge

2.2.4 Inclined Crest and/or Inclined Surcharge

For a wall with an inclined crest and/or subjected to an inclined surcharge, the failure surfaces and lateral earth pressure in active and passive conditions can readily be determined with the help of Mohr circles by the limiting equilibrium method of analysis. Failure surfaces for inclined surcharge and for inclined crest in active and passive conditions are shown in Figure 2.10. Note that failure planes for a wall with inclined surcharge will change their orientation with depth, hence result in *curved* failure planes. Failure planes for a wall with an inclined crest, however, will remain flat. The case of a wall with inclined crest *and* subjected to an inclined surcharge is given in Example 2.3.

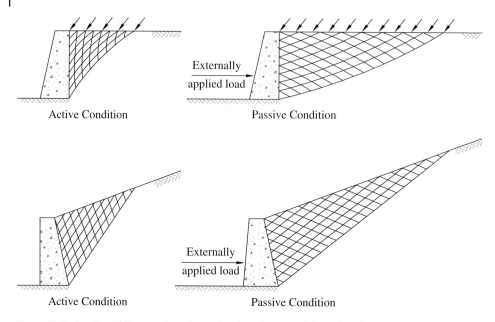

Figure 2.10 Family of failure surfaces for inclined surcharge and for inclined crest in Rankine active and passive conditions

Example 2.3 Perform Rankine analysis to determine the active lateral earth pressure and associated failure surfaces for a wall with an inclined crest and subjected to inclined surcharge, as shown in Figure Ex. 2.3(a).

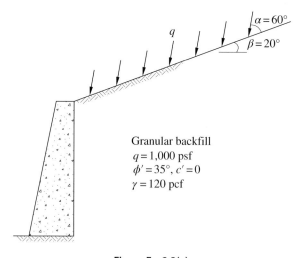

Granular backfill
$q = 1,000$ psf
$\phi' = 35°, c' = 0$
$\gamma = 120$ pcf

$\alpha = 60°$
$\beta = 20°$

Figure Ex. 2.3(a)

Solution:

At depth z, the normal and shear stresses on a plane making an angle β with the horizonal (referred to as the β-plane), due to the inclined surcharge $q = 1{,}000$ psf and the self-weight of the inclined crest are shown in Figure Ex. 2.3(b). Also shown in Figure Ex. 2.3(b) is the resulting normal and shear stresses at selected depths on the β-plane. Note that the resultant stresses are due to the combined effect of inclined surcharge and soil self-weight of the inclined crest.

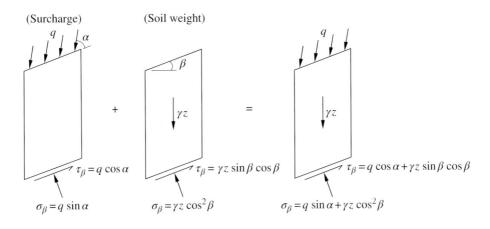

(Surcharge)   (Soil weight)

$\tau_\beta = q\cos\alpha$ $\tau_\beta = \gamma z\sin\beta\cos\beta$ $\tau_\beta = q\cos\alpha + \gamma z\sin\beta\cos\beta$

$\sigma_\beta = q\sin\alpha$ $\sigma_\beta = \gamma z\cos^2\beta$ $\sigma_\beta = q\sin\alpha + \gamma z\cos^2\beta$

Stress (psf)	Depth						
	0	2'	4'	6'	8'	10'	12'
σ_β	866	1,078	1,290	1,502	1,714	1,926	2,138
τ_β	500	577	654	731	809	866	963

Figure Ex. 2.3(b)

The Mohr circles corresponding to active failure and passive failure conditions at depths $z = 0$, 4 ft, and 8 ft are shown in Figure Ex. 2.3(c). Also given in Figure Ex. 2.3(c) are the orientation of failure planes and the values of σ and τ on the vertical plane.

The active (minimum) normal stresses and passive (maximum) normal stresses on the vertical planes at different depths determined from Figure Ex. 2.3(c) are, respectively, the active and passive earth pressures acting on the wall. The pressure profiles are plotted in Figure Ex. 2.3(d). Note that the orientation of failure planes determined from Figure Ex. 2.3(c) can be used to construct the failure surfaces. The failure planes are "curved" as the orientation changes with depth (not shown) in both active and passive conditions.

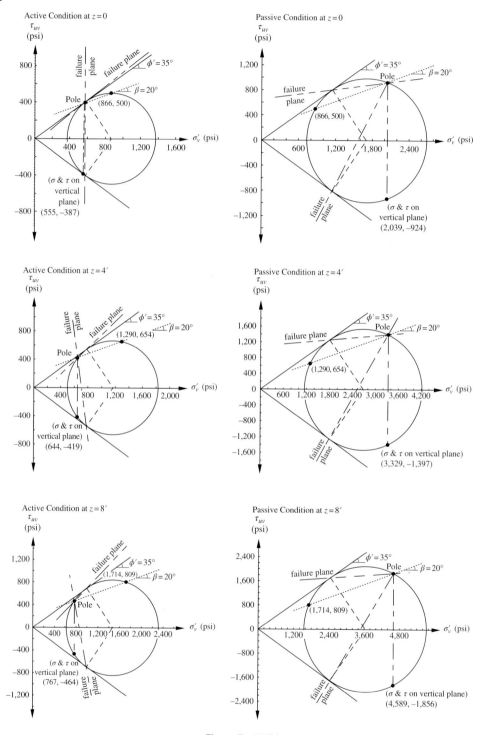

Figure Ex. 2.3(c)

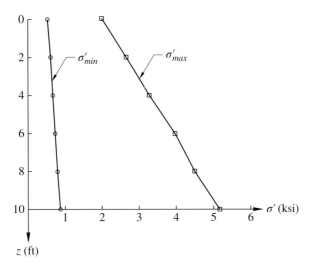

Figure Ex. 2.3(d)

2.2.5 Influence of Submergence

The influence of submergence on active and passive earth pressures under a static water level condition (i.e., without seepage) is illustrated by Example 2.4. When the backfill is entirely submerged below the free water level (e.g., due to prolonged rainfall and infiltration of rainwater), the active earth pressure at depth z is increased by approximately $(1 - K_A)\gamma_w z$, while the passive resistance at depth z is decreased by approximately $(K_P - 1)\gamma_w z$. For the conditions given in Example 2.4, due to standing water in the backfill, the active earth thrust on the wall is increased by 100% (from $300\gamma_w$ to $600\gamma_w$), and the passive resistance is decreased by 33% (from $2{,}700\gamma_w$ to $1{,}800\gamma_w$). Thus, submergence is "bad" for an earth retaining wall in both active and passive conditions, and is especially harmful for active condition. The large increase in active earth thrust has been a major cause for failure of many earth retaining walls. This is the reason why it is important to provide proper drainage in the backfill behind earth retaining structures.

Example 2.4   For the conditions given in Figure Ex. 2.4, determine earth thrusts under active and passive conditions.

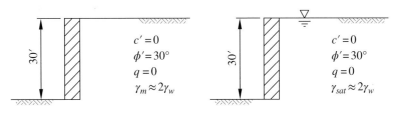

(a) Backfill is moist   (b) Backfill is submerged

Figure Ex. 2.4

Solution:

For $\phi' = 30°$, $K_A = 1/3$ and $K_P = 3$; $\gamma' = \gamma_{sat} - \gamma_w = 2\gamma_w - \gamma_w = \gamma_w$

(A) Moist:

Active: $\sigma_{x_A} = \gamma_m z K_A$

$\qquad = \dfrac{1}{3} \gamma_m z$

$\qquad = \dfrac{2}{3} \gamma_w z$

Passive: $\sigma_{x_P} = \gamma_m z K_P$

$\qquad\quad = 3 \gamma_m z$

$\qquad\quad = 6 \gamma_w z$

(B) Submerged:

Active: $\sigma_{x_A} = \gamma' z K_A + \gamma_w z$

$\qquad = \dfrac{1}{3} \gamma_w z + \gamma_w z$

$\qquad = \dfrac{4}{3} \gamma_w z$

Passive: $\sigma_{x_P} = \gamma' z K_P + \gamma_w z$

$\qquad\quad = 3 \gamma_w z + \gamma_w z$

$\qquad\quad = 4 \gamma_w z$

Active

Passive

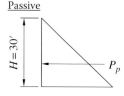

Active

Passive

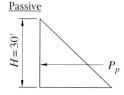

$\sigma_{x_A} = \dfrac{2}{3} \gamma_w H = 20 \gamma_w$

$P_A = \dfrac{1}{2} \left(20 \gamma_w \right) H$

$\quad = 300 \gamma_w$

$\sigma_{x_P} = 6 \gamma_w H = 180 \gamma_w$

$P_P = \dfrac{1}{2} \left(180 \gamma_w \right) H$

$\quad = 2{,}700 \gamma_w$

$\sigma_{x_A} = \dfrac{4}{3} \gamma_w H = 40 \gamma_w$

$P_A = \dfrac{1}{2} \left(40 \gamma_w \right) H$

$\quad = 600 \gamma_w$

$\sigma_{x_P} = 4 \gamma_w H = 120 \gamma_w$

$P_P = \dfrac{1}{2} \left(120 \gamma_w \right) H$

$\quad = 1{,}800 \gamma_w$

2.2.6 External Loads on Wall Crest

Increase in lateral earth pressure due to externally applied loads on the crest of a wall has commonly been determined by using influence diagrams. Two influence diagrams for lateral earth pressure due to line loads and point loads on the wall crest are given in Figure 2.11. For routine design, when external loads involve more than one surface load, the lateral earth pressures due to each surface load and due to the self-weight of the backfill (and water if present) can be superimposed to determine the lateral earth pressure due to a combination of the loads. An example illustrating how to use the influence diagrams due to multiple external loads on the wall crest is given in Section 5.4.2.

It is interesting to note that the lateral earth pressure obtained from Figure 2.11 is approximately *twice* the value obtained from the Boussinesq elastic solution. The Boussinesq elastic solution assumes that the material is homogeneous, isotropic, and of infinite depth. When a rigid wall is involved, the wall can be viewed as a *mirror* for the lateral stress in the free field, hence the lateral earth pressure on a rigid wall is theoretically twice the Boussinesq solution.

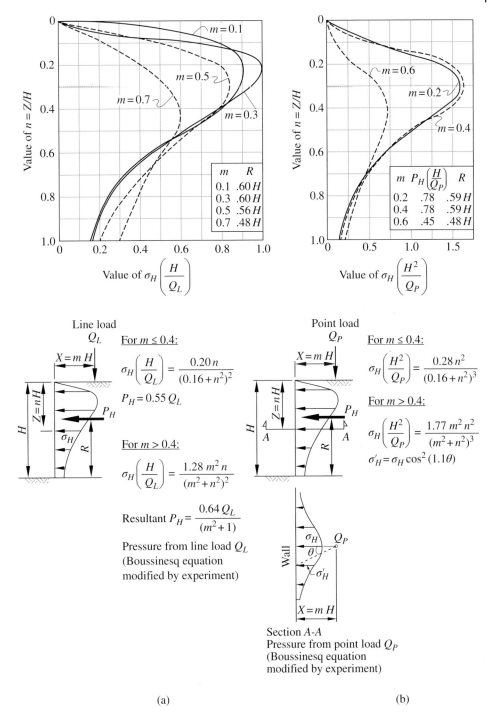

Figure 2.11 Influence diagrams for determination of change in lateral earth pressure due to (a) a line load and (b) a point load on the wall crest (NAVFAC, 1986)

There is another elastic solution for determination of lateral earth pressure, known as the *Westergaard solution*, which assumes that thin layers of a homogeneous, transversely isotropic, elastic material are sandwiched between closely spaced, infinitely thin sheets of rigid inclusion that allow compression but not lateral deformation (i.e., Poisson's ratio = 0). The Westergaard solution is arguably better suited for applications to reinforced soil walls with strong tensile inclusion. However, since the Boussinesq solution generally produces larger values of lateral earth pressure, it is often used in the design of reinforced soil for conservatism.

"Rigid" retaining walls in practice are in fact less than perfectly rigid. Measured lateral earth pressure from actual rigid retaining walls has been found to lie between the Boussinesq solution (for walls of rigidity similar to that of the soil) and twice the Boussinesq solution (for walls of perfect rigidity) (Spangler and Handy, 1982). The assumption of an unyielding rigid wall in design is conservative, and has been recommended for design of all types of rigid retaining walls.

2.2.7 Applicability of Rankine Analysis

As to be expected, the assumptions made in Rankine theory will govern the applicability of Rankine analysis. A summary of the assumptions, with a brief discussion on some, is given below.

1) *The wall is rigid.*

   Examples of rigid retaining walls include gravity walls, cantilever concrete walls, and crib walls (see Figure 2.33(a)). For flexible walls such as cantilever sheet pile walls and anchored sheet pile walls (see Figure 2.33(b)), Rankine analysis is not applicable.

2) *The wall is of infinite height.*

   In theory, the active or passive condition cannot develop in its entirety for a wall of finite height. In an actual wall, the active condition is said to develop within a wedge of soil bonded by a line passing through the heel and making an angle of $45° + \phi/2$ with the horizontal (for a vertical wall with horizontal crest and subjected to uniform vertical surcharge), as shown in Figure 2.12. The remainder of the soil behind and below the wall is said to be in a state of elastic equilibrium. This also applies to the passive condition, except the wedge is bounded by a much flatter line that makes an angle of $45° - \phi/2$ with the horizontal.

3) *There is no shear resistance along the soil–wall interface.*

   Rankine analysis assumes that the interface shear resistance between the backface of a vertical wall and the backfill is zero. Namely, the wall is either perfectly smooth, or the wall and the soil move at the same rate. The assumption of a smooth wall results in overestimation of active force and underestimation of passive force; both are *on the conservative side*. The extent of the overestimation and underestimation depends on the magnitude of the soil–wall friction and/or adhesion. The larger the interface shear resistance, the greater the overestimation or underestimation.

4) *The surcharge is uniform across the entire crest area (i.e., the top surface of a wall).*

5) *The soil is homogeneous and isotropic.*

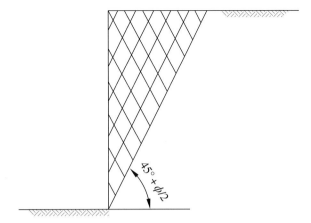

Figure 2.12 Active failure zone due to a wall being of a limited height

6) *Failure of the backfill can be described accurately by the Mohr–Coulomb failure criterion.*

   The assumption that the soil is a "Mohr–Coulomb material" has been brought into question by recent studies on soil behavior. However, use of the Mohr–Coulomb criterion is simple and also supported by an abundance of performance data.

7) *The entire soil mass reaches a state of limiting equilibrium at the same time, i.e., every point in the soil mass will experience the same deformation.*

   A uniform strain within the triangular wedge shown in Figure 2.12 can be produced by a rotational movement of the wall about its base. The lateral earth pressure for a rigid retaining wall rotating outward about its base will take on a hydrostatic distribution, as shown in Figure 2.13(a). An implicit assumption of active Rankine analysis is that the wall is rotating outward about its base. For a wall that has other modes of movement, the lateral earth pressure distribution will be quite different from being hydrostatic. Figures 2.13(b) and (c) depict typical lateral earth pressure distribution resulting from a wall rotating about the top and a wall experiencing lateral translation, respectively.

8) *There is sufficient wall movement to develop active or passive conditions.*

   In laboratory model tests performed by Terzaghi (1943) and Tschebotarioff (1973), it was found that a rotational tilt on the order of $0.1\% \times H$ (where H = wall height) at the top of the wall would be sufficient to mobilize an active condition for a cohesionless backfill. The very small movement needed to mobilize an active condition is the reason why design of gravity and semi-gravity walls usually assumes that the active condition (rather than the at-rest condition) will prevail. To fully mobilize a passive condition, however, the required wall movement in the soil is much greater. The tilt of a wall on the order of $10\% \times H$ for loose sands and $1\% \times H$ for dense sands is found required for full mobilization of a passive condition. The wall movement needed to mobilize one-half of the passive earth pressure, however, is much smaller. Thus, when resistance of soil to externally applied loads is to be considered, $(1/2)P_P$ may be used in design.

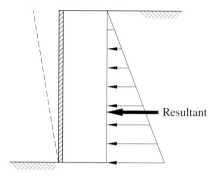

(a) Rotation about the base

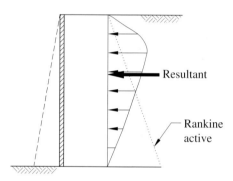

(b) Rotation about the top

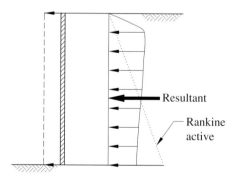

(c) Lateral translation

Figure 2.13 Active lateral earth pressure distribution and resultant thrust due to different modes of wall movement

2.3   Coulomb Analysis

Rankine analysis assumes every point in the soil mass behind a wall is in a state of limiting equilibrium. Coulomb analysis, on the other hand, assumes that a state of limiting equilibrium occurs only along a single planar failure surface passing through the heel of a wall. Coulomb analysis involves analysis of a potential failure wedge bounded by the backface of the wall, the crest of the wall, and a trial failure plane. The force exerted on a wall by a potential failure wedge is determined by the equilibrium of the forces of the failure wedge, hence Coulomb analysis is also referred to as *wedge analysis*. In Coulomb analysis, a number of trial failure surfaces need to be selected to determine the maximum (in active condition) and minimum (in passive condition) forces exerted on the wall.

The soil along the assumed trial failure plane in Coulomb analysis is assumed to reach a state of limiting equilibrium, and the stresses along trial failure planes are assumed to satisfy the Mohr–Coulomb failure criterion. To better understand Coulomb analysis, we shall first examine the forces acting on a failure plane by considering the simple test depicted in Figure 2.14(a). The test is conceptually similar to the direct shear test described in Section 1.4.1, in that the soil contained in a box is subjected to a constant vertical normal force (N) and a horizontal shear force (T). The force T is increased gradually until failure occurs along the bottom surface of the box. The forces acting on the failure plane at failure are shown in Figure 2.14(b), in which the shear resistance at failure is equal to $N \tan\phi + cL$. Forces N (normal force on the failure plane) and $N \tan\phi$ (a component of shear resistance due to N) can be combined into a single force F, which acts at an inclined angle ϕ from the vertical (on the side opposite to the direction of impending movement at failure). Therefore, there are two forces acting on the failure plane: F and cL (see Figure 2.14(c)). This is a very simple force system in which the line of action and the sense of both forces (F and cL) as well as the magnitude of cL are known, the only unknown is the magnitude of force F.

Similarly, the forces on the contact surface between the soil wedge and the backface of a wall can be combined into a single force P (inclined at an angle δ, δ = soil–wall friction angle, measured from the normal of the contact surface) and $c_a l$ (c_a = adhesion between soil and wall, and l = contact length between the soil wedge and the backface of the wall). The adhesive force $c_a l$ is usually ignored in design for conservatism because the magnitude of c_a is known to be rate-dependent. Therefore, the forces acting on a wall by the soil wedge are now reduced to a single force P of which the line of action and the sense are known, but not the magnitude. The goal of the analysis is to determine the magnitude of P in two limiting conditions: active and passive conditions. This can be accomplished by considering the equilibrium of forces for several potential failure wedges.

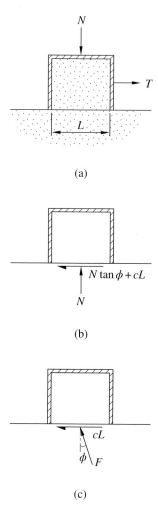

Figure 2.14 A simple test to examine forces on a failure plane at failure: (a) test setup, (b) resisting forces on the failure plane at failure, and (c) an alternate presentation of resisting forces on the failure plane at failure

2.3.1 Active Condition

Figure 2.15(a) shows a trial failure wedge for an assumed failure plane initiated at the heel and inclined at an angle θ from the horizontal. Figure 2.15(b) shows the free-body diagram of the trial failure wedge in an active condition. The wall is to rotate away from the soil, which is accompanied by a downward movement of the failure wedge with respect to the wall. This downward movement of the failure wedge is merely an assumed direction of movement. The assumption will be examined in Section 2.3.5.

There are four forces acting on the trial failure wedge (Note: The wall is assumed to be in a plane-strain condition, i.e., the wall's longitudinal dimension is much greater than other dimensions, and the wall can be analyzed by considering a "slice" of the wall

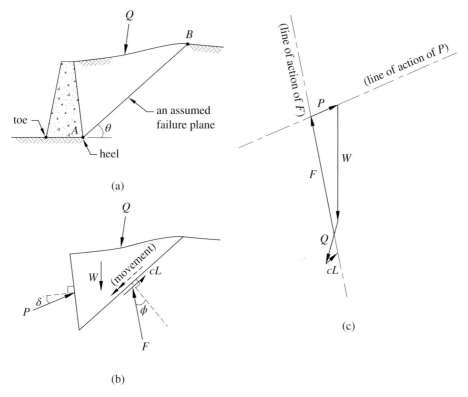

Figure 2.15 Graphical solution of Coulomb analysis in an active condition: (a) a wall with an assumed trial failure plane, (b) free-body diagram of a trial failure wedge, and (c) force polygon of forces acting on the trial failure wedge

of unit thickness in the longitudinal direction; all forces therefore are expressed in force per unit thickness):

1) *Weight of soil wedge, W*

The weight of the failure wedge per unit thickness, $W = \gamma A$, where γ is the unit weight of the soil and A is the planar area of the wedge. Since the magnitude of W is known and is acting vertically downward, the magnitude, line of action, and sense of W are known.

2) *Resultant force of all externally applied forces on the wall crest, Q*

The force Q is the resultant of all external forces applied to the crest of the failure wedge. The magnitude, line of action, and sense of Q are known. Note that the resultant force Q (units: force/length) is assumed to be applied uniformly along the longitudinal direction of the wall.

3) *Forces F and cL on an assumed failure plane* (L = length AB in Figure 2.15(a))

The magnitude, line of action, and sense of force cL are known. The line of action of force F is known to be inclined at angle ϕ measured from the normal of the assumed failure plane. The sense of force F is on the side opposite to the direction of impending movement of the failure wedge, see Figure 2.15(b). The magnitude of F, however, is not known.

4) *Interface frictional force P on the soil–wall interface*

Force P is inclined at angle δ from the normal of the wall–soil interface. Again, it is on the opposite side of the impending movement of the failure wedge (assumed to be moving downward with respect to the wall for now), see Figure 2.15(b). The magnitude of P is also not known.

Knowing the line of action and sense for all the forces acting on the trial failure wedge and all their magnitudes except P and F, a force polygon can be constructed as shown in Figure 2.15(c). Since all these forces needs to be in equilibrium, the corresponding force polygon must be a *closed* polygon. This characteristic of force polygon allows us to determine the magnitude of P. The resultant force exerted by the trial failure wedge on the wall is equal to P, with an opposite sense depicted in Figure 2.15(c).

Several trial failure planes should be examined. For each trial failure plane (i.e., for each assumed angle θ), the corresponding force P can be determined by constructing a new force polygon. The most critical failure surface is the one that results in the largest value of force P, which is the Coulomb active force, P_A. Figure 2.16 shows a plot between θ and P, and how P_A can be determined. The θ-value corresponding to P_A is θ_{cr}, the angle of the critical failure plane in an active condition.

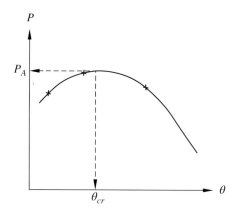

Figure 2.16 A plot of θ (angle of trial wedge) vs. P (force acting on the wall) for determination of active force P_A and angle of critical failure plane (θ_{cr})

The procedure for determination of P_A as described above is very versatile; it can accommodate walls with complex soil profiles, complex external loads on the wall crest, and complex wall configurations. The procedure, however, is rather time-consuming. For a few simple cases, analytical solutions are available for Coulomb analysis, and the time-consuming graphical procedure can be avoided. One of these cases is for a wall of simple geometry shown in Figure 2.17 for a cohesionless backfill ($c = 0$) without any external force on the crest ($Q = 0$). By denoting the angle of the sloping crest with the horizontal α, and the angle of the backface of a wall with the vertical β, the Coulomb active force is:

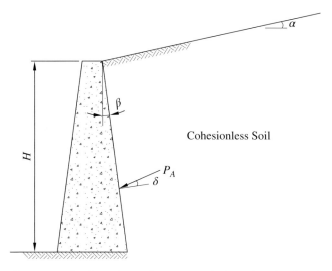

Figure 2.17 Configuration and parameters for Eqns. (2-15) to (2-18)

$$P_A = \frac{1}{2} K_A \gamma H^2 \tag{2-15}$$

in which,

$$K_A = \frac{\cos^2(\phi - \beta)}{\left(\cos^2\beta\right)\cos(\delta + \beta)\left[1 + \sqrt{\frac{\sin(\delta + \phi)\sin(\phi - \alpha)}{\cos(\delta + \beta)\cos(\beta - \alpha)}}\right]^2} \tag{2-16}$$

For $\alpha = 0$ (i.e., horizontal crest), $\beta = 0$ (i.e., vertical wall), and $\delta = 0$ (i.e., frictionless soil–wall interface), K_A in Eqn. (2-16) reduces to $(1 - \sin\phi)/(1 + \sin\phi) = \tan^2(45° - \phi/2)$, which is the same as the Rankine active earth pressure coefficient seen in Eqns. (2-12) and (2-13).

Recall in the discussion of Eqn. (2-14) that tension cracks will develop near the crest of a cohesive soil in an active condition. To account for the tension cracks in Coulomb active analysis, the trial failure wedges should have reduced lengths of failure plane, as shown in Figure 2.18.

Even though Coulomb analysis says nothing about the *location* of external resultant Q, the location of Q can be relevant in the design or analysis of earth retaining walls. An interesting case is given in Example 2.5. The example illustrates how Coulomb analysis can help us solve problems involving location of Q which seemingly is not addressed by Coulomb analysis.

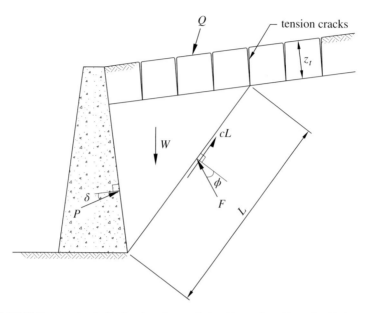

Figure 2.18 Trial failure wedge in Coulomb active analysis with a reduced length of failure surface due to tension cracks in cohesive soil

Example 2.5 For a retaining wall subject to a line load of 5,000 lb/ft on a sloping crest, as shown in Figure Ex. 2.5(a),

   a)   Determine the most critical angle λ (the angle between the line load and the sloping crest) that would produce the largest active force on the wall.

   b)   Determine the distance x beyond which the line load will have no influence on the active force on the wall.

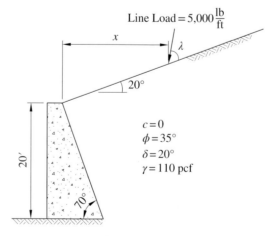

Figure Ex. 2.5(a)

Solution:

The problem can be solved by the following procedure:

1) Perform Coulomb analysis for a number of trial failure planes initiating at the heel of the wall, each making a different angle θ from the horizontal plane, Figure Ex. 2.5(b).

2) For each trial failure plane (i.e., for each assumed value of θ), construct a force polygon for forces acting on the failure wedge, Figure Ex. 2.5(c), to determine the corresponding P_A (without including Q in the failure wedge, i.e., assuming Q is applied beyond the range of the failure wedge) and $P_{A/Q}$, (with Q included in the failure wedge, i.e., assuming Q is applied within the range of the failure wedge). Note that for determination of $P_{A/Q}$, the orientation of Q that would lead to the largest $P_{A/Q}$ is when Q is acting in an orientation perpendicular to the line of action of force F, as seen in Figure Ex. 2.5(d).

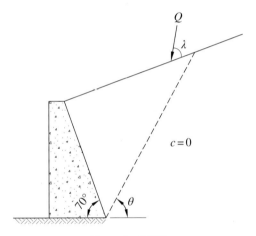

Figure Ex. 2.5(b)

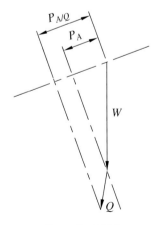

Figure Ex. 2.5(c)

3) Plot P_A and $P_{A/Q}$ vs. θ, and determine $P_{A(max)}$ (without Q), see Figure Ex. 2.5(e). This is the largest active force that must "not" be exceeded when Q is included in the failure wedge.

4) Determine θ_{cr}, the angle θ corresponding to $P_{A(max)}$ (without Q) on the $P_{A/Q}$ vs. θ curve, see Figure Ex. 2.5(e).

5) Determine distance x by drawing a straight line fom the heel of the wall making an angle θ_{cr} with the horizontal, as seen in Figure Ex. 2.5(f). The value of x is the distance beyond which the line load will have no influence on the active force.

6) For the wall geometry and material parameters given in this example, $\lambda = -16°$ and $x = 52$ ft ($P_{A(max)} = 6,500$ lb/ft, $\theta_{cr} = 39°$).

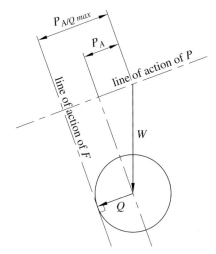

Figure Ex. 2.5(d)

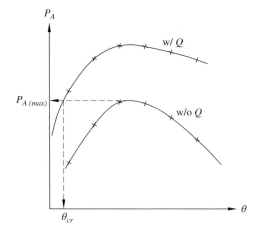

Figure Ex. 2.5(e)

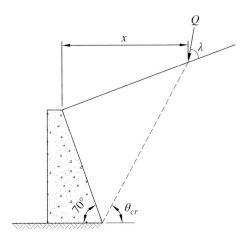

Figure Ex. 2.5(f)

Note: This example is a simplified version of an interesting project at the old Denver
Stapleton International Airport. The project was about an existing retaining wall
supporting a luggage pit at the airport. A new runway was to be added near the
luggage pit. The question at hand was: How close can the aircraft on the runway
come to the edge of the luggage pit without having to redesign and reconstruct
the retaining wall?

2.3.2 Passive Condition

For Coulomb passive analysis, a single planar failure plane is also assumed. In a
passive condition, the trial soil wedge at failure is assumed to move upward with respect
to the soil beneath the wedge and, in most cases, with respect to the wall as well (ref.
Section 2.3.5 for variation).

Figure 2.19(a) shows the free-body diagram of a trial failure wedge in a passive
condition. The free-body diagram is similar to that in an active condition (cf.
Figure 2.15(b)), except the forces P and F are now acting on *the other side* of the normal
directions. This is because the sense of the shear forces is reversed due to reversal of
the direction of the impending wedge movement in the passive condition. The force
polygon of a trial wedge for a passive condition is shown in Figure 2.19(b). The magnitude of P, which represents the resisting force provided by the soil to resist externally
applied loads, can be determined from the force polygon. Similar to the active condition, a number of trial failure planes need to be examined. The most critical failure
plane is the one that gives the smallest value of resisting force P, which is the Coulomb
passive force, P_P. Figure 2.20 shows a plot of the relationship between θ and P, and how
P_P can be determined. The plot also gives θ_{cr}, the angle of the most critical failure
plane with the horizontal.

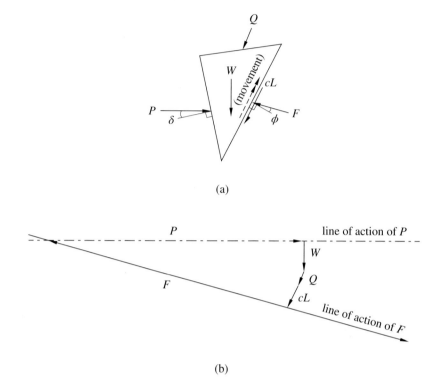

(a)

(b)

Figure 2.19 Graphical solution of Coulomb analysis in a passive condition: (a) free-body diagram of a trial failure wedge and (b) force polygon of forces acting on the trial failure wedge

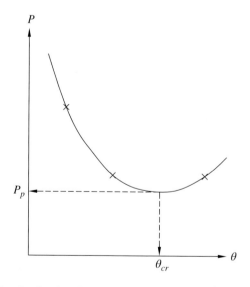

Figure 2.20 A plot of θ (angle of trial wedge) vs. P (force acting on the wall) for determination of passive force (P_P) and angle of critical failure plane (θ_{cr})

As in an active condition, analytical solutions are available for a handful of cases in a passive condition. For the same simple cases as in the active case (i.e., the simple wall geometry shown in Figure 2.17, with $c = 0$ and $Q = 0$), the Coulomb passive resistance is:

$$P_P = \frac{1}{2} K_P \gamma H^2 \tag{2 17}$$

in which,

$$K_P = \frac{\cos^2(\phi + \beta)}{\left(\cos^2 \beta\right) \cos(\delta - \beta) \left[1 - \sqrt{\dfrac{\sin(\phi - \delta) \sin(\phi + \alpha)}{\cos(\delta - \beta) \cos(\alpha - \beta)}}\right]^2} \tag{2-18}$$

Again, for $\alpha = 0$, $\beta = 0$, and $\delta = 0$, Eqn. (2-18) reduces to $K_P = (1 + \sin\phi)/(1 - \sin\phi) = \tan^2(45° + \phi/2)$, which is identical to the mathematical solution for Rankine passive analysis given in Eqns. (2-12) and (2-13).

Large-scale model tests have indicated that the curvature of the failure surface in a passive condition for higher values of soil–wall interface friction angle δ is too large to be ignored, hence the assumption of *planar* failure surface is good only for smaller δ (δ less than approximately $\phi/3$). The assumption of a planar failure surface in those cases will lead to overestimation of passive resistance (i.e., an error on the unsafe side). This point is discussed further in Section 2.4. Limiting equilibrium analysis with curved failure surfaces (such as log-spirals or ellipses) or the method of slices has been suggested for passive analysis in these cases (e.g., Shields and Tolunay, 1973).

2.3.3   Influence of Submergence

In the absence of seepage force, the influence of submergence on a retaining wall in Coulomb analysis can be evaluated by analyzing forces in trial failure wedges that account for the effect of static water. Figure 2.21(a) shows a wall with free water level (FWL) "perched" in the fill behind a wall. The free-body diagram of a trial failure wedge in terms of effective (submerged) forces is shown in Figure 2.21(b). The force W_2' in Figure 2.21(b) is the submerged weight in the trial failure wedge below the free water level, and $W_2' = \gamma' \times$ (area of trail failure wedge below the free water level). The active force P_A acting on the wall is the vector sum of P_A' and P_w. Force P_A' is due to submerged weight of the trial failure wedge, and can be obtained by constructing a force triangle of the effective (submerged) forces as shown in Figure 2.21(c). Force P_w, on the other hand, is the resultant of hydrostatic pressure in the potential failure wedge on the backface of the wall, $P_w = h^2 \times \gamma_w / (2 \sin\alpha)$, where h is the height of the free water level above the heel. The line of action of P_w is normal to the backface of the wall. The magnitude and orientation of the resultant active force P_A can then be determined as the vector sum of P_A' and P_w, as shown in Figure 2.21(d).

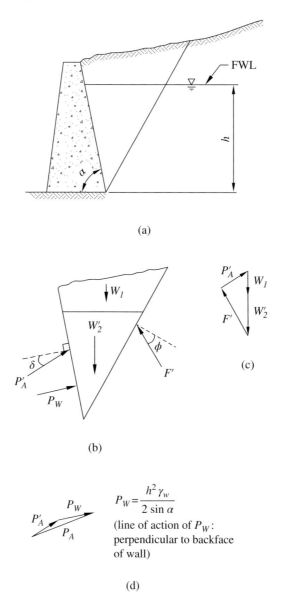

(a)

(b)

(c)

$$P_W = \frac{h^2 \gamma_w}{2 \sin \alpha}$$

(line of action of P_W: perpendicular to backface of wall)

(d)

Figure 2.21 Coulomb active analysis of a partially submerged wall: (a) a wall with free water level (FWL) perched in the fill, (b) free-body diagram of a trial failure wedge in terms of effective forces, (c) effective force triangle for the failure wedge, and (d) determination of resultant active force P_A

2.3.4 Influence of Seepage

Water is often the culprit in the failure of earth structures, including earth retaining walls. Generally speaking, seepage through the backfill behind a retaining wall will result in increased active earth pressure and decreased passive resistance. Control of problems associated with seepage and drainage of water has usually been accomplished by employing free draining backfill or by installing drains along the back of the wall or in the backfill. Figure 2.22 shows some drainage measures commonly used in rigid retaining walls.

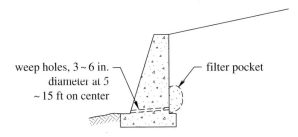

weep holes, 3 ~ 6 in.
diameter at 5
~ 15 ft on center

filter pocket

(a) Weep holes through wall
with or without pocket
(minimum drainage)

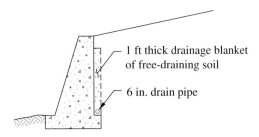

1 ft thick drainage blanket
of free-draining soil

6 in. drain pipe

(b) Vertical drainage blanket
with collector drain

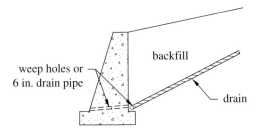

backfill

weep holes or
6 in. drain pipe

drain

(c) Inclined drain

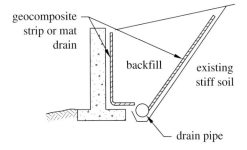

geocomposite
strip or mat
drain

backfill

existing
stiff soil

drain pipe

(d) Geocomposite drain

Figure 2.22 Common drainage systems behind a rigid retaining wall

Computation of the resultant active force acting on a wall that involves seepage can be performed with the aid of a flow net. Using a properly sketched flow net, two resultant thrusts due to seepage can be determined (see Figure 2.23(a)): (i) U_1: resultant water force along the assumed failure plane and (ii) U_2: resultant water force along the backface of a wall. Water pressure exerted on a surface is always perpendicular to the surface (water can transmit only normal stress, not shear stress). Being resultants of water pressure, both U_1 and U_2 are perpendicular to the surface under consideration. By considering all the forces acting on a trial failure wedge as shown in Figure 2.23(a), a force polygon can be constructed to determine active force P_A for the trial failure wedge, as shown in Figure 2.23(b). A number of trial wedges needs to be examined to determine the maximum $P_{A(max)}$. The total active force P_{total} is the vector sum of $P_{A(max)}$ and U_2, see Figure 2.23(c).

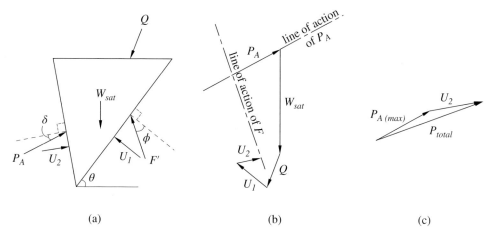

(a) (b) (c)

Figure 2.23 Graphical solution of Coulomb active analysis with influence of seepage: (a) free-body diagram of a trial failure wedge, (b) force polygon of the trial failure wedge (U_1 and U_2 are water pressure resultants on the boundaries), and (c) determination of resultant active force

At this time, you may feel a little confused about when to use total force vs. effective (submerged) force, or about when to include the resultant of water pressure and when not to. If that is the case, you will likely find the fundamental force diagram for seepage analysis shown in Figure 2.24 very useful for understanding the stability analysis of earth structures involving seepage. The force diagram indicates that the resultant force involving seepage can be determined by using one of two approaches:

a) Effective stress approach: use the submerged soil weight (γ') and the seepage force per unit volume ($i \cdot \gamma_w$) in the trial failure wedge, where i is the hydraulic gradient, for computation of the resultant force.

b) Total stress approach: use saturated soil weight $\gamma_{sat} = (\gamma' + \gamma_w)$ in the trial failure wedge and resultant fluid pressure per unit thickness *on the boundaries* of the trial failure wedge for computation of the resultant force.

Since the seepage force per unit volume needed in the effective stress approach usually changes the magnitude and line of action along the flow path, it is much easier to carry out the analysis using the total stress approach. Note that all the forces shown in Figure 2.23(b) are total forces, including W, which is the saturated unit weight of the soil wedge.

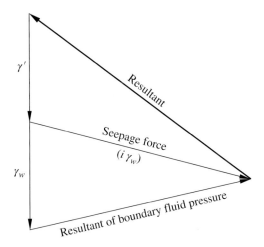

Figure 2.24 The fundamental force diagram involving seepage

A flow net for a wall constructed over an impervious foundation with a vertical drain along the backface, and the corresponding force polygon for an assumed trial failure wedge is shown in Figure 2.25(a). Note that the boundary resultant water force on the backface of the wall, $U_2 = 0$. Example 2.6 illustrates how to determine the force on the backface of a wall for a given trial failure surface with the aid of a flow net. In design, different trial failure planes need to be analyzed. A plot of P_A vs. θ (angle of an assumed failure plane with the horizontal) can be constructed to determine the maximum P_A, which is the active force of the wall (i.e., $P_{A(max)} = P_{total}$ in Figure 2.23(c), because $U_2 = 0$). For walls with a vertical drain along the backface and constructed over an impervious level ground surface, the unit boundary water force U_1 acting on an assumed failure plane can be obtained by Figure 2.26 (by determining the ordinate f first) without having to sketch a flow net.

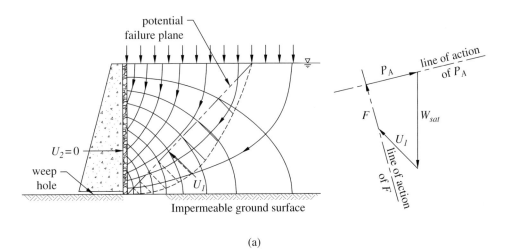

(a)

Figure 2.25 Flow nets and the associated trial wedge force polygon for a wall with (a) a vertical drain along the backface and (b) an inclined drain located below the potential failure plane, i.e., outside the failure wedge (modified after Craig, 1978)

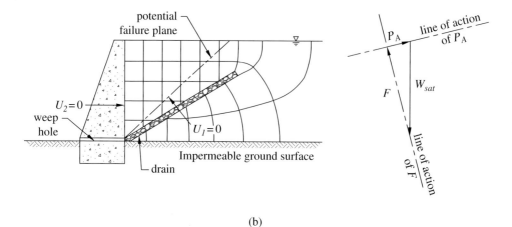

(b)

Figure 2.25 (Continued)

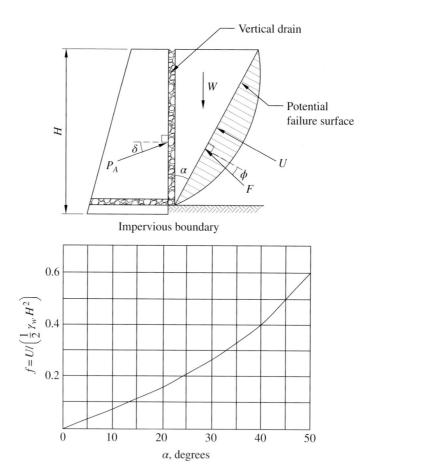

Figure 2.26 A graph for determination of seepage boundary water pressure resultant (U) for a wall with a vertical drain along the backface (NAVFAC, 1986)

Example 2.6   Determine the active force on the wall shown in Figure Ex. 2.6, due to continuous rainfall, by Coulomb analysis for a given trial failure plane inclined at $\theta = 55°$ from the horizontal plane.

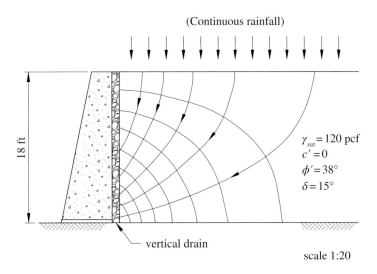

(Continuous rainfall)

$\gamma_{sat} = 120$ pcf
$c' = 0$
$\phi' = 38°$
$\delta = 15°$

18 ft

vertical drain

scale 1:20

Figure Ex. 2.6

Solution:

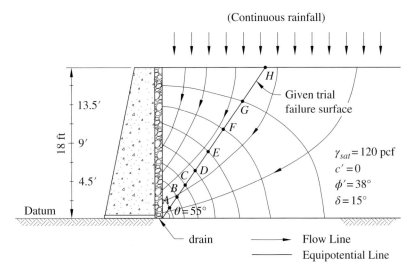

(Continuous rainfall)

13.5′

9′

4.5′

18 ft

Datum

H

G

F

E

C D

B

A

$\theta = 55°$

Given trial
failure surface

$\gamma_{sat} = 120$ pcf
$c' = 0$
$\phi' = 38°$
$\delta = 15°$

drain

Flow Line

Equipotential Line

1. Each equipotential drop $= \frac{18'}{8} = 2.25$ ft

2. Pressure Head $= H_{total}$ − Elevation Head

3.

Point	H_{total} (ft)	Elevation Head (ft)	Pressure Head (ft)
A	2.25	1.2	1.0
B	4.50	2.8	1.7
C	6.75	4.4	2.4
D	9.00	6.1	2.9
E	11.25	8.0	3.2
F	13.50	10.8	2.7
G	15.75	13.7	2.0
H	18.00	18.0	0.0

4. Pressure Diagram

5. $\cos 35° = 18$ ft$/L_1$
   $L_1 = 22$ ft

6. $\sin 35° = L_2/22$ ft
   $L_2 = 12.6$ ft

7. Area $\approx 22'(3.2') - 2\left[\frac{1}{2}(6.6')(3.2')\right]$
   Area ≈ 50 ft^2

8. $U_1 = \gamma_w A = (62.4 \text{ pcf})(50 \text{ ft}^2)$
   $U_1 = 3,120$ lb/ft

9. $W = \gamma A = (120 \text{ pcf})\left[\frac{1}{2}(18')(12.6')\right]$
   $W = 13,608$ lb/ft

10. Force Polygon

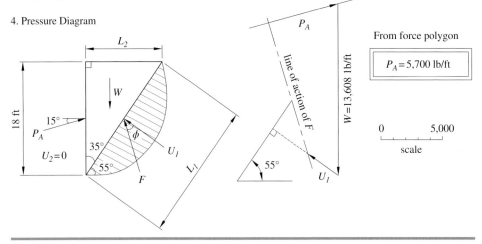

From force polygon

$P_A = 5,700$ lb/ft

0   5,000

scale

For a wall with an inclined drain in the backfill, as shown in Figure 2.25(b), the water pressure anywhere between the wall and the drain is zero; hence the resultant water forces U_1 and U_2 are both equal to zero, provided that the trial failure plane lies between the wall and the drain. The corresponding force triangle for determination of active force P_A is also shown in Figure 2.25(b). An inclined drain is a little better than a vertical drain in terms of minimizing seepage forces. Compared to the case of a wall without drainage for a 15-ft high wall, a vertical drain would typically reduce the active force by about 50%, while an inclined drain would typically reduce the active force by about 65%.

2.3.5 Influence of Relative Wall Movement

When using Coulomb analysis for analysis of a wall in an active condition, the failure wedge is commonly assumed to move down with respect to the soil beneath and with respect to the wall (see Section 2.3.1). The former is the nature of an active condition; however, the latter (that the failure wedge would move down relative to the wall) is merely an assumption. It is possible for the body of a wall to settle faster or settle more than the backfill behind it in an active condition (e.g., when the wall is heavy and/or is constructed over a weak foundation). In this situation, the failure wedge would move "up" with respect to the wall even in an active condition. The shear resistance along the wall–soil interface would then direct *downward* on the free-body diagram of the trial failure wedge, hence the line of action of force *P* will be at an angle δ on *the other side* of the normal (i.e., opposite to that shown in Figure 2.15(b)). This condition is referred to as the *negative-δ case*, while the condition shown in Figure 2.15(b) is known as the *positive-δ case*. Forces acting on a trial failure wedge and respective force polygon for both the negative-δ case and for the positive-δ case are shown in Figure 2.27.

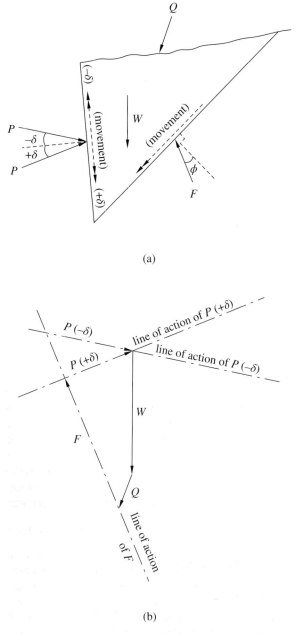

(a)

(b)

Figure 2.27 (a) Free-body diagrams of a trial failure wedge and (b) force polygons for the positive-δ and negative-δ cases (*c* = 0) (note: the double arrows indicate the sense of movement)

Although the negative-δ case is not a common occurrence in actual practice, it is important that the designer recognizes the possibility of such a condition occurring. This is because the negative-δ case will result in a "significantly larger" active force than in the positive-δ case; i.e., an error on the unsafe side when failing to recognize it, as seen in Figure 2.27(b). The difference in active forces between the negative-δ and positive-δ cases can be as large as 50%.

2.3.6 Influence of Seismic Force

During earthquakes, the active force exerted on a retaining wall will become larger, at least temporarily. For the design of walls of low to moderate height (say, 5 m or 15 ft in height or less) in low earthquake risk areas, generally no special design provision is used to account for seismic loads, although some designers would increase the active force calculated by considering only static loads by 10% to improve the safety margin. For retaining walls taller than moderate height or for walls in areas of moderate or high earthquake risks, the active force has commonly been determined by a wedge analysis method proposed by Mononobe and Okabe. Alternatively, a simple procedure proposed by Seed and Whitman (1970) for determination of active force and its location on a retaining wall may be employed for analysis and design. The Mononobe–Okabe method is briefly described in this section.

Mononobe and Matsuo (1929) and Okabe (1926) extended the Coulomb analysis and proposed a method for analysis of a rigid retaining wall subjected to seismic loads. The method accounts for earthquake-induced effects by adding pseudostatic forces on trial failure wedges. In an active condition, the pseudostatic forces take the form of a horizontal inertia force $k_h W$ (acting toward the wall) and a vertical inertia force $k_v W$ (acting upward). Figure 2.28 shows the forces on an active trail failure wedge, the associated force polygon, and a plot of θ (i.e., the assumed angle of a failure wedge) vs. the corresponding active force for determination of the active earthquake force (P_{AE}). The coefficients k_h and k_v are the ratios of the horizontal and vertical components of earthquake acceleration, respectively, to the acceleration of gravity. The values of k_h and k_v can be estimated by examining the earthquake records of the locality. A map provided by the Applied Technology Council (Richards and Elms, 1979) gives values of k_h and k_v for the United States divided into regions. The map can be used as a general guide for determination of k_h and k_v.

The free-body diagram of the failure wedge and the associated force polygon for analysis of active forces involving earthquakes, as shown in Figure 2.29, was first considered by Okabe (1926). In an independent study, Mononobe and Matsuo (1929) considered the effect of earthquakes as being equivalent to rotating the wall (along with the soil behind) by an angle θ_E (as shown in Figure 2.30 and defined in Eqn. (2-21)). The different approaches employed by Mononobe/Matuso and Okabe interestingly yield the same result. The resulting equation is often referred to as the *Mononobe–Okabe equation*, which expresses active earthquake force, P_{AE} (with combined effects of soil weight and earthquake-induced forces) as:

$$P_{AE} = \frac{1}{2} K_{AE} \gamma H^2 = \frac{K_{AE}}{K_A}(P_A)$$

(2-19)

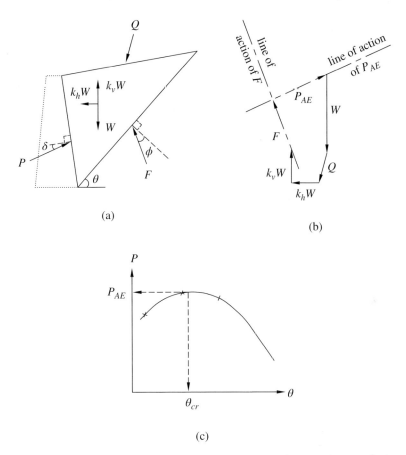

(a)

(b)

(c)

Figure 2.28 Mononobe–Okabe method for determining the active force involving an earthquake, P_{AE}: (a) free-body diagram of a trial failure wedge, (b) the associated force polygon, and (c) determination of P_{AE} from the P vs. θ curve

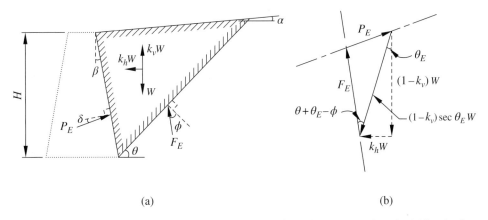

(a)

(b)

Figure 2.29 The Okabe approach for estimating the active force during an earthquake: (a) free-body diagram of a trial failure wedge and (b) the associated force polygon (modified after Okabe, 1926)

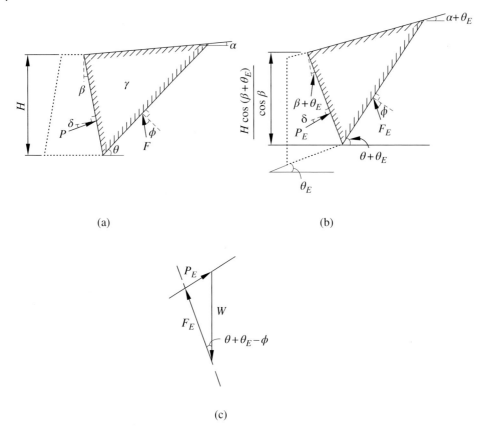

(a)

(b)

(c)

Figure 2.30 The Mononobe/Matsuo approach for estimating the active force: (a) before an earthquake, (b) during an earthquake, and (c) the associated force polygon during an earthquake (modified after Mononobe and Matsuo, 1929)

in which P_A is the Coulomb active force without consideration of seismic forces; K_A is the Coulomb active pressure coefficient, calculated using Eqn. (2-16) for selected cases; and K_{AE} is the combined active-and-earthquake coefficient, as defined below. Some design methods assume P_{AE} to act at $0.5H$ or $0.6H$ (H = wall height) above the base of the wall. In the original approaches by Mononobe and Okabe, P_{AE} is assumed to act at $1/3H$ above the wall base.

Referring to Figure 2.29(a), the combined active-and-earthquake coefficient K_{AE} in Eqn. (2-19) is

$$K_{AE} = \frac{\cos^2(\phi - \beta - \theta_E)}{\cos\theta_E \cos^2\beta \cos(\delta + \beta + \theta_E)\left[1 + \sqrt{\dfrac{\sin(\phi - \alpha - \theta_E)\sin(\phi + \delta)}{\cos(\delta + \beta + \theta_E)\cos(\alpha - \beta)}}\right]^2} \qquad (2\text{-}20)$$

in which

$$\theta_E = \tan^{-1}(k) = \tan^{-1}\left(\frac{k_h}{1-k_v}\right) \tag{2-21}$$

where k = combined seismic coefficient of acceleration $\left[k = k_h/(1-k_v)\right]$, k_h = maximum seismic coefficient of horizontal acceleration, and k_v = maximum seismic coefficient of vertical acceleration. Table 2.3 shows k_A, k_{AE}, and k_{AE}/k_A for $k_h = 0.2$, $k_v = 0.1$, $\beta = 0°$, $\delta = 0$, and $\alpha = 0°$. It is seen that the earthquake active force varies from 1.43 P_A to 1.77 P_A (i.e., an increase of 24%) as ϕ varies from 20° to 50° for the prescribed condition.

Table 2.3 Active earth pressure coefficients before and during an earthquake for different angles of internal friction of soil: for $k_h = 0.2$, $k_v = 0.1$, $\beta = 0°$, $\delta = 0$, and $\alpha = 0$

Angle of internal friction, ϕ	$\phi = 20°$	$\phi = 30°$	$\phi = 40°$	$\phi = 50°$
Before earthquake, K_A	0.49	0.33	0.22	0.13
During earthquake, K_{AE}	0.70	0.49	0.34	0.23
K_{AE}/K_A ratio	1.43	1.48	1.55	1.77

Example 2.7 illustrates how to calculate active thrust in "ordinary time" and during an earthquake event with a given value of the combined seismic coefficient of acceleration. Note that for backfills that are sloped at 3H:1V or steeper, it may not be possible to evaluate P_{AE} by Eqn. (2-20) because the term $\left[\sin(\phi - \alpha - \theta_E)\right]$ in the equation may become negative and P_{AE} will become unrealistically large when this condition is approached.

For the passive condition, the Mononobe–Okabe passive earthquake coefficient, K_{PE}, is

$$P_{PE} = \frac{1}{2}K_{PE}\gamma H^2 = \frac{K_{PE}}{K_P}(P_P) \tag{2-22}$$

and

$$K_{PE} = \frac{\cos^2(\phi + \beta - \theta_E)}{\cos\theta_E \cos^2\beta \cos(\delta - \beta + \theta_E)\left[1 - \sqrt{\frac{\sin(\phi + \alpha - \theta_E)\sin(\phi + \delta)}{\cos(\delta - \beta + \theta_E)\cos(\beta - \alpha)}}\right]^2} \tag{2-23}$$

Example 2.7 For the retaining wall shown in Figure Ex. 2.7, determine the active lateral force acting on the wall (a) in ordinary time and (b) during an earthquake event with a combined seismic coefficient of acceleration, $k = 0.2$.

(modified after Suzuki, 2008)

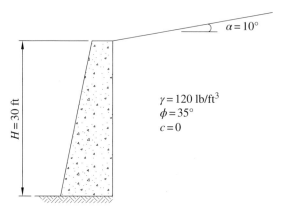

Figure Ex. 2.7

Solution:

a) In ordinary time

To obtain a conservative estimate of active force, the angle of friction at the soil–wall interface is assumed to be zero, $\delta = 0$. From Eqn. (2-16), for $\alpha = 10°$ and $\beta = 0$, the coefficient of active earth pressure K_A is

$$K_A = \frac{\cos^2 35°}{\left[1+\sqrt{\dfrac{\sin 35° \sin(35°-10°)}{\cos(-10°)}}\right]^2} = \frac{0.671}{\left[1+\sqrt{\dfrac{0.574\times0.423}{0.985}}\right]^2}$$

$$= 0.300$$

Hence the active force, P_A, is

$$P_A = \frac{1}{2}K_A \gamma H^2 = \frac{1}{2}(0.300)(120)(30)^2 = 16,200\left(\frac{\text{lb}}{\text{ft}}\right) = 16.2\left(\frac{\text{kip}}{\text{ft}}\right)$$

The active force P_A is to act at $H/3$ (or 10 ft) from the base of the wall, and oriented at 10° from the horizontal plane.

b) During an earthquake event

$$\theta_E = \tan^{-1}(k) = \tan^{-1}(0.2) = 11.3°$$

Assume a frictionless soil–wall interface, $\delta = 0$

Vertical backface, $\beta = 0$

From Eqn. (2-20),

$$K_{AE} = \frac{\cos^2(\phi-\beta-\theta_E)}{\cos\theta_E \cos^2\beta \cos(\delta+\beta+\theta_E)\left[1+\sqrt{\dfrac{\sin(\phi-\alpha-\theta_E)\sin(\phi-\delta)}{\cos(\delta+\beta+\theta_E)\cos(\alpha-\beta)}}\right]^2}$$

$$= \frac{\cos^2(35°-11.3°)}{\cos^2(11.3°)\left[1+\sqrt{\dfrac{\sin(35°-10°-11.3°)\sin(35°)}{\cos(11.3°)\cos(10°)}}\right]^2} = \frac{0.838}{1.818} = 0.461$$

The active earthquake force P_{AE} is determined from Eqn. (2-19) as

$$P_{AE} = \frac{K_{AE}}{K_A} P_A = \frac{0.461}{0.300}(16,200) = 24,900\left(\frac{\text{lb}}{\text{ft}}\right) = 24.9\left(\frac{\text{kip}}{\text{ft}}\right)$$

The active earthquake force P_{AE} may be assumed to be acting at $0.5H$ (or 15 ft) from the base of the wall, and oriented at $10°$ from the horizontal plane.

2.4 Rankine Analysis versus Coulomb Analysis

Coulomb theory was developed well over two centuries ago in 1776, and Rankine theory was developed 81 years afterwards. Selection of Rankine theory or Coulomb theory for analysis and design of a retaining wall is not a matter of personal preference or convenience. As noted at the beginning of Section 2.3, Coulomb analysis is more versatile than Rankine analysis with respect to geometric and loading conditions. We shall recall the applicability of Rankine analysis discussed in Section 2.2.7. We shall also recall an important point noted in Section 2.3 that Rankine and Coulomb analyses will give identical resultant lateral forces for both active and passive conditions for a vertical wall with a horizontal crest and frictionless soil–wall interface.

For walls that can be analyzed by both methods of analysis, Coulomb analysis has been found to give a smaller active force (i.e., is less conservative) than Rankine analysis. This is due to the fact that the frictional resistance at the soil–wall interface can be accounted for in Coulomb analysis, but not in Rankine analysis. The general applicability of Coulomb analysis (with a single planar failure surface) for cohesionless backfill walls that are free to rotate about their toe is given in Table 2.4. It is seen that the applicability depends largely on the value of soil–wall interface friction angle δ.

Table 2.4 Applicability of Coulomb analysis for walls with cohesionless backfill and free to rotate about their toe

Active condition	$\delta = 0$	Coulomb's planar slip surface is correct
	$\delta > 0$	Coulomb's planar slip surface is approximately correct
Passive condition	$\delta = 0$	Coulomb's planar slip surface is correct
	$\delta \leq \phi/3$	Coulomb's planar slip surface is subject to small errors
	$\delta > \phi/3$	Calculated passive resistance based on Coulomb's planar slip surface is unconservative (i.e., calculated value is too large); use curved slip surface

Common values of wall friction for various soil–wall interface conditions are given in Table 2.5. When using Coulomb analysis, it is recommended that $\delta = 0$ be used for routine design of rigid retaining walls so that a conservative estimate of active force is obtained. When there is a good reason that the actual wall friction should be counted on in a design, direct shear tests should be performed to evaluate the actual value of δ, otherwise $\delta = 0$ should be used.

Table 2.5 Soil–wall interface friction angle and adhesion (NAVFAC, 1986)

Structural Material	Rock or Soil Material	Friction Factor, tan δ	Interface Friction Angle δ, Degrees
Mass concrete or masonry	Clean sound rock	0.7	35
	GW, GP	0.55–0.60	29–31
	GM, GC, SW, SP	0.45–0.55	24–29
	SM, SC	0.35–0.45	19–24
	ML (non-plastic)	0.30–0.35	17–19
	CL (stiff to hard)	0.40–0.50	22–26
	CL (medium stiff to stiff)	0.30–0.35	17–19
Steel sheet piles	Well-graded rock fill with spalls	0.40	22
	GW, GP, SW, SP	0.40	22
	Hard rock fill (single size), GM, SW	0.30	17
	GM, GC, SM	0.25	14
	ML (non-plastic)	0.20	11
Formed concrete or concrete sheet piling	Welled-graded rock fill with spalls	0.40–0.50	22–26
	GW, GP, SW, SP	0.40–0.50	22–26
	Hard rock fill (single size), GW, SW	0.30–0.40	17–22
	GM, GC, SM	0.30	17
	ML (non-plastic)	0.25	14
Various	Masonry on masonry: crystalline rocks		
	Dressed soft rock on dressed soft rock	0.70	35
	Dressed hard rock on dressed soft rock	0.65	33
	Dressed hard rock on dressed hard rock	0.55	29
	Masonry on wood (cross grain)	0.50	26
	Steel on steel at sheet-pile interlocks	0.30	17

Cohesive soil s_u, tsf	Adhesion c_a, tsf
Very soft (0–0.125)	0–0.125
Soft (0.125–0.250)	0.125–0.250
Medium stiff (firm) (0.25–0.50)	0.250–0.375
Stiff (0.50–1.0)	0.375–0.475
Very stiff (1.0–2.0)	0.475–0.650

The applicability of Rankine analysis versus Coulomb analysis depends also on wall geometry and soil strength properties. As shown in Figure 2.31(a), line AB in the backfill is uninterrupted by the backface or stem of the wall, Rankine analysis would apply. On the other hand, if the development of the shear zone is restricted by the backface or stem of a wall, e.g., the cantilever wall in Figure 2.31(b), Coulomb analysis would apply.

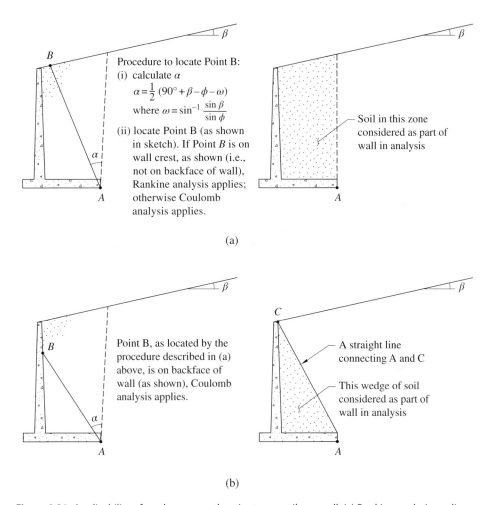

Procedure to locate Point B:

(i) calculate α

$$\alpha = \frac{1}{2}(90° + \beta - \phi - \omega)$$

where $\omega = \sin^{-1}\frac{\sin\beta}{\sin\phi}$

(ii) locate Point B (as shown in sketch). If Point B is on wall crest, as shown (i.e., not on backface of wall), Rankine analysis applies; otherwise Coulomb analysis applies.

Soil in this zone considered as part of wall in analysis

(a)

Point B, as located by the procedure described in (a) above, is on backface of wall (as shown), Coulomb analysis applies.

A straight line connecting A and C

This wedge of soil considered as part of wall in analysis

(b)

Figure 2.31 Applicability of earth pressure theories to a cantilever wall: (a) Rankine analysis applies and (b) Coulomb analysis applies (modified after Teng, 1962)

For walls constructed in front of a *constrained fill zone*, i.e., constructed in front of a firm surface of existing stiff clay or rock slope or a firm slope face formed by soil nailing, which obstructs the development of a full active condition (see Figure 2.32 as an example), neither Rankine nor Coulomb analyses would apply. In these situations, the design lateral earth pressure should correspond to at least the at-rest condition, especially when the fill is compacted by machine in small lifts (say, 8–10 in. or 20–25 cm lifts).

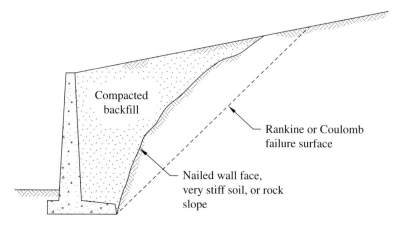

Figure 2.32 A wall with "constrained fill," i.e., with a firm limit obstructing full development of an active failure plane

2.5   Additional Topics Regarding the Design of Rigid Retaining Walls

Retaining walls that support a vertical or near-vertical soil mass by providing external support are referred to as externally supported earth retaining walls. Figure 2.33 shows schematic drawings of common types of externally supported retaining walls. The first five wall types (see Figure 2.33(a)) are considered *rigid retaining* walls, of which the deformation in the soil is small enough that they do not influence the magnitude and distribution of earth pressure acting on the wall. By contrast, the other three wall types (see Figure 2.33(b)) are considered *flexible* retaining walls, of which the deformation is sufficiently large that the magnitude and distribution of the earth pressure are influenced significantly by the extent of deformation and the interaction among the soil, wall, and internal/external restraints on the wall.

Rankine and Coulomb analyses are applicable only to rigid retaining walls. Lateral earth pressure acting on flexible retaining walls, on the other hand, can be determined by other theories that take wall yielding into account (e.g., Hansen, 1953). Empirical and semi-empirical methods incorporating variations of Rankine and/or Coulomb theories have been used for the design of flexible retaining walls. A number of these methods have been presented by Teng (1962) and Hunt (1986). Over the past two decades, the finite element methods of analysis have gradually replaced these methods for the analysis and design of flexible retaining walls.

In this section we shall discuss some additional topics related to the design of rigid retaining walls, including common proportions of walls, design charts for estimation of active force, equivalent fluid density, compaction-induced stress, evaluation of wall stability, and selection of shear strength parameters of fill material in the design of rigid retaining walls.

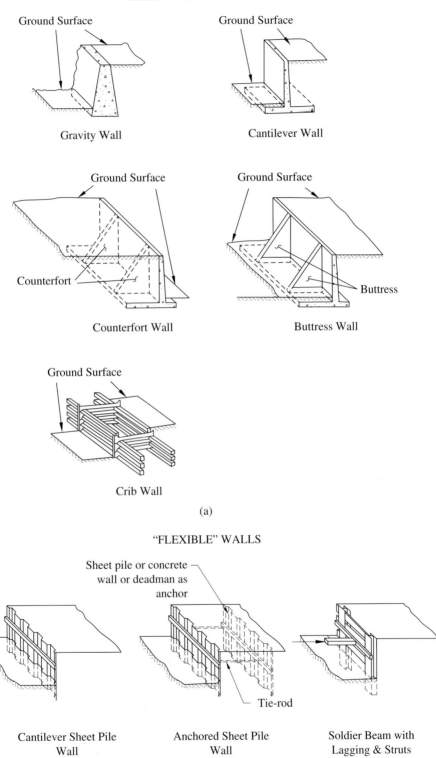

Figure 2.33 Common types of earth retaining walls: (a) rigid retaining walls and (b) flexible retaining walls (modified after Perloff and Baron, 1976, and McCarthy, 2007)

2.5.1 Common Proportions of Rigid Retaining Walls

As in the design of most civil engineering structures, the detailed design of an earth retaining wall typically starts with a preliminary choice of wall type and tentative dimensions. Upon checking the stability and structural adequacy, the tentative dimensions are revised to obtain final dimensions under prescribed design criteria. Figure 2.34 shows common proportions for three types of rigid retaining walls. They are useful for the selection of tentative dimensions.

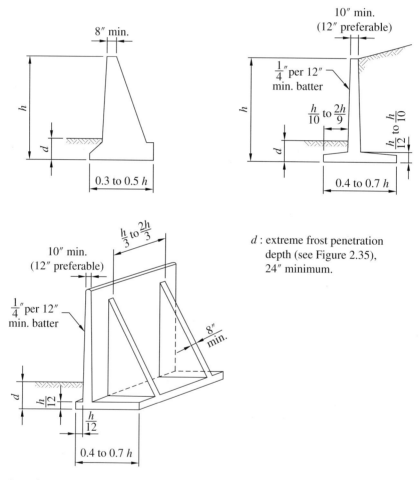

Figure 2.34 Common proportions of three types of rigid retaining walls (modified after Teng, 1962 and Hunt, 1986)

Most earth retaining walls have been designed with a minimum embedment of 0.6 m (2 ft) or the maximum frost penetration depth on the wall site, whichever is greater. The embedment may involve a substantial volume of excavation and backfill. The contours of extreme frost penetration depths in the United States as shown in

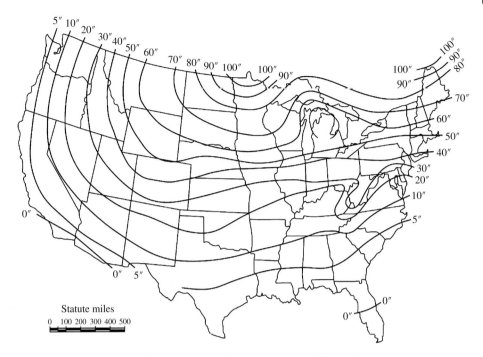

Figure 2.35 Map of extreme frost penetration depths in the United States, based on state average, given by the National Oceanic and Atmospheric Administration

Figure 2.35 can be used as guide. Note that the actual depths of extreme frost penetration may vary considerably depending on cover, soil type, moisture, and weather.

2.5.2 Design Charts for Estimation of Active Force

For rigid retaining walls not taller than 20 ft (6 m) in height, conservative values of active force can be obtained from simple design charts. A series of design charts is given in Figure 2.36 for walls with straight slope backfill (including horizontal crest) and for walls with broken slope backfill. For design of gravity walls with a sloping back, cantilever walls, and counterfort walls, the soil wedge projected through the heel is commonly considered an added weight to the wall. The active forces determined from Figures 2.36 are assumed to act on a vertical plane projected through the heel of the wall.

2.5.3 Equivalent Fluid Density

The concept of *equivalent fluid density* has been used widely for the design of rigid retaining walls. Despite its widespread use in practice, the concept may be an abomination of earth pressure theories. The concept corresponds strictly to Rankine earth pressure theory for a unique case of walls with a horizontal crest without surcharge. The equivalent fluid density γ_f is in fact the product of K_A and γ. A specific value of γ_f is often

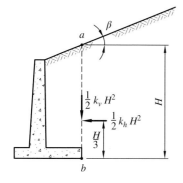

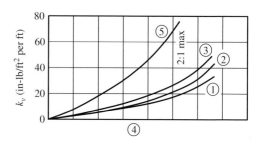

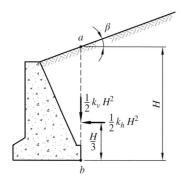

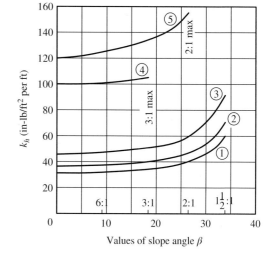

Notes:

Numerals on curves indicate soil types as described below.

For material of type 5, computations should be based on value of H four feet less than actual value.

<u>Types of backfill for retaining walls</u>

① Coarse-grained soil without admixture of fine soil particles, very free-draining (clean sand, gravel or broken stone).

② Coarse-grained soil of low permeability due to admixture of particles of silt size.

③ Fine silty sand; granular materials with conspicous clay content; or residual soil with stones.

④ Soft or very soft clay; organic silt; or soft silty clay.

⑤ Medium or stiff clay that may be placed in such a way that negligible amount of water will enter the spaces between the chunks during floods or heavy rains.

(a)

Figure 2.36 Charts for conservative estimates of active force for wall height not more than 20 ft (6 m): (a) walls with straight slope backfill and (b) walls with broken slope backfill (NAVFAC, 1986)

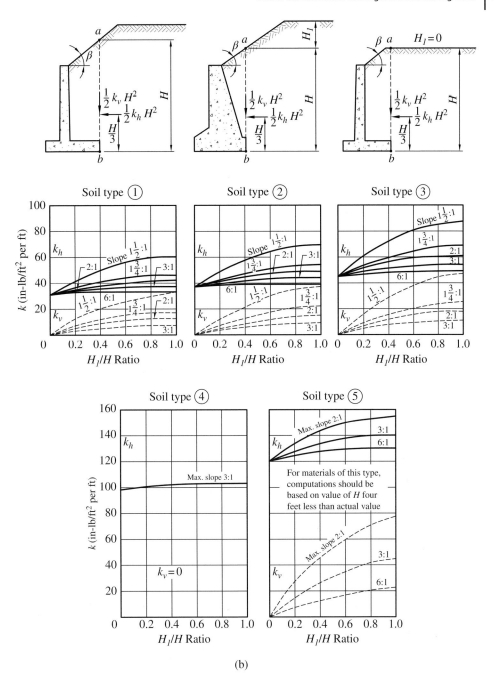

Figure 2.36 (Continued)

used as an *office standard* without consideration of other factors in some consulting firms. It has been a serious issue that the concept is misused in cases where the crest is not horizontal, as shown in Figure 2.37. Note that the misuse will result in an unconservative estimate of active force.

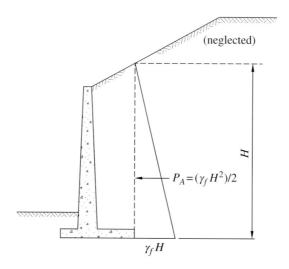

Figure 2.37 Misuse of the equivalent fluid density concept

2.5.4 Compaction-Induced Stress

When a soil mass is subjected to an increase in vertical load, there will generally be an increase in vertical and horizontal stresses in the soil mass. The increase in vertical stress has commonly been evaluated by the theory of elasticity. The increase in horizontal stress, however, depends largely on the constraint to lateral deformation of the soil mass. The higher the constraint, the larger the increase in horizontal stress. If the increase in vertical load is subsequently removed, the increased vertical stress will reduce to a very small value or even zero; however, the increase in horizontal stress may only reduce slightly. The *net* increase in horizontal stress that remains in the soil mass when it is subject to a loading–unloading sequence is commonly referred to as the *residual* or *lock-in* lateral stress. An increase in vertical load followed by subsequent removal of the load is typical in fill compaction operation during wall construction. The residual lateral stress due to compaction is referred to as *compaction-induced stress*.

Compaction-induced stress in a soil mass has been the subject of study by many researchers, including Rowe (1954), Broms (1971), Aggour and Brown (1974), Seed (1983), Seed and Duncan (1986), and Duncan et al. (1991). These studies have indicated that compaction-induced stress will result in a significant increase in the lateral stress of a soil mass if there is sufficient constraint to the lateral movement of the soil. Figure 2.38 shows a design earth pressure diagram due to different types of compaction equipment. The earth pressure diagram was suggested by the U.S. Army Corps of Engineers and has been used in some design communities. In other design communities, compaction-induced stress is either ignored or not addressed explicitly. Compaction-induced stress in reinforced soil is addressed in Section 3.3.3.

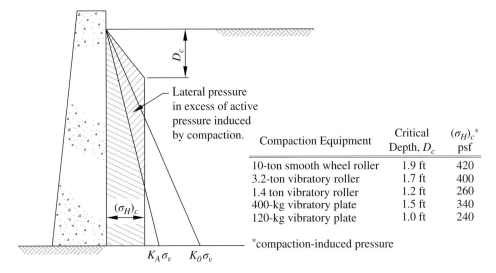

Figure 2.38 Design earth pressure diagrams resulting from different types of compaction equipment, given by the U.S. Army Corps of Engineers

2.5.5 Evaluation of Wall Stability

The following criteria are routinely checked for evaluation of stability in the design of a rigid retaining wall:

- Structural components of the retaining wall itself must be capable of resisting internal shears, bending moments, and tensions developed as a result of lateral earth pressure and other loads, including surcharge and seepage forces (see Figure 2.39(a)).

- The retaining wall must be safe against overturning about the toe, as shown in Figure 2.39(b). A minimum safety factor against overturning failure of 1.5 for granular backfill and 2.0 for cohesive backfill is generally required.

- The retaining wall must be safe against forward sliding at the base (see Figure 2.39(c)). If only the resistance along the base is considered, the safety factor against sliding failure should be at least 1.5 in most design criteria. Passive resistance P_p is often ignored or $0.5P_p$ is used instead of P_p. If passive resistance of the soil in front of the wall is included in the analysis, the minimum safety factor is often raised to 2.0.

- The bearing capacity of the foundation soil supporting the retaining wall must be sufficient to prevent bearing failure (see Figure 2.39(d)). Some design criteria require a minimum safety factor against bearing failure of 2.0 for granular foundation and 3.0 for cohesive foundation.

- The soil mass around the retaining wall must be safe against an overall slope failure (see Figure 2.39(e)). The minimum safety factor for this mode of failure is usually set as 1.3 or 1.5.

(a) Structural Stability

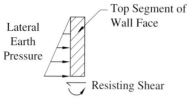

*Check structural strength of
each component of the wall*

(b) Overturning

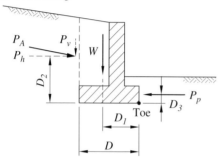

$$F_s = \frac{Resisting\ Moment\ about\ the\ Toe}{Overturning\ Moment\ About\ the\ Toe}$$

$$= \frac{WD_1 - P_v D + P_p D_3}{P_h D_2}$$

(c) Sliding

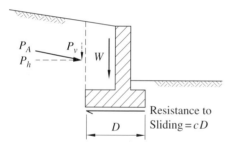

$$F_s = \frac{Resisting\ Force}{Driving\ Force}$$

$$= \frac{(W + P_v)\ f + cD}{P_h}$$

$c = $ *Cohesion of the soil at wall base*
$f = $ *Friction factor between wall
base and soil*

(d) Bearing

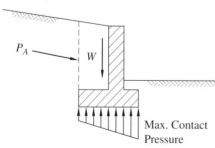

$$F_s = \frac{Ultimate\ Bearing\ Capacity}{Max.\ Base\ Contact\ Pressure}$$

Figure 2.39 Failure modes and evaluation of factors of safety for design of rigid retaining walls

(e) Overall Slope Stability

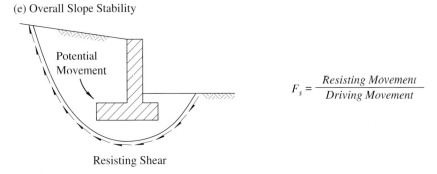

$$F_s = \frac{Resisting\ Movement}{Driving\ Movement}$$

Figure 2.39 (Continued)

In addition, total and differential settlements of the retaining wall due to compression of the foundation soil and the backfill must be limited to tolerable values to eliminate damage to the wall and wall-supported structures, such as pavements, buildings, machines, etc.

2.5.6 Selection of Shear Strength Parameters in Design

Whenever feasible, free draining materials (coefficient of permeability no less than 10^{-4} cm/sec, e.g., clean granular soils) should be employed as backfill for the design and construction of earth retaining walls. With a free draining backfill, the analysis of wall stability involving the backfill should be based on the effective strength parameter ϕ', with $c' = 0$.

Backfill used in wall construction needs to be substantially free of clay, silt, or other poor-durability particles (e.g., shale, mica, gypsum) and usually should contain no more than 1% organic material because such soils tend to deform significantly with time. Studies of existing retaining walls (Peck, 1948) have indicated that a majority of failures of rigid retaining walls are associated with walls supported on, and/or backfilled with clay, silt, or clayey soils. If only such soils are available for construction, they should be carefully compacted by using sheepsfoot rollers so that all the chunks are broken up and no conspicuous voids are left in place. To avoid wetting-induced movement, it is recommended that a clayey soil is compacted in a wetter state as it will be less likely to attract moisture after fill placement.

With a cohesive backfill, the selection of strength parameters for a retaining wall depends on the time element. If short-term stability controls, undrained strength parameters (c_u and ϕ_u) should be used. On the other hand, drained strength parameters (c' and ϕ') may be used for conditions controlled by long-term stability, as discussed in Section 1.5.2. For a curved Mohr–Coulomb failure envelope, a secant line over the range of interest in stress should be used for determination of the strength parameters, see Section 1.5.1.

References

Aggour, M.S. and Brown, C.B. (1974). The Prediction of Earth Pressure on Retaining Walls Due to Compaction. *Geotechnique*, 24(4), 489–502.

Bishop, A.W. (1958). Test Requirements for Measuring the Coefficient of Earth Pressure at Rest. *Proceedings, Brussels Conference on Earth Pressure Problems, Volume 1*, pp. 2–14.

Bowles, J.E. (1996). *Foundation Analysis and Design*, 5th edition, McGraw-Hill, New York, 1175 pp.

Brooker, E.W. and Ireland, H.O. (1965). Earth Pressure at Rest Related to Stress History. *Canadian Geotechnical Journal*, 2(1), 1–15.

Broms, B. (1971). Lateral Earth Pressure due to Compaction of Cohesionless Soils. *Proceedings, 4th Budapest Conference on Soil Mechanics and Foundation Engineering*, pp. 373–384.

Craig, R.F. (1978). *Soil Mechanics*, 2nd edition. Van Nostrand Reinhold Company, New York, 318 pp.

Duncan, J.M., Williams, G.W., Sehn, A.L., and Seed R.M. (1991). Estimation Earth Pressures due to Compaction. *Journal of Geotechnical Engineering, ASCE*, 117(12), 1833–1847.

Hansen, J.B. (1953). *Earth Pressure Calculation*. Danish Technical Press, Institute of Danish Civil Engineering, Copenhagen.

Henry, F.D.C. (ed.) (1986). *Design and Construction of Engineering Foundations*. Chapman & Hall, 600 pp.

Hunt, R.E. (1986). *Geotechnical Engineering Analysis and Evaluation*. McGraw-Hill, New York, 729 pp.

Jaky, J. (1948). Pressure in Silos. *Proceedings, 2nd International Conference of Soil Mechanics and Foundation Engineering, Rotterdam, Volume I*, pp. 103–107.

Kulhawy, F.H., Jackson, C.S., and Mayne, P.W. (1989). First-Order Estimation of K_0 in Sands and Clays. *Foundation Engineering: Current Principles and Practices, Volume 1*, Kulhawy (ed.). ASCE, New York, pp. 121–134.

Mayne, P.W. and Kulhawy, F.H. (1982). K_0-OCR (At rest Pressure – Overconsolidation Ratio) Relationships in Soil. *Journal of the Geotechnical Engineering Division, ASCE*, 108(GT6), 851–872.

McCarthy, D.F. (2007). *Essentials of Soil Mechanics and Foundations, Basic Geotechnics*, 7th edition. Prentice Hall, Upper Saddle River, New Jersey, 850 pp.

Mononobe, N. and Matsuo, H. (1929). On the Determination of Earth Pressure during Earthquakes. *Proceedings, the World Engineering Conference, Volume 9*, str. 176.

NAVFAC (1986). *Design Manual 7.02 Foundations and Earth Structures*. Bureau of Yards and Docks, U.S. Navy.

Okabe S. (1926). General Theory of Earth Pressure. *Journal of the Japanese Society of Civil Engineers*, 12(1).

Peck, R.B. (1948). *History of Building Foundations in Chicago*. Bulletin 373, University of Illinois Engineering Experimental Station, 64 pp.

Perloff, W.H. and Baron, W. (1976). *Soil Mechanics – Principles and Applications*. John Wiley and Sons, 745 pp.

Richards, R.J. and Elms, D. (1979). Seismic Behavior of Gravity Retaining Walls. *Journal of Geotechnical Engineering Division, ASCE*, 105(GT4), 449–464.

Rowe, P.W. (1954). A Stress-Strain Theory for Cohesionless Soil with Applications to Earth Pressures at Rest and Moving Walls. *Geotechnique*, 4(2), 70–88.

Seed, H.B. and Whitman, R.V. (1970). Design of Earth Retaining Structures for Dynamic Loads. In *Lateral Stresses in the Ground and Design of Earth Retaining Structures*. ASCE, New York, pp. 103–107.

Seed, R.M. (1983). *Compaction-induced Stresses and Deflections on Earth Structure*. Doctoral Dissertation, Department of Civil Engineering, University of California, Berkeley, 447 pp.

Seed, R.M. and Duncan, J.M. (1986). FE Analyses: Compaction-Induced Stresses and Deformations. *Journal of Geotechnical Engineering, ASCE*, 112(1), 23–43.

Shields, D.H. and Tolunay, A.Z. (1973). Passive Pressure Coefficients by Method of Slices. *Journal of the Soil Mechanics and Foundations Division, ASCE*, 99(SM12), Paper 10221, 1043–1053.

Sowers, G.B. and Sowers, G.F. (1979). *Introductory Soil Mechanics and Foundations*, 4th edition. Macmillan, New York, 621 pp.

Spangler, M.G., and Handy, R.L. (1982). *Soil Engineering*, 4th edition. Harper and Row, New York, 819 pp.

Suzuki, O. (2008). *Geotechnical Engineering Example Exercise* (in Japanese), 3rd edition. Toyo Publisher, Tokyo, 424 pp.

Swedish Road Administration (2008). *Bridge Design Code for New Constructions – Bro 2004* (in Swedish).

Teng, W.C. (1962). *Foundation Design*. Prentice Hall, Englewood Cliffs, New Jersey, 466 pp.

Terzaghi, K. (1943). *Theoretical Soil Mechanics*. John Wiley & Sons, New York.

Tschebotarioff, G.P. (1973). *Soil Mechanics, Foundations and Earth Structures*, 2nd edition. McGraw-Hill, New York, 642 pp.

Yasukawa, I., Imanishi, K., and Tateishi, Y. (2012). *Soil Mechanics with Sketches* (in Japanese), 2nd edition, Ohmsha Publisher, Tokyo, 216 pp.

3

Reinforced Soil and Geosynthetic Reinforced Soil (GRS) Walls

The idea of incorporating tensile inclusion to reinforce a soil mass is not at all new; it is known to have existed for at least 6,000 to 7,000 years. Modern reinforced soil technology was reintroduced in the early 1960s. Since then, many reinforced soil wall systems have been developed for earthwork construction where a sudden change in grade is necessary or desired. To date, about 200,000 reinforced soil walls have been constructed worldwide, and the number is growing at an increasing rate. These reinforced soil wall systems can be grouped into two categories: externally stabilized reinforced walls and internally stabilized reinforced walls. For the former, the tensile inclusion embedded in the fill material serves primarily as frictional quasi-tiebacks, and the fill is *retained* by facing that is connected to the tensile inclusion to form a stable system. If facing fails, failure of an externally stabilized reinforced wall would generally be imminent. For the latter, on the other hand, the tensile inclusion embedded in fill material serves to improve the properties of the soil so that and the soil–geosynthetic composite would be stable by itself. Facing, in this case, is mostly just a façade.

The underlying design concept of externally stabilized reinforced walls is somewhat similar to that of conventional earth retaining walls (such as gravity or cantilever reinforced concrete walls), except that instead of relying on the weight of wall structure to stabilize the backfill, the external support is derived from facing (serving to retain the backfill) and tensile reinforcement (serving to stabilize the facing). These externally stabilized reinforced walls are often referred to as mechanically stabilized earth (or MSE) walls. MSE walls with geosynthetics as reinforcement are sometimes referred to as GMSE walls.

The fundamental design concept for an internally stabilized reinforced walls is quite different from that of externally stabilized reinforced walls. The backfill in internally stabilized reinforced walls is "reinforced" by geosynthetic inclusion in such a way that the soil–geosynthetic composite needs to be sufficiently stable under its self-weight and upon load applications. Internally stabilized soil walls rely on soil–reinforcement interaction to generate different reinforcing mechanisms to achieve a stable system. Facing in internally stabilized reinforced walls is not a load-carrying component in design.

This chapter focuses on internally stabilized walls. We shall begin with a discussion of the theory of *reinforced soil* in general and *geosynthetic reinforced soil* (GRS) in

Geosynthetic Reinforced Soil (GRS) Walls, First Edition. Jonathan T.H. Wu.
© 2019 John Wiley & Sons Ltd. Published 2019 by John Wiley & Sons Ltd.

particular, followed by a discussion of various reinforcing mechanisms of internally stabilized soil walls, and conclude with a description of some common types of GRS walls. It should be noted that because GRS walls (walls reinforced with closely spaced geosynthetic reinforcement) do not count on external support to achieve stability, they are referred to as *soil walls*, as opposed to *soil (or earth) retaining walls*.

3.1 Reinforced Soil and GRS

Soil is relatively strong in resisting shear stress and compressive normal stress, but is very weak in resisting tensile normal stress. The concept of reinforced soil in some respects is quite similar to that of reinforced concrete. In reinforced soil, tensile reinforcement embedded in a soil mass plays the role of resisting tensile stress, much like the role of steel in reinforced concrete. There are, however, significant differences between reinforced soil and reinforced concrete. Other than the obvious difference in parent material (unbonded soil vs. bonded aggregate and cement) and reinforcement material (geosynthetics or other high tension-resistance materials vs. rebar), a more important difference is in the reinforcing mechanism. In reinforced concrete, the reinforcement (rebar) bonds with aggregate when the cementitious material in concrete hardens. The bonding allows the reinforcement and the concrete to deform together as long as strain compatibility is maintained. In reinforced soil, however, the interface bonding between the fill material and the geosynthetic reinforcement derives primarily from friction, which is strongly dependent on the normal stress. More importantly, the role of reinforcement in reinforced soil goes far beyond resisting tensile stress. There are many other mechanisms by which reinforcement brings about reinforcing effects in a reinforced soil mass. These mechanisms are discussed in Section 3.3.

Tensile reinforcement (also referred to as *tensile inclusion*) in reinforced soil may take a few different forms, such as planar sheets, strips, mats, mesh, or three-dimensional cells. This book focuses on planar sheet reinforcement; the design principles for other forms of reinforcement are essentially the same as that for planar sheet reinforcement. The tensile reinforcement may also be made of different materials. Natural materials, such as reeds, straws, and bamboos, were used as reinforcement in ancient times and are still being used in some developing countries. Since the early 1960s, however, the reinforcement has almost exclusively been man-made materials, including metallic and polymeric materials, with the latter being referred to as *geosynthetics*.

The use of metallic members (in the form of strips, wires, cables, or plates) as reinforcement is attractive to civil engineers because they have been used in earthwork construction for centuries and with very good success. Polymeric material, on the other hand, is relatively new in earthwork construction. Polymeric reinforcement, however, offers several distinct advantages over metallic reinforcement, including lower costs, easier to transport to construction site, fewer life-cycle maintenance activities, and higher resistance to environmental attack, including attack by water. The tradeoffs between metallic and polymeric reinforcements are discussed in Section 4.3.

Figure 3.1 shows a set of stress–strain curves of reinforced and unreinforced sands obtained from a series of triaxial compression tests (Broms, 1977). For the reinforced sands, sheets of a geotextile are cut into small disks and placed horizontally inside soil specimens in different layouts, as seen in Figure 3.1. Tests were performed at low (20 kPa) and high (200 kPa) confining pressures. It is seen that, other than at very small strains, the reinforced sand exhibits appreciably higher stiffness (i.e., steeper slopes on stress–strain plots) and higher shear strength than the unreinforced sand. Note that the behavior of reinforced soil with geotextile disks at the top and bottom of the test specimens is nearly the same as that of unreinforced soil because these geotextile disks hardly play a role in resisting tensile stress.

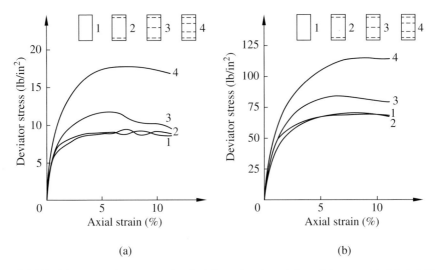

(a)   (b)

Figure 3.1 Triaxial compression test results of reinforced and unreinforced dense sands: (a) at 20 kPa (3 psi) confining pressure and (b) at 200 kPa (30 psi) confining pressure (Broms, 1977, taken after Koerner, 1994)

The stress–strain curves in Figure 3.1 indicate that the presence of reinforcement has little effect under very small strains. Results of other laboratory triaxial tests have shown a similar behavior, in that a threshold reinforcement strain of about 0.5% is needed to begin mobilizing the reinforcing effect (e.g., Gray and Al-Refeai, 1986; Tafreshi and Asakereh, 2007). Recent studies using field-scale specimens (see Section 3.2), however, have disputed the magnitude or even the existence of such a threshold strain.

Some laboratory tests have suggested that a reinforced soil may even be less stiff than an unreinforced soil at axial strains between 0 to 1–2% (e.g., Gray and Al-Refeai, 1986; Tafreshi and Asakereh, 2007). This behavior would have been a serious issue, had it been true. After all, maximum geosynthetic reinforcement strains under typical service loads in reinforced soil walls have been found to be usually lower than 1.0%, and almost always less than 2.0%. Fortunately, the low stiffness of the reinforced soil at lower strains has been found to be due entirely to compression of the geotextile layers in the test specimens (Liu, 1987). The relative thickness of the reinforcement layers in those laboratory tests is much larger than that in actual construction, therefore the effect of compression in reinforcement thickness is much exaggerated in those laboratory tests. In other words, the reinforced soil being weaker than unreinforced soil at low strains is due to compression of geosynthetic disks. The problems with most reduced-scale experiments for geosynthetic reinforced soil are discussed in greater detail in Sections 3.2.3 and 3.3.2. This suggests strongly the need for field-scale experiments (where reinforcement spacing in field installation is properly simulated) when studying reinforced soil behavior.

Figure 3.2 shows the distribution of vertical and lateral stresses in an unreinforced soil mass and a reinforced soil mass subjected to a vertical load (6 kN) on the top surface, as obtained from plane-strain finite element analysis. The two soil masses have identical geometry and properties except that the reinforced soil mass contains three sheets of geosynthetic reinforcement, equally spaced at the top, mid-height, and bottom, on 0.3 m spacing. As seen in Figure 3.2(a), there is little difference in vertical stress between the unreinforced and reinforced soils. The lateral stress in reinforced soil, however, is significantly higher, as seen in Figure 3.2(b). The increase in lateral stress over unreinforced soil is the highest immediately adjacent to the reinforcement and diminishes with distance away from the reinforcement. The increase in lateral stress due to reinforcement is only significant within about 0.1 m from the reinforcement and becomes fairly small at about 0.15 m from the reinforcement. This suggests that when reinforcement spacing is greater than about 0.2–0.3 m, the interaction between soil and reinforcement will become much less significant under the conditions of this example.

Geosynthetic reinforced soil (GRS) refers to a reinforced soil mass with geosynthetic inclusion on tight vertical spacing, generally not more than 0.2 m (8 in.) and never greater than 0.3 m (12 in.). In the literature, the term "GRS" has sometimes been used to describe a soil mass with geosynthetic reinforcement on *any* spacing. In light of the significant benefits of tight reinforcement spacing, however, GRS herein refers specifically to geosynthetic reinforced soil on small reinforcement spacing. Unless otherwise noted, this definition is used throughout this book.

When new earthwork construction technologies are being developed, field practice often leads the way, then research follows. The benefit of closely spaced reinforcement was first realized by wall builders in actual construction. The finding was later validated by a number of field-scale experiments (see Section 3.2) and analytical studies. Now, many reinforcing mechanisms by which soil is reinforced by geosynthetic reinforcement have been identified, and close reinforcement helps to activate or enhance most of these mechanisms (see Section 3.3). It is true that tight reinforcement means a greater number of layers in a wall. This, however, does not necessarily mean higher costs because reinforcement of lower stiffness/strength can be used with closer spacing. The material cost of geogrid reinforcement in a geosynthetic reinforced soil wall typically ranges between 15% and 20% of the total project cost; the percentage is slightly smaller when woven geotextiles are used. The total cost of geosynthetic reinforcement in GRS walls, with a greater number of layers but lower stiffness/strength values, is typically comparable to that of reinforced soil walls with fewer layers of reinforcement but higher stiffness/strength values.

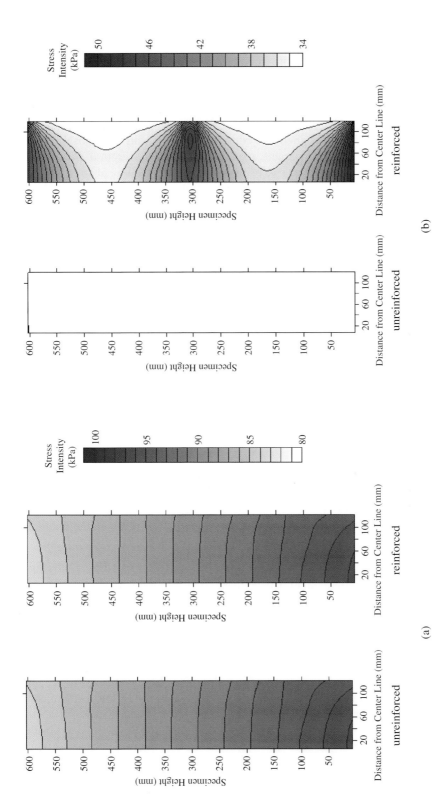

Figure 3.2 Stress distribution in an unreinforced soil mass vs. a reinforced soil mass in plane-strain condition: (a) vertical stress distribution and (b) lateral stress distribution (Ketchart and Wu, 2001)

In actual construction, GRS walls have demonstrated many distinct advantages over conventional earth retaining walls, including the following:

- GRS walls are more flexible, hence inherently more tolerant to foundation movement.

- GRS walls are more adaptable to lower quality backfill.

- When properly designed and constructed, GRS walls with closely spaced geosynthetic reinforcement are remarkably stable. From 1990 to 2010, researchers were not able to bring about complete failure of any field-scale GRS until the Federal Highway Administration (FHWA) was finally able to bring about failure of closely spaced GRS columns using a unusual loading setup, and the load-carrying capacities were over 1,000 kPa or 20,000 lb/ft^2 (Adams et al., 2011b; Nicks et al., 2013).

- Construction of GRS walls is rapid and requires only ordinary construction equipment.

- GRS walls do not require embedment into the foundation soil to maintain stability. This is especially important when the project involves excavation of contaminated foundation soil.

- When used in conjunction with "deformable" facing (say, dry-stacked light-weight concrete blocks with friction-only connection between blocks), the lateral earth pressure exerted on the wall face is very small.

- GRS walls typically involve reduced life-cycle maintenance costs and activities.

- GRS walls are generally less expensive (typical savings of about 20–30%) compared with conventional earth retaining walls.

With these advantages of GRS walls, however, designers should be cautioned not to take a lackadaisical attitude toward investigation of site conditions. Unfavorable site conditions can often lead to problems associated with external stability (i.e., stability surrounding the reinforced soil mass, involving the soil beneath and behind the reinforced soil mass, see Section 5.1).

One common concern of reinforcing a soil wall, including both GRS and GMSE, with geosynthetics as reinforcement is that their *system stiffness* may not be as high as reinforced earth walls with metallic strip reinforcement. However, studies have clearly demonstrated that the stiffness of GRS and GMSE wall systems is much more than needed to fulfill their intended functions as long as the fill is well compacted (see Section 3.2). The not-as-high stiffness of geosynthetic reinforced soil walls is only relevant in applications where wall deformation needs to be kept to an absolute minimum, e.g., walls for high-speed railways. Tatsuoka et al. (2005) showed that using cement-mixed gravelly soil with geogrid reinforcement can produce a soil–reinforcement composite of extremely high stiffness. In many parts of Japan, Japan Railway has used cement-mixed gravels successfully in construction of reinforced soil walls and bridge abutments for *shinkansen* (bullet-train railway lines) where tolerance of deformation is very small.

Berg et al. (2009) list potential disadvantages of reinforced soil walls as: (i) require-ment of a relatively large space for excavation, (ii) use of select granular free-draining backfill, and (iii) often needing a shared design responsibility between geosynthetic material suppliers and owners. While (i) is debatable (a *true* GRS wall system with tight reinforcement spacing of 0.2 m or less requires little to no excavation unless the founda-tion soil is too weak to support the wall, see Step 12, Section 5.7.1), (ii) is shared by all earth retaining structures; however, (iii) is indeed a distinct disadvantage of GRS walls. The GRS wall components manufacturer/supplier is usually responsible for the internal stability (see Section 5.1) of the system (as they are usually intimately familiar with the system components, hence able to optimize the design of the system), while the respon-sibility for external stability (see Section 5.1) typically falls on the owner or his/her rep-resentative. Shared responsibility, when needed, requires good coordination for a successful project.

3.2   Field-Scale Experiments of GRS

This section describes and discusses three sets of experiments conducted to investigate the load–deformation behavior of GRS with closely spaced reinforcement. These experiments are of *field-scale* in that the reinforcement spacing in the experi-ments are typical of what has been used in actual construction of GRS walls in the field. The three sets of experiments are: (i) unconfined "mini pier" experiments by Adams et al. (2002, 2007), (ii) unconfined compression experiments by Elton and Parawaran (2005), and (iii) generic soil-–geosynthetic composite (SGC) plane-strain experiments by Wu et al. (2010).

3.2.1   "Mini Pier" Experiments (Adams et al., 2002, 2007)

Five "mini pier" experiments were conducted and reported by Adams and his col-leagues at the Turner-Fairbank Highway Research Center, Federal Highway Administration (Adams et al., 2002, 2007). The test specimens were cuboidal in shape (referred to as "mini piers"), measuring 1.0 m by 1.0 m in cross-section and 2.0 m in height. The mini piers were subjected to incrementally increasing vertical loads on the top surface under an unconfined condition. Figure 3.3(a) shows a mini pier experiment after it has been loaded to failure. Of the five experiments, one was conducted without reinforcement, two were on reinforcement spacing of 0.2 m, and the other two on mixed reinforcement spacing of 0.4 m to 0.6 m. Two woven geotextiles of different stiffness/strength were used as reinforcement, with wide-width strengths (T_f) of 70 kN/m and 21 kN/m.

The stress–strain curves of the five mini piers are shown in Figure 3.3(b). It is seen that the reinforced piers, especially Piers C and D on reinforcement spacing of 0.2 m, exhibit much higher stiffness and strength than the unreinforced pier (Pier A). The relative influence of reinforcement spacing vs. reinforcement strength on stress–strain behavior can be evaluated by inspecting carefully the difference in stress–strain

(a)

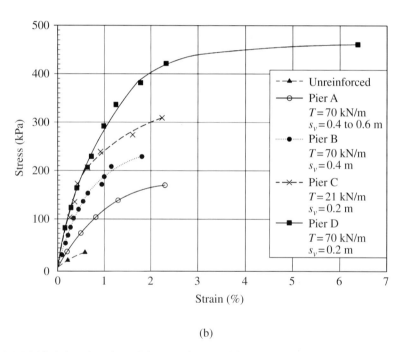

(b)

Figure 3.3 (a) A "mini pier" reaches a failure condition and (b) stress–strain relationships of all five "mini piers" (modified after Adams et al., 2007)

relationships between Pier D (on 0.2 m spacing) and Pier B (on 0.4 m spacing), as compared to the difference between Pier D (T_f = 70 kN/m) and Pier C (T_f = 21 kN/m). It is obvious that the former is much greater than the latter, indicating that the influence of reinforcement spacing is greater than that of reinforcement strength. For a quantitative evaluation, the applied vertical pressures needed to bring about a global vertical strain of 1.0% are 190 kPa, 245 kPa, and 300 kPa for Piers B, C, and D, respectively. Considering Pier D as a baseline for comparisons, the required vertical pressure to reach 1.0% strain for Piers B and C is, respectively, 1.6 times and 1.2 times that for Pier D, while the reinforcement strength of Pier C is 3.5 times that of Pier D, yet the spacing of Pier B is only twice that of Pier D. This series of experiments provide solid evidence that reinforcement spacing plays a more important role than reinforcement strength in the load–deformation behavior of reinforced soil.

3.2.2   Unconfined Compression Experiments (Elton and Patawaran, 2005)

Elton and Patawaran (2005) conducted seven field-scale unconfined compression experiments on 0.76 m diameter, 1.5 m high cylindrical specimens (see Figure 3.4). Reinforcement spacing was 0.15 m for all the experiments, except one with 0.3 m. The reinforcements were needle-punched nonwoven geotextiles of different weights, designated TG500, TG600, TG700, TG800, TG1000, and TG028, with respective ultimate strength of 9, 14, 15, 19, 20, and 25 kN/m, as shown in Table 3.1. The soil was a poorly graded sandy soil with less than 5% fines and a maximum grain diameter of 12.7 mm. Modified Proctor compaction tests (ASTM D1557) of the soil gave an optimum moisture content of 9.3% and a maximum dry unit weight of 19.0 kN/m^3. Direct shear tests on the sand in a moist compacted state (at a wet unit weight of 18.5 kN/m^3) gave an angle of internal friction (ϕ) of 40° with cohesion (c) of 29 kPa.

(a)   (b)

Figure 3.4 Soil–geosynthetic composite specimen in a unconfined compression test (a) before loading and (b) at failure (Elton and Patawaran, 2005)

Table 3.1 Test conditions and failure loads of unconfined compression tests (Elton and Patawaran 2005)

Designation		TG 500	TG 500-12	TG 600	TG 700	TG 800	TG 1000	TG 028
Reinforcement strength, T_f (kN/m)	(Machine)	9	9	14	15	19	20	25
	(Cross-Machine)	14	14	19	20	20	23	22
Reinforcement spacing, s_v (m)		0.15	0.30	0.15	0.15	0.15	0.15	0.15

The stress–strain curves of the seven soil–geosynthetic composites are shown in Figure 3.5. It is seen that reinforcement strength influences the stress–strain behavior appreciably. The curves follow a general trend, namely, the stronger the reinforcement, the higher the strength of the soil–geosynthetic composite. The experiment designated TG500-12, the only one on reinforcement spacing of 0.3 m, shows much lower composite strength than TG500 which has the same reinforcement as TG500-12, but on half the reinforcement spacing. The influence of reinforcement spacing is seen to be very significant.

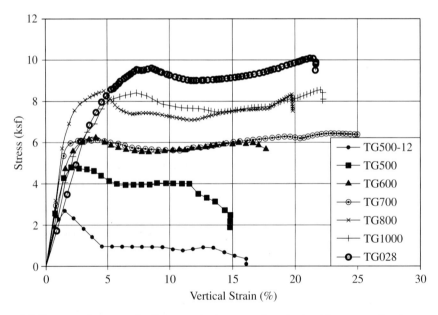

Figure 3.5 Stress–strain curves of soil–geosynthetic composites obtained from unconfined compression tests (Elton and Patawaran, 2005)

3.2.3 Generic Soil–Geosynthetic Composite Plane-Strain Experiments (Wu et al., 2010, 2013)

Wu et al. (2010, 2013) conducted and reported the results of four generic plane-strain soil–geosynthetic composite (SGC) experiments and, for comparison purposes, one soil-only experiment (see Figure 3.6). Since a soil mass with layers of geosynthetic

Figure 3.6 A plane-strain soil–geosynthetic composite (SGC) experiment conducted at TFHRC of the Federal Highway Administration

inclusion is far from being *uniform*, it was deemed necessary that the test specimen be sufficiently large to provide "representative" behavior of soil–reinforcement composites. The finite element method of analysis with an elasto-plastic hardening soil model was used to determine the specimen dimensions needed to provide representative stress–strain–volume change behavior of the soil–geosynthetic composites. Specimens of different dimensions were analyzed, and the global (overall) stress–strain and volume change relationships are shown in Figure 3.7. It was concluded that a specimen 2.0 m in height and 1.4 m in depth on 0.2 m spacing under a confining pressure of approximately 30 kPa would give an adequate representation of 7.0-m high GRS on 0.2 m reinforcement spacing. The conditions of the five experiments are summarized in Table 3.2. Among the five SGC experiments, all except one were conducted under a confining pressure of 34 kPa, and the experiments varied in reinforcement stiffness/strength and reinforcement spacing.

The backfill used in the experiments was a well-graded gravelly sand. The Mohr–Coulomb strength parameters for as-compacted soil are: $c = 70$ kPa and $\phi = 50°$ for confining pressure between 0 and 200 kPa; $c = 240$ kPa and $\phi = 38°$ for confining pressure between 200 and 750 kPa. The reinforcement was a woven geotextile with a wide-width strength of 70 kN/m, except for Test 3, where the reinforcement was formed by gluing two sheets of the same geotextile together, producing twice the stiffness and

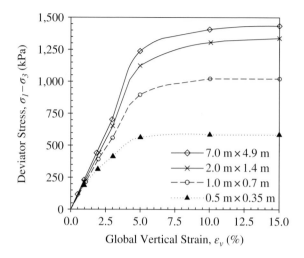

(a)

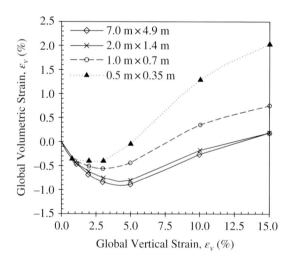

(b)

Figure 3.7 (a) Deviator stress vs. strain relationships, and (b) axial strain vs. volumetric strain relationships for soil–geosynthetic composites of different specimen sizes under a confining pressure of 30 kPa (Wu et al., 2013)

Table 3.2 Test conditions of the SGC experiments (Wu et al., 2013)

Designation	Confining Pressure (kPa)	Reinforcement Strength, T_f (kN/m)	Reinforcement Spacing, s_v (m)
Test 1	34	None	N/A
Test 2	34	70	0.2
Test 3	34	140	0.4
Test 4	34	70	0.4
Test 5	0	70	0.2

the wide-width strength of a single sheet (as verified by wide-width tensile tests) while maintaining the same soil–reinforcement interface conditions as for a single sheet.

The measured load–deformation and volume change relationships of the soil–geosynthetic composites obtained from the SGC experiments are shown in Figure 3.8. A summary of the test conditions and ultimate vertical load-carrying pressure of the SGC experiments is given in Table 3.3. The following observations are made:

- Using vertical strains at failure as an indicator for the ductility of the soil–geosynthetic composites, all soil–geosynthetic composites (Test 2 to Test 5) exhibit much higher ductility than the unreinforced soil mass (Test 1). The vertical strain at failure of Test 2 (the baseline) is more than 200% of Test 1.

- All the soil–geosynthetic composites (Test 2 to Test 5) have significantly higher load-carrying capacity than the unreinforced soil (Test 1). The ultimate load-carrying capacity of Test 2 is about 350% of Test 1.

- Test 2 ($s_v = 0.2$ m) and Test 3 ($s_v = 0.4$ m) have the same T_f/s_v ratio, yet the load-carrying capacity of Test 3 is only 65% of Test 2.

- The influence of reinforcement spacing can be seen by comparing the load–deformation curves of Test 4 and Test 2. By reducing reinforcement spacing by 50% (and maintaining the same reinforcement strength), the load-carrying capacity is seen to increase about 100%. The ductility is also seen to reduce significantly with the increase in reinforcement spacing.

- The influence of reinforcement strength can be seen by comparing the load–deformation behavior of Test 3 and Test 4. By reducing the reinforcement strength by 50% (and maintaining same reinforcement spacing), the load-carrying capacity is seen to decrease only about 25%. Comparing this to the influence of reinforcement spacing (see the bullet immediately above), it is evident that reinforcement spacing has a much stronger influence than reinforcement strength on the load-carrying capacity and load–deformation behavior of GRS.

- The load-carrying capacity of Test 5 (unconfined) is about 30% lower than that of Test 2. This shows that confining pressure can have a significant influence on the load–deformation behavior of GRS.

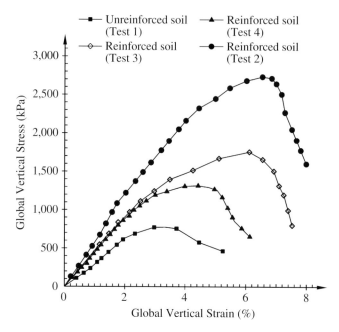

Figure 3.8 Measured load–deformation relationships of soil–geosynthetic composites obtained from the SGC experiments (Wu and Pham, 2013)

Table 3.3 Summary of measured results of the SGC experiments (Wu et al., 2013)

Parameter	Test 1 (unreinforced)	Test 2 (T, s_v)	Test 3 ($2T$, $2s_v$)	Test 4 (T, $2s_v$)	Test 5 (T, s_v)
Ultimate strength of reinforcement, T_f (kN/m)	NA	70	140	70	70
Reinforcement spacing, s_v (m)	NA	0.2	0.4	0.4	0.2
Confining pressure, σ_c (kPa)	34	34	34	34	0
Ultimate pressure at failure, q_{ult} (kPa)	770	2,700	1,750	1,300	1,900
Vertical strain at failure	3.0%	6.5%	6.1%	4.0%	6.0%

3.3 Reinforcing Mechanisms of GRS Walls

This section discusses the reinforcing mechanisms by which layers of planar tensile inclusion improve the load–deformation behavior of a soil mass. Reinforcing mechanisms have been the subject of study by numerous researchers and for a few decades (e.g., Schlosser and Long, 1974; Yang, 1972; Hausmann, 1976; Bassett and Last, 1978; Ingold, 1982; Gray and Al-Refeai, 1986; Maher and Woods, 1990; Athanasopoulos,

1993; Elton and Patawaran, 2005; Pham, 2009, etc.). Many reinforcing mechanisms have been identified. The operative mechanism(s) in a reinforced soil mass depends on the type of structure, reinforcement spacing, loading conditions, etc. The reinforcing mechanisms of GRS walls include:

i)   increase lateral confinement of soil

ii)   effect cohesion in granular fill (without introducing the undesirable behavior of a cohesive soil)

iii)   suppress dilation of soil

iv)   increase compaction-induced stress of soil

v)   restrain lateral deformation of soil

vi)   stabilize potential failure wedge of a soil structure (i.e., serving as quasi-tiebacks)

vii)   preserve integrity of soil by preventing loss of soil particles

viii) accelerate dissipation of porewater pressure in low permeability fill

ix)   improve ductility of soil

In the following, each of the above mechanisms is described, with greater detail given to the first three mechanisms for which the extent of reinforcing has been expressed in a quantitative manner.

3.3.1   Mechanisms of Apparent Confining Pressure and Apparent Cohesion

If we have a bucketful of dry clean sand and wish to make it into the steepest possible pile without external support or compaction, we can perhaps accomplish it by pouring the sand onto a flat surface very slowly. We will then likely find the steepest angle to be between 30° and 40°. This angle is commonly referred to as the *angle of repose*. Without cohesion between dry granular particles, the steepest angle that any granular fill can assume is less than about 50°. Only with cohesion can a granular fill stand at a 90° angle without lateral support.

Figure 3.9(a) shows a clean dry sand at its angle of repose, and Figure 3.9(b) shows the same sand reinforced by strips of plain paper embedded horizontally inside the sand. By incorporating layers of reinforcement in the sand, the sand is *internally reinforced* so that it can assume a 90° angle. The embedded paper strips are said to generate an *apparent cohesion* in the dry sand and allow the sand to behave as if it possesses a certain amount of cohesion. Note that the paper is wrapped at the face to prevent loss of dry sand particles (sloughing); it otherwise assumes little structural function at the face.

The mechanism of *apparent cohesion* as a reinforcing mechanism for reinforced soil was first identified by Schlosser and Long (1974). In this mechanism, the interaction between soil and tensile reinforcement is said to bring about an *apparent cohesion* in the soil without introducing any appreciable time-dependent deformation commonly associated with cohesive soils. As seen in the Mohr circles shown in

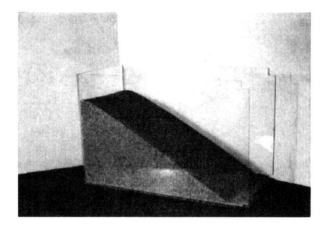

(a)

(b)

Figure 3.9 Unreinforced vs. reinforced dry sands: (a) a unreinforced sand at the angle of repose and (b) the same sand reinforced internally by layers of paper strips (Mitchell and Villet, 1987)

Figure 3.10(a), under a confining pressure σ_{3c}, the *apparent cohesion* acquired by tensile reinforcement, c_R, will lead to an increase in shear strength, i.e., an increased major principal stress at failure from σ_1 to σ_{1R}. In the figure, it is tacitly assumed that the internal friction angle of the soil ϕ will not be affected by the presence of tensile reinforcement. Limited data available to date suggest that the internal friction angle of geosynthetic reinforced soil will indeed stay approximately the same as that of the unreinforced soil (see Figure 3.11). Some recently published data seems to contradict this assumption. Additional reliable test data from field-scale experiments (such as those described in Section 3.2) are needed before a conclusion regarding the assumption can be reached.

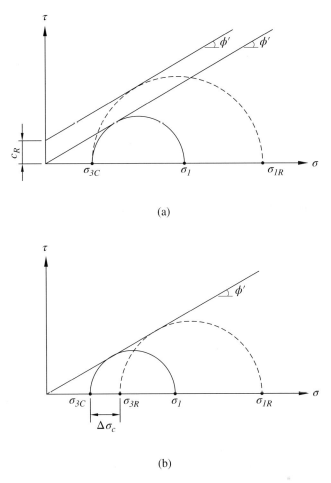

(a)

(b)

Figure 3.10 Two mechanisms for the increase in shear strength of granular soil due to geosynthetic inclusion: (a) mechanism of apparent cohesion (generation of c_R) and (b) mechanism of apparent confining pressure (generation of $\Delta\sigma_c$)

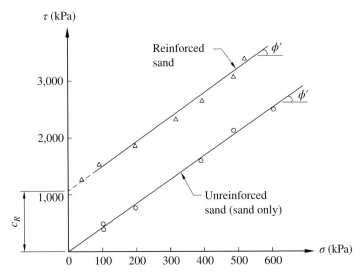

Figure 3.11 Strength envelopes of unreinforced and reinforced sands (modified after Mitchell and Villet, 1987)

Yang (1972) proposed a reinforcing mechanism, known as the mechanism of *apparent confining pressure*, which is somewhat parallel to the mechanism of apparent cohesion. In the mechanism of apparent confining pressure, the reinforcement is said to generate an added confining pressure in the direction of the reinforcement layers. The added confining pressure can be regarded as an increase in the minor principal stress, $\Delta\sigma_{3R}$, and will lead to a larger major principal stress at failure, σ_{1R}, as seen in Figure 3.10(b).

The mechanisms of apparent cohesion and apparent confining pressure are inter-related. By equating the shear strength of the two mechanisms (i.e., setting σ_{1R} of a given σ_{3c} to be equal for the two mechanisms), see Figure 3.10, the apparent cohesion of a reinforced soil can be determined from the apparent confining pressure, and vice versa. This is addressed below.

For a reinforced soil mass of which the reinforcement has a tensile rupture resistance of T_f, and vertical spacing between adjacent layers of reinforcement of s_v, it has been *assumed* that the increase in confining pressure due to tensile inclusion, $\Delta\sigma_{3R}$, is:

$$\Delta\sigma_{3R} = \frac{T_f}{s_v} \tag{3-1}$$

By equating the σ_{1R} in Figure 3.10(a) to the σ_{1R} in Figure 3.10(b), and making use of Rankine's passive pressure theory, the apparent cohesion, c_R, can be expressed as:

$$c_R = \frac{\Delta\sigma_{3R}\sqrt{K_P}}{2} \tag{3-2}$$

Substituting Eqn. (3-1) into Eqn. (3-2), the apparent cohesion becomes

$$c_R = \frac{0.5 \cdot T_f \sqrt{K_P}}{s_v} \tag{3-3}$$

where K_P is the coefficient of Rankine passive earth pressure. Note that Eqn. 3-3 is for a cohesionless soil, $c = 0$.

Eqn. (3-3), if valid, can be very useful for the stability analysis of earth structures involving reinforced soil. Given the strength parameters of soil in the unreinforced zone ($c = 0$ and $\phi > 0$) and T_f and s_v of the reinforcement in the reinforced zone, Eqn. (3-3) can be used to determine the strength parameter c_R of the reinforced soil as a function of T_f and s_v. By assuming $\phi_R = \phi$ (i.e., the same friction angle for unreinforced and rein-forced soils), the stability analysis of a reinforced soil structure can be carried out the same way as for soil-only earth structures by using $c = 0$ and ϕ for soil in the unrein-forced zone, and $c = c_R$ and ϕ for soil in the reinforced zone.

Moreover, with the aid of Eqn. (3-3), the ultimate load-carrying capacity, q_{ult}, of a reinforced granular soil mass ($c = 0$) becomes:

$$q_{ult} = \sigma_{1R} = \sigma_3\, K_P + 2c_R\sqrt{K_P} = \left(\sigma_c + \frac{T_f}{s_v}\right)K_P \tag{3-4}$$

In recent years, however, the validity of Eqns. (3-1) to (3-4) has been brought into question. The seemingly correct expression of $\Delta\sigma_{3R} = T_f/s_v$ (i.e., Eqn. (3-1)) suggests that the reinforcing effect is reflected completely by the simple ratio of T_f and s_v. It implies that the resulting reinforcing effect is the same for any two reinforcing schemes of the same T_f/s_v ratio. In other words, reinforced soil walls of any value of reinforcement spacing will behave the same as long as the reinforcement strength values are proportional to the reinforcement spacing. This *assumption* has encouraged designers to use larger reinforcement spacing with proportionally higher strength reinforcement because the cost of higher strength geosynthetic products does not differ very significantly, yet placing reinforcement on larger spacing would save considerable time in field installation. Consequently, reinforcement spacing as large as 0.6–1.0 m (or even larger) has routinely been employed in geosynthetic reinforced soil walls since manufacturers have little difficulty producing ever stronger reinforcement. It has been common practice for GMSE walls (MSE walls with geosynthetics as reinforcement) to use reinforcement on 0.3–0.9 m or 12–36 in. spacing (with 0.6 m or 24 in. being typical) to reduce construction time.

Field-scale experiments, such as the FHWA mini pier experiments (see Section 3.2.1) and the SGC experiments (see Section 3.2.3), have demonstrated conclusively that reinforcement spacing plays a more significant role in the load–deformation behavior of a reinforced soil mass than reinforcement strength. Wu et al. (2010) and Wu and Pham (2013) have presented an analytical model which introduces a *W-factor* in the expression for $\Delta\sigma_c$ to reflect more accurately the relative roles of reinforcement spacing (s_v) and reinforcement strength (T_f) as:

$$\Delta\sigma_c = W\left(\frac{T_f}{s_v}\right) = \left[0.7^{\left(\frac{s_v}{6d_{max}}\right)}\right]\left(\frac{T_f}{s_v}\right) \tag{3-5}$$

where d_{max} = maximum particle size of the soil. Eqn. (3-5), referred to as the *W-equation*, leads to a revised expression for apparent cohesion (c_R):

$$c_R = \frac{\Delta\sigma_c}{2}\sqrt{K_P} + c = \left[0.7^{\left(\frac{s_v}{6d_{max}}\right)}\right]\frac{T_f}{2s_v}\sqrt{K_P} + c \tag{3-6}$$

where c = cohesion of unreinforced soil. It follows that the ultimate load-carrying capacity of a soil–geosynthetic composite, q_{ult}, becomes:

$$q_{ult} = \sigma_{1R} = \left[\sigma_c + 0.7^{\left(\frac{s_v}{6d_{max}}\right)}\frac{T_f}{2s_v}\right]K_P + 2c\sqrt{K_P} \tag{3-7}$$

Eqn. (3-7) is referred to as the *load-carrying capacity equation* of soil–geosynthetic composites.

For GRS walls with dry-stacked concrete block facing, the value of σ_c can be estimated as $\sigma_c = \gamma_b D \tan \delta$, where γ_b is the *bulk* unit weight of the facing block, D is the depth of the facing block unit (in the direction perpendicular to the wall face), and δ is the friction angle between vertically adjacent facing blocks (or between the geosynthetic reinforcement and the facing block, if the geosynthetic is sandwiched between adjacent blocks). The value of σ_c is very small for light-weight dry-stacked concrete blocks without any connection enhancement elements (e.g., lips, keys, or pins), and can be conservatively assumed to be zero.

The load-carrying capacity equation, Eqn. (3-7), has been shown to give drastically improved prediction of load-carrying capacity of soil–geosynthetic composites than Eqn. (3-4) (Wu et al., 2010, 2013). The Federal Highway Administration (FHWA) compared all available measured data where a GRS mass is loaded to failure with those calculated by the load-carrying capacity equation (Eqn. 3-7), as seen in Figure 3.12. The GRS included in the comparison are field-scale experiments with soils ranging from fine sand to crushed gravel, geosynthetic reinforcement varying from light-weight nonwoven to heavy-weight woven geotextiles, and reinforcement spacing ranging from 0.15 m to 0.3 m. It is seen that the load-carrying capacity equation, Eqn. (3-7), provides rather accurate predictions of the load-carrying capacity of GRS for all the cases. The FHWA (Adams et al., 2011a,b) has incorporated Eqn. (3-7) in the design guidance of a GRS bridge system called the geosynthetic reinforced soil–integrated bridge system (GRS-IBS).

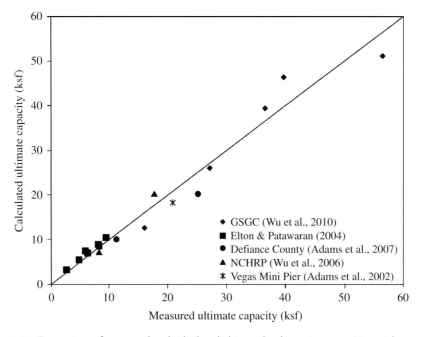

Figure 3.12 Comparison of measured and calculated ultimate load-carrying capacities, with calculated values determined by the load-carrying capacity equation, Eqn. (3.7), data compiled by TFHRC of the FHWA (Adams et al., 2011b)

3.3.2   Mechanism of Suppression of Soil Dilation

When subject to shear stress, a dense granular soil tends to dilate if volume change is permitted. Leonards (1962) used a set of simple drawings (Figure 3.13(a)) to explain the dilative behavior. In a dense granular soil, soil particles packed tightly and will need to roll over past other particles (without breaking) to produce shear-induced deformation and lead to an increase in volume, i.e., dilation. If dilation is suppressed, i.e., if shear deformation is prohibited, the soil stiffness and strength will increase.

Suppression of soil dilation in reinforced soil may be a result of two effects. The first is illustrated conceptually in Figure 3.13(b) where the reinforcement in tension contains soil particles to stay between reinforcement layers and prevent them from rolling over past neighboring soil particles, hence suppress the increase in volume. The second effect is related to restraint of lateral deformation due to soil–reinforcement interface friction (see Section 3.3.3). The latter effect is explained below.

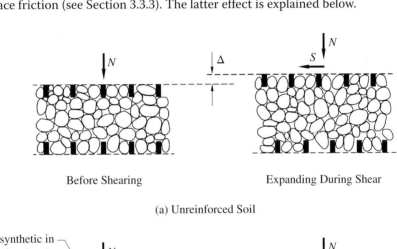

(a) Unreinforced Soil

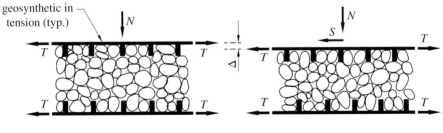

(b) Reinforced Soil

Figure 3.13 Schematic volume change behavior of a densely packed granular soil subject to shear stress for (a) unreinforced soil and (b) reinforced soil with closely spaced reinforcement (Note: the small dark rectangles show the relative positions of the top and bottom) (Wu et al., 2014)

When a soil mass is subjected to vertical loads, there will be a reduction in volume in the vertical direction (i.e., contraction) and an increase in volume in the lateral direction (i.e., expansion). In a dense granular soil, the latter typically overcomes the former, and results in a net increase in volume, hence dilation. For reinforced soil, however, the increase in volume in the lateral direction is significantly smaller than that in unreinforced soil, which can result in smaller dilation. For reinforced soil with closely spaced reinforcement, the dilative lateral volume change can be less than the compressive vertical volume change. This can lead to complete suppression of dilation or even a net decrease in the total volume.

Suppression of soil dilation by the presence of reinforcement layers can be seen by inspecting Figure 3.14, which shows measured stress–stain and volume change

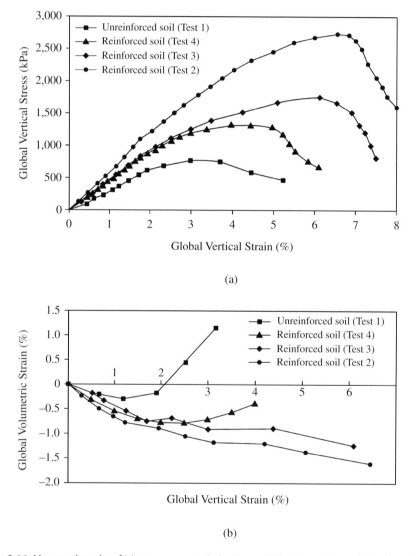

(a)

(b)

Figure 3.14 Measured results of (a) stress vs. strain behavior and (b) volume change behavior of unreinforced and reinforced soil masses, as obtained from the SGC experiments (Wu et al., 2014)

relationships of unreinforced and reinforced soils obtained from the field-scale SGC plane-strain experiments described in Section 3.2.3. In the unreinforced soil, dilation (i.e., increasing volumetric strain due to shear), after a certain amount of initial contraction, is seen from measured data. In the reinforced soils, on the other hand, the dilative behavior is clearly seen to be suppressed due to the presence of geosynthetic reinforcement.

Depending on soil type, reinforcement spacing, reinforcement stiffness, and, to a lesser degree, confining pressure, the presence of geosynthetic reinforcement may or may not be able to fully suppress soil dilation. The volume change behavior due to change in reinforcement spacing is shown in Figure 3.15. The influence of reinforcement spacing is obvious. The unreinforced soil is seen to exhibit significant dilation. For reinforced soil with $s_v = 1.0$ m, soil dilation still occurs, but is much less than that of unreinforced soil; with $s_v = 0.5$ m, the soil–geosynthetic composite becomes nearly incompressible (i.e., no net volume change) once the axial strain exceeds about 2%; with $s_v = 0.2$ m, dilation of the composites is suppressed over the entire range of stress and strain; the tendency for suppression of dilation is even more pronounced for $s_v = 0.1$ m than for $s_v = 0.2$ m.

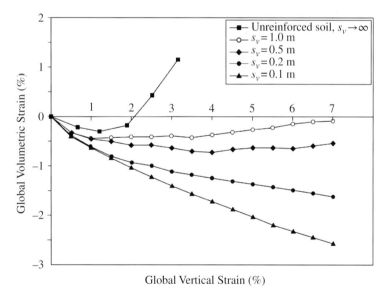

Figure 3.15 Volume change behavior of soil–geosynthetic composites for different values of reinforcement spacing (Wu et al., 2014)

The dilation angles (denoted by ψ) of the soil–geosynthetic composites for reinforcement spacing of 0.1 m, 0.2 m, 0.5 m, 1.0 m, and unreinforced soil are about $-10°$, $-3.5°$, $0°$, $+3°$, and $+20°$, respectively (a smaller negative ψ-value indicates more contraction, implying less tendency to dilate). It is conceivable that ψ-value may be used to characterize the extent of suppression due to geosynthetic reinforcement. As reinforcement spacing decreases from ∞ (unreinforced) to 0.1 m, the corresponding dilation angle ψ gradually decreases from $+20°$ to $-10°$, with $s_v = 0.5$ m showing nearly incompressible behavior.

It is interesting to note that tests of small-size specimens have been seen to produce entirely different behavior. Figure 3.16 shows triaxial test results of a sand with zero, one, two, and three layers of geosynthetic reinforcement. The test specimens used in the tests were quite small (100 mm in diameter and 200 mm in height), and reinforcement spacing ranged from 50 mm to 100 mm. The presence of reinforcement is seen to increase the stiffness and strength of the soil, but does not offer any suppression effect on the dilative behavior observed in the unreinforced soil. In fact, the tests show that the dilative behavior is more prevalent with decreasing reinforcement spacing. Similar findings of triaxial tests performed on small size specimens have been reported, including the tests conducted by Broms (1977), which have been cited frequently in the literature. When the material in a test specimen is non-uniform, as in a soil–geosynthetic composite, it is important to bear in mind that the test specimen needs to be large enough to accurately account for the interaction among the different components and to give *representative* composite global behavior, otherwise it may lead to misleading results, as exemplified by the behavior seen in Figure 3.16.

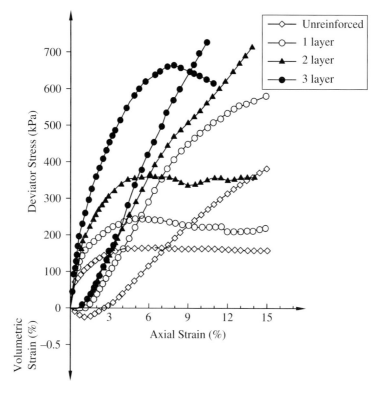

Figure 3.16 Stress–strain and volume change relationships of unreinforced and reinforced soils with different numbers of reinforcement layers, as obtained from using 100 mm diameter test specimens (Chandrasekaran et al., 1989)

3.3.3 Other Reinforcing Mechanisms

Increase in Compaction-Induced Stress
When a laterally confined soil mass is subjected to external vertical loading, there will generally be an increase in both vertical and lateral stresses in the soil mass. If the loading is subsequently removed, the increase in vertical stress will essentially vanish,

but not the increase in lateral stress. The net increase in lateral stress is commonly known as the *residual lateral stress* or *lock-in lateral stress*. For the same vertical loading, the stronger the restraint to lateral deformation, the higher the residual lateral stress. The residual lateral stress resulting from fill compaction operations, which typically involves a series of loadings and unloadings onto the top surface of a compaction lift, is referred to as *compaction-induced stress* (CIS).

Compaction-induced stress is an important factor to consider in terms of the load–deformation behavior of a compacted fill. Compaction-induced stress in unreinforced soil has been the subject of many studies, including Rowe (1954), Broms (1971), Aggour and Brown (1974), Seed (1983), Bolton (1990), and Duncan et al. (1991). These studies have indicated that compaction-induced stress can be rather significant if there is sufficient restraint to lateral movement of the soil, and will lead to increased stiffness and strength.

When a soil mass is reinforced by layers of geosynthetic reinforcement, compaction-induced stress is more pronounced than in an unreinforced condition because a reinforced soil has a higher degree of restraint to lateral deformation of the soil due to soil–reinforcement interface friction. For the same reason, closer reinforcement spacing tends to result in higher compaction-induced stress. Closer reinforcement spacing also "compels" the compactor operator to keep compaction lift thickness small (no more than reinforcement spacing) and improves the efficiency of fill compaction.

Figure 3.17 shows a simplified stress path at a given depth in the fill due to multiple passes of compaction. The stress path assumes that the slopes of unloading and reloading are the same. The slopes of the stress path can be calibrated by data obtained from well-controlled field-scale experiments of GRS, and the value of CIS can then be evaluated. Wu and Pham (2010) have provided an example.

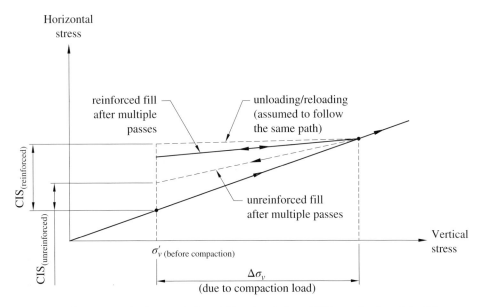

Figure 3.17 The stress path of an unreinforced fill and a reinforced fill for evaluation of the CIS (Wu and Pham, 2010)

Restrain Lateral Deformation of Soil

When a soil mass is compressed in one direction (usually the vertical direction), tensile strains tend to develop in the orthogonal direction (usually the horizontal direction). Reinforcement placed in the expected direction of tensile strains of a GRS mass will help restrain the tensile deformation of the soil mass, hence bringing about greater compaction-induced stress, and improve stiffness and strength of the soil. The restraint of tensile strain is derived from interface bonding between the soil and reinforcement. This restraining behavior can be seen by comparing the deformation of two triaxial test specimens when subject to vertical loads: soil only vs. soil with one layer of reinforcement at the mid-height (see Figure 3.18). The restraining effect of lateral deformation by reinforcement is evident in the photos.

(a)   (b)

Figure 3.18 Deformation of test specimens in triaxial compression tests performed on (a) unreinforced soil and (b) the same soil with one layer of reinforcement at the mid-height (courtesy of Cho-Sen Wu)

Knowing the distribution of lateral strains in an unreinforced earth structure can be helpful in determining locations where reinforcement will produce the greatest improvement. This can be accomplished by performing finite element analysis on the unreinforced earth structure to identify the locations of tensile strains. The most beneficial orientation of geosynthetic reinforcement may not be the horizontal direction. However, placing geosynthetic sheets horizontally is usually least time-consuming in actual field installation.

Stabilization of Potential Failure Wedge

If and when a potential failure wedge is to develop within the reinforced zone of a GRS wall, the geosynthetic reinforcement extending beyond the failure wedge is said to act as frictional quasi-tiebacks to help stabilize the potential failure wedge, as seen in

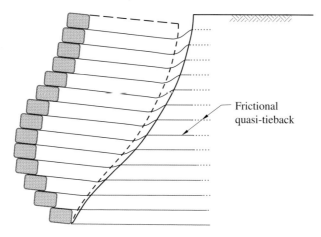

Frictional
quasi-tieback

Figure 3.19 Stabilization of a potential failure wedge by geosynthetic reinforcement extending beyond a potential failure surface where the reinforcement serves as quasi-tiebacks

Figure 3.19. This failure mechanism has been considered as a major mode of failure for checking the internal stability (i.e., failure occurred within reinforced zone) of GMSE walls, and has been used for the determination of the required reinforcement length of GMSE walls.

Preserve Soil Integrity

Many nonwoven and woven geotextiles can serve the function of filtration well, and allow water to pass through without compromising the integrity of the soil. This mechanism can help maintain the stability of GRS walls in certain situations. As an example, wrapped-face geotextile reinforcement (see Figure 3.20) has been found to help alleviate *bridge bumps* (differential settlement found at the junction between a bridge deck and the fill behind) by the Wyoming Department of Highways (Edgar et al., 1989). A major reason for the ability to eliminate bridge bumps by geosynthetic inclusion was not the geotextile's reinforcing function, rather it was attributed to the capability of the wrapped-face geotextile to contain soil particles in the approach fill. The wrapped-face geotextile was able to prevent soil particles from falling into the *gap* formed between the abutment wall and the fill due to shrinkage of the bridge deck in cold seasons (Wu and Helwany, 1990; Monley and Wu, 1993).

Acceleration of Porewater Pressure Dissipation

Some geosynthetics, notably heavier needle-punched geotextiles, can serve the function of transmitting water in the plane of their matrices in addition to reinforcing. These geosynthetics, when used in clayey fill, can facilitate drainage and accelerate dissipation of excess porewater pressure, hence allowing the soil to gain strength with time more rapidly. In Japan, for example, clayey soil is occasionally used as backfill for construction of embankments and walls due to the wide availability of some clayey soils such as a volcanic ash soil known as Kanto loam; the mechanism of accelerating porewater pressure dissipation is important in this application.

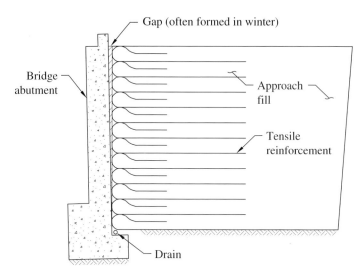

Figure 3.20 Wrapped-face geotextile to prevent soil particles from falling into a *gap* formed between an abutment wall and reinforced soil mass due to seasonal change (Monley and Wu, 1993)

Figure 3.21 shows a clay embankment reinforced by a nonwoven needled geotextile with different layout on the two sides. It is seen that the side with more layers of geotextile settles visibly less (i.e., stronger) than the other side with fewer geotextile layers. Measurement of porewater pressure has verified the capability of the geotextile to accelerate dissipation of the porewater pressure (Yamauchi, 1987). Since the stiffness of most nonwoven needled geotextiles with good drainage capability is usually too low to serve effectively as reinforcement, geotextile manufacturers in Japan have produced woven–nonwoven composite geotextiles to combine the advantages of the two types of geotextile, i.e., good capability to dissipate the porewater pressure of a nonwoven needled geotextile, coupled with the higher tensile stiffness of a woven geotextile.

Improve Ductility of Soil Mass

Figure 3.22 shows the stress–deformation relationships of a ductile material versus a brittle material with about the same strength. *Ductility* is a term used to denote a material's ability to undergo significant deformation before rupture. The term "ductility" also applies to a structure. For design of any structure constructed in public places, it is highly desirable that the structure possesses sufficiently high ductility so that if the structure is to fail unexpectedly (due to unexpected external influence or deficiencies in design or construction), failure will not be sudden catastrophic collapse.

Unreinforced soil structures with granular soils often exhibit low ductility. The lack of ductility, however, can be improved significantly with the inclusion of closely spaced geosynthetic reinforcement. This is seen in full-scale laboratory experiments and in actual earth structures. The increase in ductility can be evaluated by examining the stress–deformation curves of field-scale experiments of reinforced versus unreinforced soil. In the SGC plane-strain experiments, for example, the strain at failure of GRS is about 200% of that in an unreinforced condition (see Figure 3.8). In the unconfined FHWA mini pier experiments, the ratio of failure strains between GRS and unreinforced soil is much higher, between 4.5 and 12.5 (see Figure 3.3(b)).

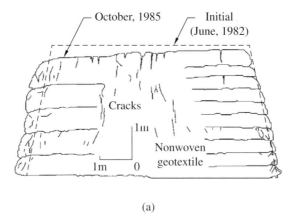

Figure 3.21 A clayey embankment reinforced by different number of layers of a needled-punched geotextile on two sides of an embankment: (a) layout of reinforcement and deformation, and (b) photo of the embankment (courtesy of Fumio Tatsuoka)

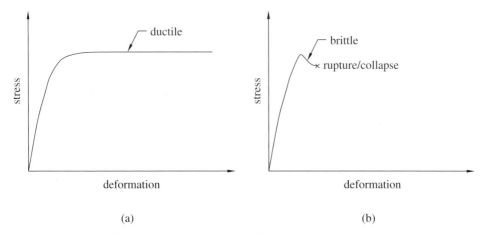

Figure 3.22 Stress–deformation relationships of (a) a ductile material/structure and (b) a brittle material/structure

3.4   Geosynthetic Reinforced Soil (GRS) Walls

A geosynthetic reinforced soil (GRS) wall comprises two major components: geosynthetic reinforced soil mass and facing. Some common types of GRS walls are shown in Figure 3.23. Note that these walls differ primarily in the facing and the connection between reinforced soil mass and facing. This section gives a brief description of major types of GRS walls, including wrapped-face GRS walls, concrete block GRS walls, cast-in-place rigid full-height facing GRS walls, full-height precast panel facing GRS walls, and timber facing GRS walls. GRS walls of other types of facing are also noted briefly.

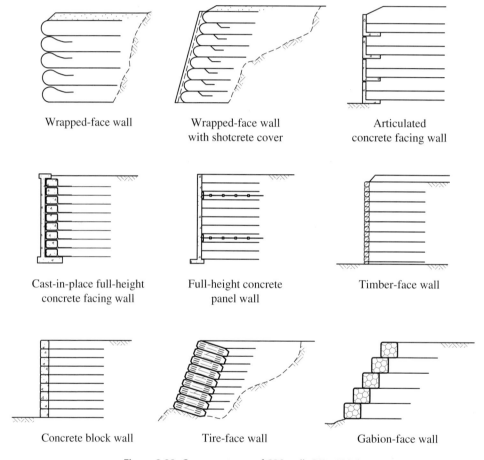

<table>
<tr><td>Wrapped-face wall</td><td>Wrapped-face wall with shotcrete cover</td><td>Articulated concrete facing wall</td></tr>
<tr><td>Cast-in-place full-height concrete facing wall</td><td>Full-height concrete panel wall</td><td>Timber-face wall</td></tr>
<tr><td>Concrete block wall</td><td>Tire-face wall</td><td>Gabion-face wall</td></tr>
</table>

Figure 3.23 Common types of GRS walls (Wu, 1994)

3.4.1   Wrapped-Face GRS Wall

A geotextile reinforced soil wall with a wrapped face was first constructed in Siskiyou National Forest in Oregon in 1974, and soon after in Olympic National Forest in Shelton, Washington in 1975 by the U.S. Forest Service (Steward and Mohney, 1982). The satisfactory performance and low cost of these walls provided an impetus for many wrapped-face GRS walls subsequently constructed in the U.S. and around the world. Most wrapped-face GRS walls have used geotextiles as reinforcement and as facing.

More recently, use of geogrid as reinforcement with a pre-seeded mat or geocomposite as facing has seen applications.

Wrapped-face GRS walls constructed to date are of two types: (a) geotextile wrapped-face GRS walls and (b) wire-mesh wrapped-face GRS walls. Figure 3.24 shows

(a)

(b)

Figure 3.24 (a) A geotextile wrapped-face GRS wall (courtesy of Bob Barrett), (b) a wire-mesh wrapped-face GRS wall (courtesy of Strata Systems, Inc.), and (c) a wire-mesh wrapped-face GRS wall (courtesy of Calvin VanBuskirk)

(c)

Figure 3.24 (Continued)

three completed wrapped-face GRS walls; the first is a geotextile wrapped-face wall, and the other two are wire-mesh wrapped-face walls. The differences between the two wall types are described in Section 6.1.2. The face of wrapped-face walls is sometimes covered with shotcrete, gunite, or asphalt emulsion (see Figure 3.25) to prevent deterioration (due to extended UV exposure) and vandalism. The construction procedure for wrapped-face GRS walls is given in Section 6.1.2.

Figure 3.25 A wrapped-face geotextile reinforced soil wall being covered with shotcrete (courtesy of Bob Barrett)

The backfill used in the construction of wrapped-face GRS walls is typically granular soil, ranging from silty sand to coarse gravel. Compacted cohesive backfill has also been used, especially in Asia and South America. Both nonwoven needle-punched and heat-bonded geotextiles and woven geotextiles have been employed as reinforcement.

Wrapped-face GRS reinforced soil walls have been constructed mostly in remote areas or as temporary walls; however, some have been for permanent applications in an urban setting. They have also been used commonly as the "stage-1" wall in construction of a "two-stage wall", where the 2nd stage involves installing a permanent-looking wall façade. Wrapped-face GRS walls generally range in height from 1 m to 6 m, with taller walls being built mostly in Asia, where their faces are often vegetated. Figure 3.26 shows a very tall (21-m high) vegetated face GRS wall in Japan and a partially vegetated face in Canada.

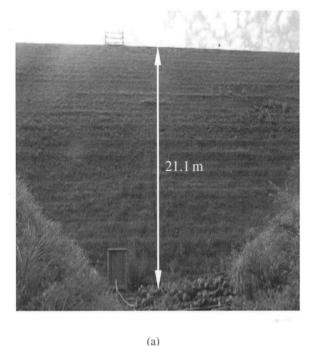

21.1 m

(a)

(b)

Figure 3.26 (a) A 21-m high vegetated face GRS wall in Japan (courtesy of Fumio Tatsuoka) and (b) a partially vegetated face GRS wall in Canada (courtesy of Calvin VanBuskirk)

Experiences distilled from construction of wrapped-face GRS walls include the following:

- Wrapped-face GRS walls are very economical and can be built by a general contractor.

- Wrapped-face GRS walls do not exhibit any appreciable creep as long as the backfill is predominantly granular and is well-compacted.

- The geosynthetic reinforcement in a wrapped-face GRS wall can effectively generate apparent cohesion (see Section 3.3.1) for a cohesionless fill to assume a vertical slope while subject to common traffic loads. Figure 3.27 shows a photo of an unprotected vertical cut in a wrapped face wall five years after the cut was made.

- Wrapped-face GRS walls can tolerate large differential settlement without visible distress.

- The facing of wrapped-face GRS walls may experience significant deformation, especially if a weak reinforcement is employed or if the fill is not compacted well.

- A wrapped GRS wall face covered with asphalt products or shotcrete is not aesthetically appealing compared to other types of wall face.

Figure 3.27 A cut in the Glenwood Canyon geotextile wrapped-face GRS test wall ($s_v = 9$ in.); it remained unprotected five years after the cut was made

Most wrapped-face GRS walls constructed to date have been designed based on the U.S. Forest Service method (see Section 5.4). Observation and field measurements of wall performance reveals that the design method is very conservative.

3.4.2 Concrete Block GRS Wall

Since being introduced in the mid-1980s, concrete block reinforced soil walls have quickly become the most popular type of reinforced soil walls in North America. The configuration of a concrete block geosynthetic reinforced wall is seen in Figure 3.23. Figure 3.28 shows four completed concrete block GRS walls, including (a) a single tier

(a)

(b)

Figure 3.28 Examples of completed concrete block GRS walls: (a) a single-tier concrete block wall transitioned into a three-tier concrete block wall (courtesy of Bob Barrett), (b) a concrete block wall with the highest point standing 16.7 m tall (courtesy of Bob Barrett), (c) a concrete block wall in a rural setting, and (d) a curved-face concrete block wall with SRW units in an urban setting

(c)

(d)

Figure 3.28 (Continued)

concrete block wall transitioned into a three-tier concrete block wall in DeBeque Canyon, Colorado, (b) a concrete block wall with its highest point standing 16.7 m (55 ft) tall in Grand County, Colorado, (c) a concrete block wall of multiple block colors in a rural setting, Winter Park, Colorado, and (d) a concrete block wall with a curved face in an urban setting, Centennial, Colorado.

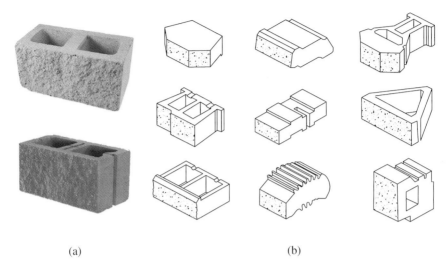

(a)   (b)

Figure 3.29 Concrete facing blocks: (a) examples of split face concrete masonry units (CMUs) and (b) examples of segmental retaining wall (SRW) units ((b) was modified after Bathurst and Simac, 1994)

The facing of concrete block GRS walls is typically formed by dry-stacking concrete blocks, which are usually made light and small enough to be handled by a worker with bare hands. The blocks may be dry- or wet-cast machine molded, with the former being far more common. A variety of split face concrete masonry units (CMUs) and segmental retaining wall (SRW) units varying in size, shape, weight, color and texture are available commercially (see Figure 3.29). The most common dimensions of split face CMUs used in the construction of GRS walls are $410 \times 200 \times 200$ mm ($16 \times 8 \times 8$ in.). SRW units, on the other hand, are concrete blocks specifically made for segmental retaining walls, typically ranging from 100 to 200 mm (4 to 8 in.) in height and 200 to 750 mm (8 to 30 in.) in width. CMU blocks cost about \$1.50 per block, depending on the region and the cement content. SRW blocks, on the other hand, typically cost between \$4.50 and \$6.50 per block.

Concrete blocks are either stacked vertically or at a small batter (usually less than 10° degrees from the vertical) to form the face of a concrete block GRS wall. Most concrete blocks have built-in hollow cores/cells which may be filled with crushed stone or sand/gravel during construction to increase block weights. Some blocks have cast-in lips or keys to help with facing alignment and to increase the bond strength between blocks. Geosynthetic reinforcement is usually sandwiched between vertically adjacent blocks through interface friction to connect the reinforced soil mass with the wall face. Some wall builders also install mechanical shear pins between blocks to help maintain facing alignment. The facing connection stability of concrete block walls is discussed in detail in Section 5.3.5.

When concrete block walls were first introduced to the wall industry, they were used mostly for landscaping, with typical wall height less than 2 m (6 ft). The use of geosynthetic inclusion, however, took the wall system to new heights because geosynthetic inclusion (especially when deployed on tight spacing) strengthens the backfill and reduces the lateral earth pressure. Concrete blocks, while providing a certain degree of local bending resistance, can be merely an architectural façade in an internally stabilized wall system (see opening statement of this chapter). Concrete block GRS walls typically range from 2 to 9 m (6 to 30 ft) in height, although walls over 20 m (65 ft) in height have been constructed without problems. Concrete block GRS walls are commonly built with a slightly battered face, although some wall builders consider the batter of little practical value. When the total height of a wall exceeds about 6 m (20 ft), most concrete block GRS walls are built in tiers.

In addition to the advantages for all GRS walls noted in Section 3.1, concrete block GRS walls enjoy the distinct advantage of being rather aesthetically pleasing. A combination of a concave face and the rough textured finish ("split face") of concrete blocks (see Figure 3.28(a) to (c)) can help *mask* minor construction misalignment. Both CMUs and SRW units of different colors, textures, shapes, and sizes can be used to improve the aesthetic appeal of a GRS wall (see Figure 3.28). CMU blocks can be cut to fit or the block tails broken off to accommodate tight curves. Specially-made narrow SRW units also serve the same purpose.

Material and engineering costs for concrete block GRS walls typically range from $5 to $10 per square foot of wall face (or $50 to $100 per square meter of wall face), excluding the cost of fill material. Not unlike other GRS walls, judicious use of on-site soils may be permitted for construction of concrete block GRS walls, although it is generally not recommended especially if the on-site soil is cohesive or poorly drained. Construction of concrete block GRS walls is addressed in Section 6.1.1.

3.4.3   Cast-in-Place Full-Height Facing GRS Wall

Beginning in the mid-1980s, Fumio Tatsuoka at the University of Tokyo and researchers at the Japan Railway Technical Research Institute jointly developed a GRS wall system with cast-in-place full-height rigid facing (Tatsuoka et al., 1992, 1997). This system is herein referred to as the cast-in-place full-height facing (CIP-FHF) GRS wall system.

Figure 3.30 shows the detailed construction sequence of a CIP-FHF system. The construction involves two major stages. In the 1st stage, a wrapped-face GRS wall, with layers of geosynthetic inclusion approximately $0.3H$ to $0.4H$ (H = wall height) in length on 0.3 m spacing (typical), is erected using stacked gravel-filled gabions as temporary facing. This temporary wrapped-face GRS wall is allowed to deform under its own weight (Step 5, Figure 3.30). In the 2nd stage, full-height rigid concrete facing (cast-in-place) is constructed over the temporary face of the GRS wall (Step 6, Figure 3.30). The rigid concrete facing is securely attached to the reinforced soil mass via protruded steel anchors that are embedded in the reinforced soil mass during the 1st stage construction

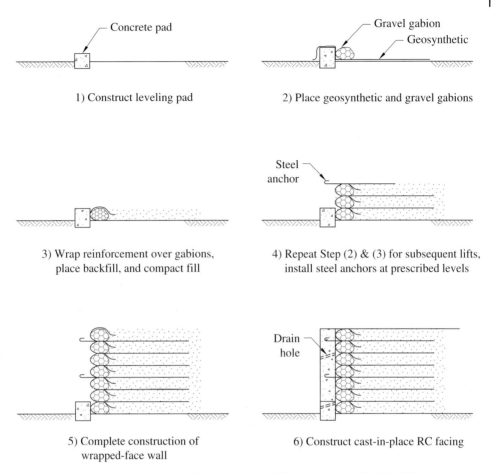

1) Construct leveling pad

2) Place geosynthetic and gravel gabions

3) Wrap reinforcement over gabions, place backfill, and compact fill

4) Repeat Step (2) & (3) for subsequent lifts, install steel anchors at prescribed levels

5) Complete construction of wrapped-face wall

6) Construct cast-in-place RC facing

Figure 3.30 Construction sequence of the cast-in-place full-height facing (CIP-FHF) GRS wall (modified after Tatsuoka, 2008)

(Step 4, Figure 3.30). Figure 3.31 shows photos of CIP-FHF systems soon after the 1st stage construction, in-service condition, and 15 years after being in service.

When rigid facing is employed in a GRS wall, a major problem is with mobilization of loads in the reinforcement, as the rigid facing will restrain lateral deformation and inhibit the reinforcing function. This problem is resolved by the CIP-FHF system, which involves a two-stage construction procedure. Reinforcement loads in the system are mobilized during 1st stage construction before installation of cast-in-place facing in the 2nd stage.

Since the temporary GRS wall in the 1st stage construction has deformed under its self-weight, the lateral earth pressure exerted on a completed cast-in-place full-height facing is negligible. This allows the wall system to withstand much greater subsequent external loads, including seismic loads due to earthquakes. The rigid facing, in

(a)   (b)

(c)

Figure 3.31 The CIP-FHF GRS wall system: (a) completion of the 1st stage of construction (stacked gravel-gabions temporary GRS wall with hooked steel bars protruding from wall face), (b) in-service condition (carrying the Yamanote-sen, a busy train line in Tokyo, Japan), and (c) a CIP-FHF GRS wall 15 years after putting into service (courtesy of Fumio Tatsuoka)

combination with the tight connection between the reinforced soil mass and the facing, also serves to increase lateral *stress* in the soil mass hence increase stiffness and strength of the soil behind the facing.

Due to its superior capability to withstand earthquakes and prolonged heavy rainfalls, the CIP-FHF system became the default retaining wall type in Japan Railway's wall repertoire. Figure 3.32 shows a CIP-FHF GRS wall before and after the 1995 Kobe

GRS RW with a FHR facing
for a rapid transit at Tanata

(a)

The wall survived!

24 Jan. 1995

(b)

Figure 3.32 A cast-in-place full-height facing (CIP-FHF) GRS wall: (a) in 1992, after the wall was put into service, and (b) in 1995, after the Kobe earthquake, which measured 7.9 on the Richter scale (courtesy of Fumio Tatsuoka)

earthquake, which measured 7.9 on the Richter scale. The wall's capability to survive major earthquakes is evident. As of 2008, over 120 km of CIP-FHF walls had been built for railway embankments in Japan with great success (Tatsuoka, 2008). In other parts of the world, however, the CIP-FHF system has seen few applications, primarily because of cost considerations and longer construction time.

3.4.4 Precast Full-Height Panel Facing GRS Wall

In 1996, the Colorado Department of Transportation (CDOT) developed a new reinforced soil wall system with full-height precast panel facing, referred to as an independent full-height facing (IFF) reinforced soil wall. An IFF reinforced soil wall system, as shown schematically in Figure 3.33, comprises three major components: (i) precast full-height reinforced concrete facing panel, (ii) reinforced soil mass, and (iii) flexible face anchor. The face anchor serves to connect the first two components, i.e., facing panels and reinforced soil mass, into an integral system. In this system, the connection needs to allow outward movement of the facing panel to take place during construction (hence, referred to as "flexible" face anchors). In the absence of this feature, the reinforcement in the reinforced soil mass would hardly be able to mobilize any loads, and would serve hardly any reinforcing function. Also, since outward movement of the facing panel is allowed during construction, a mechanism for adjusting the final positions of the panels must be in place so that the panels can be aligned in the final phase of the construction.

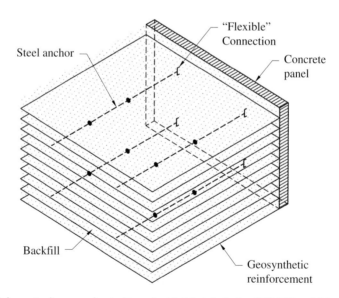

Figure 3.33 Schematic diagram of an independent full-height facing (IFF) GRS wall (Ma and Wu, 2004)

An IFF reinforced soil wall, locally known as the Fox wall, was constructed in 1996 along the entrance ramp from Interstate-25 to Interstate-70. The wall was an extension of an existing cantilever reinforced concrete retaining wall, of which the facing had the appearance of full-height grooved concrete. For aesthetic reasons, the GRS wall must bear the same appearance as the existing wall, hence full-height facing was needed. The total cost of an IFF reinforced soil wall system was estimated to be about half the cost of a conventional reinforced concrete cantilever wall on piles. The decision to go with the IFF reinforced soil wall system, however, was not based on cost, rather it was because little or no excavation would be needed for construction of an IFF wall system. This is important for two reasons. First, the wall was situated along a very busy section of Interstate-25, so minimal excavation meant minimal traffic disruption. Second, the sub-soil on the project site was found to have been contaminated in the past. Excavation of the contaminated soil, if needed, would have involved laborious laboratory testing and securing permits, and would likely have resulted in undue delay. Since construction, the Fox wall has performed very satisfactorily.

To construct an IFF reinforced soil wall, the facing panels are first erected at a slight batter in a shallow trench with the aid of temporary bracing. The reinforced soil mass is then constructed behind the facing with inclusion of tensile reinforcement layers on prescribed spacing. Face anchors are installed at pre-selected elevations to attach facing panels to the reinforced soil mass. The temporary bracing is removed after the facing panels have been securely attached to the reinforced soil mass. The final step of construction involves adjusting the positions of the facing panels to achieve proper alignment. Construction of IFF reinforced soil walls is discussed in more detail in Section 6.1.3. Figure 3.34 shows two photos of the Fox wall during construction. Figure 3.35 shows the Fox wall in a service condition.

The IFF reinforced soil wall system is under-used in the repertoire of reinforced soil walls. The IFF system offers a few distinct advantages over conventional cantilever reinforced concrete retaining walls, including the following:

- The full-height facing gives the appearance of a strong and permanent structure.
- Construction of IFF reinforced soil walls is relatively rapid; it does not require any on-site concrete formwork or piles.
- Construction of IFF reinforced soil walls requires no over-excavation, hence little traffic disruption during construction.
- IFF reinforced soil walls are low in cost. The total cost of an IFF GRS wall is comparable to that of a concrete block GRS wall, and is about half the cost of a conventional cantilever reinforced concrete retaining wall on piles.

3.4.5   Timber Facing GRS Wall

A GRS wall system with timber facing was developed in the early 1980s by the U.S. Forest Service. In areas where timber logs can be obtained economically, timber facing GRS walls may be the least expensive wall type. Figure 3.36 shows two completed timber facing GRS walls, one in an urban setting and the other in a rural setting. The facing of the GRS wall is made of wooden timber logs that serve as temporary forming and also as final facing. All the timber logs need to be preservative-treated, and are interconnected behind the facing by forming elements (e.g., plywood boards) with nails, as

(a)

(b)

Figure 3.34 An IFF reinforced soil wall, known locally as the "Fox wall": (a) during construction, with facing panels positioned in a shallow trench, and (b) temporary bracing in front of the facing panels; note that traffic remained open during construction (Ma and Wu, 2004)

Figure 3.35 The completed Fox wall, an IFF reinforced soil wall along Interstate-25 (Ma and Wu, 2004)

(a)

(b)

Figure 3.36 Timber facing GRS walls: (a) in an urban setting (courtesy of Bob Barrett) and (b) in a rural setting (courtesy of Gordon Keller)

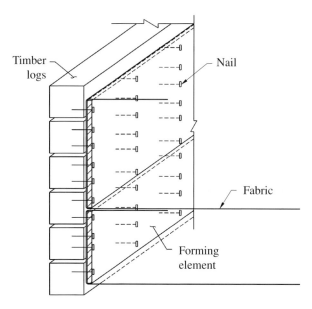

Figure 3.37 Inter-connected timber logs of a timber facing GRS wall

shown in Figure 3.37. Geosynthetic reinforcement is attached to the facing between forming elements and timber logs, and with a "tail" folding flat toward the back of wall. The inter-connected timber facing offers both local and (small) global bending stiffness. Construction of timber facing GRS walls is discussed in Section 6.1.4.

Timber facing GRS walls that have been constructed to date are typically less than 6 m (20 ft) in height. Higher walls are commonly built in tiers, which allow room for planting in the setback area and enhance aesthetics. A wide variety of soils, ranging from granular soil to low-quality clayey soil, have been used as backfill. Various geosynthetics, including woven geotextiles, nonwoven geotextiles, and geogrids, have been used as reinforcement.

Timber facing GRS walls have been researched rather extensively. In 1991, two 2.8-m (9.2-ft) high timber facing geotextile reinforced soil walls (known as the *Denver test walls*), one with a cohesive backfill and the other a granular backfill, were constructed and load-tested under well-controlled conditions in the laboratory. The reinforcement was a light-weight (3 oz/yd^2) nonwoven heat-bonded polypropylene geotextile on 0.28-m (11-in.) vertical spacing (Wu, 1992a,b). The test walls were featured in the International Symposium on Geosynthetic-Reinforced Soil Retaining Walls, in which researchers were invited to predict wall behavior. The material properties, construction, measured behavior, and predictions of the two walls were documented (Wu, 1992c). A few highlights of the test wall are as follows:

- With a very weak geotextile as reinforcement, the surcharge pressure at failure for the granular-backfill wall was 200 kPa (29 psi), and for the cohesive-backfill wall was slightly over 230 kPa (33 psi). Available design methods for GRS walls predicted surcharge pressures at failure (i.e., for factor of safety = 1.0) ranging from 0 (i.e., the wall cannot be erected) to 50 kPa (7.3 psi) for the granular-backfill wall, much lower than the measured value (Claybourn and Wu, 1992). None of the design methods allows cohesive backfill. Additional details about how measured behavior compared to design methods as well as their implications are given in Section 5.2.

- The cohesive-backfill wall is at least as strong as the granular-backfill wall provided that the moisture of the soil is kept relatively constant (at 2% wet of optimum in this case).
- Neither wall exhibited appreciable creep deformation under 103 kPa (15 psi) surcharge, typical traffic load used in design.

Timber facing geosynthetic reinforced walls have shown satisfactory performance in actual construction. Wu (1994) reported applications of timber facing GRS walls in Colorado, including the use of double-sided walls as rock fall barriers developed by wall builder Bob Barrett.

Timber facing GRS walls offer a number of advantages over other types of GRS walls, including the following:

- With timber being a *natural* material, timber facing GRS walls have the distinct advantage of a natural appearance, as compared to GRS walls with concrete facing.
- Where there is abundant supply of timber logs, a timber facing GRS wall is usually the least expensive wall type.
- Construction of timber facing GRS walls is simple and rapid; there is no requirement for an external forming system for construction. The wall can be built by a general contractor.
- Timber facing GRS walls can withstand fairly large settlement and differential settlement without hindering their function.
- The lifespan of the timber facing tends to dictate the service life of the wall system. When the timber is properly treated, the design life of a timber facing wall can be around 50 years.

3.4.6 Other Types of GRS Walls

In addition to the five wall types described in Sections 3.4.1 to 3.4.5, a number of reinforced soil walls (both GMSE walls and GRS walls) with different facing systems have been constructed. Examples of such walls are rock facing GRS walls, articulated panel facing GRS walls, tire facing GRS walls, welded wire mesh facing GRS walls, and gabion facing GRS walls, as shown in Figures 3.38 to 3.42, respectively. Note that the latter two wall types are sometimes constructed as the first-stage wall of a two-stage wall system, with the second stage often being concrete panel facing.

(a)

(b)

(c)

Figure 3.38 Rock facing GRS: (a) bridge abutment, (b) limestone block facing wall (courtesy of Geosolutions, Inc.), and (c) rockery

(a)

(b)

Figure 3.39 (a) An articulated panel facing reinforced soil wall during construction and (b) articulated panel facing as the façade of a gabion facing wall during construction (courtesy of Tensar International Corporation)

Figure 3.40 Tire facing wall (courtesy of Gordon Keller)

Figure 3.41 Welded wire mesh facing wall (courtesy of Hilfiger Retaining Walls).

Figure 3.42 Gabion facing walls with metal wire baskets (courtesy of NAUE, Germany)

3.5 Advantages and Disadvantages of Different Types of GRS Walls

In this section we shall discuss the advantages and disadvantages of wrapped-face geotextile reinforced soil walls, concrete block GRS walls, and timber facing GRS walls. Note that the terms "advantages" and "disadvantages" have to be addressed in a relative sense. We will begin by addressing the advantages and disadvantages of wrapped-face GRS walls relative to conventional earth retaining walls. Wrapped-face GRS walls are then used as a baseline when addressing the advantages and disadvantages of GRS walls with other types of facing.

3.5.1 Wrapped-Face GRS Walls

<u>Advantages</u>

Wrapped-face GRS walls have demonstrated the following advantages over conventional reinforced concrete walls:

- Construction of wrapped-face GRS walls is rapid and requires no heavy construction equipment.
- Wrapped-face GRS walls are inherently flexible, hence can tolerate larger differential settlement without visible distress.
- When properly designed and constructed and with well-compacted granular fill, wrapped-face GRS walls have high load-carrying capacity (see Figure 3.43, where a wrapped-face GRS composite with bed sheets as reinforcement and a road base as fill material is seen to be able to carry 22 "Jersey barriers" made of solid concrete; no time-dependent deformation

was observed), and exhibit high ductility (i.e., can deform considerably without sudden collapse).

- Wrapped-face walls do not require embedment into the foundation soil, thus eliminate the need for sub-excavation. This feature is especially important when environmental constraints are present. Potential damage to adjacent steams and root systems due to excavation can also be avoided to protect natural environments.

- High-quality granular backfill, although highly preferred, is not mandatory for the construction of wrapped-face GRS walls.

- A wide variety of geosynthetics with different stiffness/strength and costs are readily available and can usually be easily transported to remote construction sites; geosynthetics are free from corrosion problems and generally have strong resistance to bacterial action.

- Other than reinforcing, geotextiles can serve multiple functions, e.g., facilitate drainage of fill material, preserve soil integrity, and maintain separation of different soils during construction and under repeated external loading (see Section 4.1 for further discussion).

- The overall cost of wrapped-face GRS walls is low. In many applications, it has proven to be the lowest cost of all reinforced soil walls.

Figure 3.43 A GRS mass with bed sheets (purchased from J.C. Penny) as reinforcement was able to carry 22 Jersey barriers; no noticeable time-dependent deformation was observed (courtesy of Bob Barrett)

Disadvantages

Wrapped-face GRS walls have a number of disadvantages compared to conventional reinforced concrete retaining walls:

- Wrapped-face GRS walls lack the appearance of a permanent structure. When the face is covered with asphalt products or shotcrete, it is less than aesthetically pleasing.

- Wrapped-face GRS walls may experience appreciable deformation, especially if a weak geotextile reinforcement is employed or if the fill is not well compacted.

- Some geotextiles are susceptible to chemical degradation and may deteriorate when exposed to UV lights for an extended period of time (say, more than a few days).

3.5.2   Concrete Block GRS Walls

Advantages

In addition to all the advantages of wrapped-face geotextile reinforced walls described in Section 3.5.1, concrete block GRS walls offer a few advantages:

- A concrete block wall face is durable and can be aesthetically appealing. Concrete block GRS walls generally give the perception of a more "permanent" structure.

- A wide variety of concrete blocks with different sizes, shapes, weights, textures, and colors are readily available as facing units of GRS walls (see Figure 3.29).

- Concrete block GRS walls can be easily adapted to situations where fairly sharp curves in the longitudinal direction are needed or desired.

- Since concrete blocks offer high *local* compressive and some bending resistance, the deformation of concrete block GRS walls is typically much smaller than that of wrapped-face GRS walls.

- When very small reinforcement spacing (say 0.1 m or 4 in.) is used, it promotes better compaction near the face of the wall hence improved performance near the wall face, which is especially important for GRS bridge abutments (Adams, personal communication).

Disadvantages

Concrete block GRS walls can overcome most of the disadvantages of wrapped-face GRS walls described in Section 3.5.1, especially problems with aesthetics and deformation. However, close attention and control of facing block placement is required for concrete block GRS walls to avoid misalignment or uneven batter of the wall face. Effloresence (mineral salts leached from soil and concrete to the surface; see Figure 3.44) will leave stains on a concrete wall face if not properly treated. In addition, concrete block facing may experience problems with deterioration within a few years after it is put into service. Figure 3.45 shows an example of severe block deterioration. The cause of such deterioration, which has been seen to occur in both SRW blocks and CMU blocks, may be multi-fold, including insufficient cement content, poor quality control,

Figure 3.44 Mineral salts leached onto the surface of concrete blocks

Figure 3.45 Deterioration of concrete block facing

reactive aggregates, etc. Adherence to specifications for the use of concrete blocks as facing units of GRS walls is needed. Moreover, proper control of surface runoff to direct the runoff away from the wall face is a key requisite.

3.5.3 Timber Facing GRS Walls

Advantages

In addition to all the advantages of wrapped-face GRS walls described in Section 3.5.1, timber facing GRS walls offer a few advantages, including the following:

- Construction is simple and rapid as there is no requirement for external forming work for wall construction.
- Timber facing has a natural appearance and can be very aesthetically pleasing.

- Where timbers are readily available, timber facing GRS walls are inexpensive.
- Since timber facing offers a moderate degree of local and global compressive and bending resistance, the deformation of timber facing GRS walls is typically smaller than that of wrapped-face GRS walls.

Disadvantages

Timber-facing GRS walls share the disadvantages of wrapped-face GRS walls described in Section 3.5.1. In addition, they also suffer from the following drawbacks:

- Timber must be acquirable at a reasonable cost to make it a cost-effective wall system.
- The lifespan of the timber facing tends to dictate the wall's service life. When timber is not treated properly, the facing can deteriorate within only a few years.

References

Adams, M.T., Lillis, C.P., Wu, J.T.H., and Ketchart, K. (2002). Vegas Mini Pier Experiment and Postulate of Zero Volume Change. *Proceedings, Seventh International Conference on Geosynthetics, Nice, France*, pp. 389–394.

Adams, M.T., Ketachart, K., and Wu, J.T.H. (2007). Mini Pier Experiments – Geosynthetic Reinforcement Spacing and Strength as Related to Performance. *Proceedings, Geo-Denver 2007, ASCE, Denver*.

Adams, M.T., Nicks, J., Stabile, T., Wu, J.T.H., Schlatter, W. and Hartmann, J. (2011a). *Geosynthetic Reinforced Soil Integrated Bridge System Interim Implementation Guide*. Report No. FHWA-HRT-11-026, Federal Highway Administration, McLean, Virginia, 169 pp.

Adams, M.T., Nicks, J., Stabile, T., Wu, J.T.H., Schlatter, W. and Hartmann, J. (2011b). *Geosynthetic Reinforced Soil Integrated Bridge System Synthesis Report*. Report No. FHWA-HRT-11-027, Federal Highway Administration, McLean, Virginia.

Aggour, M.S. and Brown, C.B. (1974). The Prediction of Earth Pressure on Retaining Walls Due to Compaction. *Geotechnique*, 24(4), 489–502.

Athanasopoulos, G.A. (1993). Effect of Particle Size on the Mechanical Behavior of Sand-Geotextile Composite. *Geotextiles and Geomembranes*, 12, 255–273.

Bassett, A.K. and Last, N.C. (1978). Reinforcing Earth below Footings and Embankments. *Proceedings, ASCE Spring Convention and Exhibit, Pittsburgh, Pennsylvania*.

Bathurst, R.J. and Simac, M.R. (1994). Geosynthetic Reinforced Segmental Retaining Wall Structures in North America. *Proceedings, Fifth International Conference on Geotextiles, Geomembranes and Related Products, Singapore*, pp. 1–41.

Berg, R.R. Christopher, B.R., and Samtani, N.C. (2009). *Design of Mechanically Stabilized Earth Walls and Reinforced Soil Slopes, Design & Construction Guidelines*. Report No. FHWA-NHI-00-043. Federal Highway Administration, 394 pp.

Bolton, M.D. (1991). Geotechnical Stress Analysis for Bridge Abutment Design. *Transportation Research Laboratory Contractor Report*, London, 270 pp.

Broms, B.B. (1971). Lateral Earth Pressure due to Compaction of Cohesionless Soils. *Proceedings, 4th Budapest Conference on Soil Mechanics and Foundation Engineering*, pp. 373–384.

Broms, B.B. (1977). Triaxial Tests with Fabric-Reinforced Soil. *Proceedings, International Conference on Use of Fabrics in Geotechnics, L'Ecole Nationale des Ponts et Chaussees, Volume Ill, Paris, France*, pp. 129–133.

Chandrasekaran, B., Broms, B.B., and Wong, K.S. (1989). Strength of Fabric Reinforced Sand under Axisymmetric Loading. *Geotextiles and Geomembranes*, 8(4), 293–310.

Claybourn, A. and Wu, J.T.H. (1992). Failure Loads of the Denver Walls by Current Design Methods. In *International Symposium on Geosynthetic-Reinforced Soil Retaining Walls*. A.A. Balkema, Rotterdam, pp. 61–77.

Duncan, J.M., Williams, G.W., Sehn, A.L., and Seed R.M. (1991). Estimation of Earth Pressures due to Compaction. *Journal of Geotechnical Engineering*, 117(12), 1833–1847.

Edgar, T.V., Puckett, J.A., and D'Spain, R.B. (1989). Effects of Geotextiles on Lateral Pressure and Deformation in Highway Embankments. *Geotextiles and Geomembranes*, 8(4), 275–306.

Elton, D.J. and Patawaran, M.A.B. (2005). *Mechanically Stabilized Earth (MSE) Reinforcement Tensile Strength from Tests of Geotextile Reinforced Soil*. Technical Report, Alabama Highway Research Center, Auburn University.

Gray, D.H. and Al-Refeai, T. (1986). Behavior of Fabric- versus Fiber-Reinforced Sand. *Journal of Geotechnical Engineering*, 112(8), 804–820.

Hausmann, M.R. (1976). Strength of Reinforced Soil. *Proceedings, 8th Australian Road Research Conference, Volume 8, Section 13*, pp. 1–8.

Ingold, T.S. (1982). *Reinforced Earth*. Thomas Telford Ltd, London.

Ketchart, K. and Wu, J.J.H. (2001). *Performance Test for Geosynthetic-Reinforced Soil Including Effects of Preloading*. FHWA-RD-01-018, Turner-Fairbank Highway Research Center, Federal Highway Administration, McLean, Virginia, 270 pp.

Koerner, R.M. (1994). *Designing with Geosynthetics*, 3rd edition. Prentice Hall, Upper Saddle River, New Jersey, 783 pp.

Leonards, G.A. (1962). Engineering Properties of Soils. In *Foundation Engineering*. McGraw-Hill, New York.

Liu, H.C. (1987). *Dynamic Behavior of a Geotextile Reinforced Sand*. M.S. Thesis, Department of Civil Engineering, University of Colorado Denver.

Ma, C. and Wu, J.T.H. (2004). Performance of an Independent Full-Height Facing Reinforced Soil Wall. *Journal of Performance of Constructed Facilities*, 18(3), 165–172.

Maher, M.H. and Woods, R.D. (1990). Dynamic Response of Sand Reinforced with Randomly Distributed Fibers. *Journal of Geotechnical Engineering*, 116(7), 1116–1131.

Mitchell, J.K. and Villet, W.C.B (1987). *Performance of Earth Slopes and Embankments*. NCHRP Report No. 290, Transportation Research Board, Washington, D.C.

Monley, G.J. and Wu, J.T.H. (1993). Tensile Reinforcement Effects on Bridge-Approach Settlement. *Journal of Geotechnical Engineering*, 119(4), 749–762.

Nicks, J.E., Adams, M.T. and Ooi, P.S.K. (2013). *Geosynthetic Reinforced Soil Performance Testing – Axial Load Deformation Relationships*. Report No. FHWA-HRT-13-066, Federal Highway Administration, McLean, Virginia, 169 pp.

Pham, T.Q. (2009). *Investigating Composite Behavior of Geosynthetic-Reinforced Soil (GRS) Mass*. Doctoral Dissertation, Department of Civil Engineering, University of Colorado Denver.

Rowe, P.W. (1954). A Stress-Strain Theory for Cohesionless Soil with Applications to Earth Pressures at Rest and Moving Walls. *Geotechnique,* 4(2), 70–88.

Schlosser, F. and Long, N.T. (1974). Recent Results in French Research on Reinforced Earth. *Journal of the Construction Division,* 100(CO3), 223–237.

Seed, R.M. (1983). *Compaction-induced Stresses and Deflections on Earth Structure.* Ph.D. Dissertation, University of California, Berkeley, California, 447 pp.

Steward, J.E. and Mohney, J. (1982). Trial use Results and Experience for Low Volume Forest Roads. *Proceedings, 2nd International Conference on Geotextiles, Las Vegas, Volume II,* pp. 335–340.

Tafreshi, S.N.M. and Asakereh (2007). Strength Evaluation of Wet Reinforced Silty Sand by Triaxial Test. *International Journal of Civil Engineering,* 5(4), 274–283.

Tatsuoka, F. (2008). Recent Practice and Research of Geosynthetic-Reinforced Earth Structures in Japan. *Journal of GeoEngineering,* 3(3), 77–100.

Tatsuoka, F., Murata, O., and Tateyama, M. (1992). Permanent Geosynthetic-Reinforced Soil Retaining Walls Used for Railway Embankments in Japan. In *Geosynthetic-Reinforced Soil Retaining Walls, Denver, Colorado, 8–9 August 1991,* Wu (ed.). A.A. Balkema, Rotterdam, pp. 101–130.

Tatsuoka, F., Tateyama, M., Uchimura, T., and Koseki, J. (1997). Geosynthetic-reinforced Soil Retaining Walls as Important Permanent Structures. In *Mechanically-Stabilized Backfill,* Wu (ed.). A.A. Balkema, Rotterdam, pp. 3–24.

Tatsuoka, F., Tateyama, M., Aoki, H., and Watanabe, K. (2005). Bridge Abutment Made of Cement-Mixed Gravel Backfill. In *Ground Improvement – Case Histories.* Elsevier Geo-Engineering Book Series, Volume 3, pp. 829–873.

Wu, J.T.H. (1992a), Predicting Performance of the Denver Walls: General Report. In *Proceedings, International Symposium on Geosynthetic-Reinforced Soil Retaining Walls,* Wu (ed.), A.A. Balkema, Rotterdam, pp. 3–20.

Wu, J.T.H. (1992b). Construction and Instrumentation of the Denver Walls. In *Proceedings, International Symposium on Geosynthetic-Reinforced Soil Retaining Walls,* Wu (ed.), A.A. Balkema, Rotterdam, pp. 21–30.

Wu, J.T.H. (1992c). Measured Behavior of the Denver Walls. In *Proceedings, International Symposium on Geosynthetic-Reinforced Soil Retaining Walls,* Wu (ed.). A.A. Balkema, Rotterdam, pp. 31–42.

Wu, J.T.H. (1994). *Design and Construction of Low Cost Retaining Walls: The Next Generation in Technology.* Report No. CTI-UCD-1-94, Colorado Transportation Institute, Colorado Department of Transportation, Denver, Colorado, 162 pp.

Wu, J.T.H. and Helwany, H. (1990). Alleviating Bridge Approach Settlement with Geosynthetic Reinforcement. In *Proceedings, 4th International Conference on Geotextiles, Geomembranes and Related Products, The Hague, Volume 1,* A.A. Balkema, Rotterdam, pp. 107–111.

Wu, J.T.H. and Pham. Q. (2010). An Analytical Model for Evaluation of Compaction-Induced Stresses in Geosynthetic-Reinforced Soil (GRS) Mass. *International Journal of Geotechnical Engineering,* 4(4), 549–556.

Wu, J.T.H. and Pham, T.Q. (2013). Load Carrying Capacity and Required Reinforcement Strength of Closely Spaced Soil-Geosynthetic Composites. *Journal of Geotechnical and Geoenvironmental Engineering,* 139(9), 1468–1476.

Wu, J.T.H., Pham, T.Q., and Adams, M.T. (2010). *Composite Behavior of Geosynthetic-Reinforced Soil (GRS).* Technical Report, Civil Engineering Department, University of Colorado Denver, 277 pp.

Wu, J.T.H., Pham, T.Q., and Adams, M.T. (2013). *Composite Behavior of Geosynthetic Reinforced Soil Mass.* Publication No. FHWA-HRT 10-077, Federal Highway Administration, McLean, Virginia. 211 pp.

Wu, J.T.H., Yang, K.-H., Mohammed, S., Pham, T.Q., and Chen, R.-H. (2014). Suppression of Soil Dilation – A Reinforcing Mechanism of Soil-Geosynthetic Composites. *Transportation Infrastructure Geotechnology*, 1(1), 68–82.

Yamauchi, H. (1987). *Use of Nonwoven geotextiles for Reinforcing Clayey Embankment* (in Japanese). Doctoral Dissertation, University of Tokyo, Japan, 507 pp.

Yang, Z. (1972). *Strength and Deformation Characteristics of Reinforced Sand.* Doctoral Dissertation, Department of Civil Engineering, University of California at Los Angeles.

4

Geosynthetics Reinforcement

The concept of reinforced soil originated at least 6,000 to 7,000 years ago. Natural materials such as reeds, twigs, cotton, jute, straw, wood, etc. were first used as tensile inclusion in soil to construct ziggurats, dwellings, roadways, and even the Great Wall of China (Jones, 1985). A common problem for using natural materials as reinforcement in earthwork construction is with biodegradation of these materials due to microorganisms. With the advent of polymers in the early 20th century, a more stable and durable engineered material became available as tensile inclusion for earthwork construction. When properly formulated, polymers with half-lives of hundreds of years or more can be stable even under harsh environments.

The term *geosynthetics* refers to polymeric materials that are manufactured and used to help solve civil engineering problems. As proclaimed by Robert M. Koerner, founder of the Geosynthetic Research Institute, "Geosynthetics are bona fide engineering materials and must be treated as such."

In this chapter, we shall discuss the engineering behavior and properties of geosynthetics that are relevant to the analysis and design of GRS walls, including (i) load–deformation behavior, (ii) creep and relaxation behavior, (iii) soil–geosynthetic interface behavior, and (iv) hydraulic properties. We shall also discuss the advantages and disadvantages of using geosynthetics as reinforcement.

4.1 Geosynthetics as Reinforcement

The American Society for Testing and Materials (ASTM) D4439 defines *geosynthetics* as "a planar product manufactured from polymeric material used with soil, rock, earth, or other geotechnical engineering related material as an integral part of a manmade project, structure, or system." Since the late 1970s, the applications of geosynthetics in earth structure construction have grown several hundred folds. In 2003, for example, annual global usage of 1,400 million square meters of geosynthetics was reported (Cook, 2003). Today, the global market of geosynthetics is over 5,000 million square meters annually.

Geosynthetic Reinforced Soil (GRS) Walls, First Edition. Jonathan T.H. Wu.
© 2019 John Wiley & Sons Ltd. Published 2019 by John Wiley & Sons Ltd.

Geosynthetics originate from natural gas which reacts to form resin in the form of flakes and mixes with additives into different formations to produce various types of polymeric material known as geosynthetics. Different geosynthetic products have been developed to serve five different functions, including,

- filtration (to allow sufficient fluid flow across the plane of geosynthetics while maintaining the integrity of the soil without appreciable loss of soil particles)
- reinforcement (to improve the stiffness and strength of the soil by forming a soil–geosynthetic composite, or to serve as tiebacks/anchors to stabilize potential failure wedges of an earth structure)
- drainage (to allow sufficient fluid flow to occur within the plane of geosynthetics)
- separation (to set apart dissimilar materials so that the functioning of each material is maintained, e.g., geotextiles used below rail track ballast for separation of subgrade from ballast)
- liquid or soil containment (to contain liquid or soil particles).

To fulfill the various functions noted above, different types of geosynthetics have been manufactured, including

- geotextiles
- geogrids
- geonets
- geomembranes
- geocells
- geosynthetic clay liners
- geocomposites.

The common functions of each type of geosynthetics are summarized in Table 4.1. Detailed descriptions and applications of geosynthetics have been given in many excellent books, notably the books by Koerner (1994, 2005), Van Santvoort (1994), Ingold and Miller (1988), and Holtz et al. (1997).

There are currently more than 600 geosynthetics products available in North America. Among different types of geosynthetics, geotextiles, geogrids, geocells, and geocomposites have been used as reinforcement in reinforcing applications. A brief description of these four types of geosynthetics is given in this section. Note that the function of reinforcement is sometimes a result of *synergistic* improvement of system performance. In other words, functions of separation, filtration, drainage, and soil grain containment may work synergistically in conjunction with the reinforcing function to improve the performance of an earth structure. Different functions of geosynthetics often behave in such a manner that the total effect is greater than the sum of the individual effects. The term *geosynthetic reinforcement* is used in this book interchangeably with *geosynthetic inclusion*, although the former is more appropriate when the primary function of a geosynthetic is reinforcing, while the latter is more appropriate for situations where the synergistic effect of a geosynthetic is emphasized.

Table 4.1 Functions of different types of geosynthetics (modified after Koerner, 1994)

Type of Geosynthetic	Separation	Reinforcement	Filtration	Drainage	Liquid/Soil Grain Containment
Geotextiles	√	√	√	√	√
Geogrids		√			
Geonets				√	
Geomembranes					√
Geosynthetic clay liners					√
Geocells	√	√			
Geocomposites	√	√	√	√	√

Synthetic polymers used for producing geosynthetics include (in descending order of volume of usage) polypropylene (PP), polyester (PET), polyethylene (PE), polyamide (PA), and polyvinyl chloride (PVC). Generally speaking, polyester (PET) has higher strength and stronger resistance to creep than polypropylene (PP) and polyethylene (PE); and polypropylene (PP) and polyethylene (PE) (both belong to the *polyolefins* family) have tougher resistance to organic and acid attacks than polyesters.

4.1.1 Geotextiles

Geotextiles, when used independently, are the most versatile among all types of geosynthetics, taking up about 3/4 of the geosynthetics market. As seen in Table 4.1, geotextiles are capable of performing one or more of the five functions of geosynthetics: reinforcement, filtration, drainage, separation, and soil containment (also liquid containment, when impregnated).

Like all other types of geosynthetics, geotextiles are made of synthetic polymers. The polymers, in pellets or granules, are heated and converted into synthetic fibers. Most geotextile fibers are made from polypropylene (about 80%) and polyester (about 15%). Only about 5% of geotextile fibers are manufactured from polyethylene, polyamide (nylon), and polyvinyl chloride (PVC).

Geotextile fibers may take one of three forms, namely, filaments, staple fibers, and slit films:

- *Filaments* are produced by extruding melted polymers through dies or spinnerets. Since this process is continuous, filaments are often called continuous filaments. After the extrusion, filaments are usually drawn to straighten them.
- *Staple fibers* are obtained by cutting filaments to shorter lengths, typically 20–100 mm (0.8–4.0 in.) in length.

- *Slit films* are flat, tape-like, 10–30 mm (0.4–1.2 in.) wide fibers, produced by slitting an extruded plastic film with blades. After slitting, the tape-like fibers are often drawn to increase their strengths.

The synthetic geotextile fibers are subsequently made into flexible and generally porous fabrics, called *geotextiles*, by a few different methods. Depending on the method by which the fibers (or yarns) are converted into fabrics, geotextiles are grouped into:

- woven geotextiles
- nonwoven geotextiles
- knit geotextiles

Woven Geotextiles

Woven geotextiles comprise two sets of parallel yarns in two directions (referred to as *warp* and *weft* directions, or *machine* and *cross-machine* directions, respectively) by interlacing the yarns in a systematic manner. The two sets of yarns are typically weaved in perpendicular directions. Woven geotextiles are manufactured with one or a combination of the following types of yarns:

- *monofilament* yarn, made of a single filament
- *multifilament* yarn, made of fine multiple filaments aligned together
- *slit film* (or *tape*) yarn, made of a single slit film fiber
- *spun* yarn, made of staple fibers interlaced and twisted together to form a single yarn
- *fibrillated* yarn, made of strands of tape-like fibers partially attached to one another.

Some examples of woven geotextiles made of different types of yarns are shown in Figure 4.1. Monofilament yarns are often made from polyethylene (PE), multifilament yarns from polyester (PET), and slit film yarns from polypropylene (PP). This allows an *informed guess* of the polymer type in a woven geotextile.

Nonwoven Geotextiles

Nonwoven geotextiles are formed by bonding fibers together in an oriented or random pattern. Due to the difference in bonding process, a nonwoven geotextile may be of one of three types:

- *needle-punched* (mechanical bonding), see Figure 4.2(a)
- *heat-bonded* (thermal bonding), see Figure 4.2(b)
- *resin-bonded* (chemical bonding), see Figure 4.2(c).

Needle-punched geotextiles are formed by punching and withdrawing a bank of thousands of small barbed needles to entangle the fibers of a fibrous web (see Figure 4.3); thermal boned geotextiles are formed by heating and adhering partially melted fibers at

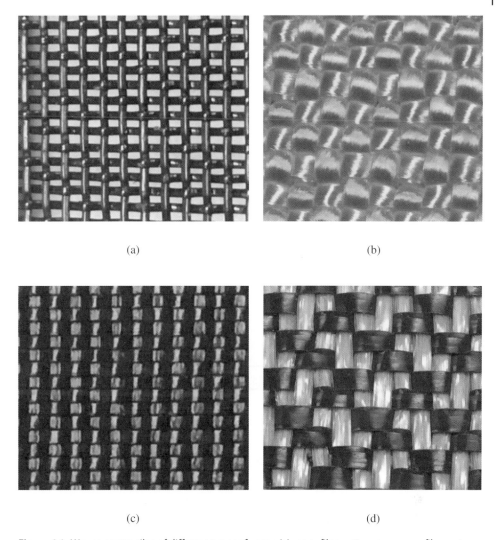

(a)

(b)

(c)

(d)

Figure 4.1 Woven geotextiles of different types of yarns: (a) monofilament yarn on monofilament yarn, (b) multifilament yarn on multifilament yarn, (c) slit film yarn on slit film yarn, and (d) multifilament yarn on slit film yarn

crossover points; whereas chemically bonded geotextiles are formed by adding cementing agents (such as acrylic resin) to join the fibers together.

Needle-punched geotextiles, the most common type of nonwoven geotextile, vary widely in weight (100–550 g/m^2 or 3–16 oz/yd^2) and thickness (varies with normal compressive pressure). Needle-punched geotextiles have seen applications in filtration, drainage, separation, protection, and as pavement overlay fabrics. Heat-bonded geotextiles are usually much thinner and denser than needle-punched geotextiles; hence have

(a)   (b)

(c)

Figure 4.2 Types of nonwoven geotextiles: (a) needle-punched, (b) heat-bonded, and (c) resin-bonded (side view, courtesy of Fiber Bond)

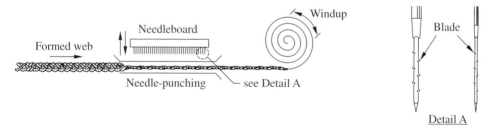

Figure 4.3 Diagram of manufacturing needle-punched geotextile (INDA, after Koerner, 1994)

higher stiffness at small strains, but with lower drainage capability than needled geotextiles. Resin-bonded geotextiles are usually much thicker (typically 10–25 mm or 0.4–1.0 in.), stiffer, more expensive, more permeable, and more compressible than other nonwoven geotextiles. Resin-bonded geotextiles have seen limited applications as liners of golf course bunkers (sand-traps), as they can provide high permeability and good long-term durability in a moist and abrasive environment even under low surcharge (Mills, 2005).

Knitted Geotextiles

Knitted geotextiles are formed by knitting loops of one or more nonwoven yarns (other than slit film yarn) to form a planar fabric. Knitted geotextiles are usually of very low stiffness, and have seen limited applications as filters in pipe wrap applications.

4.1.2   Geogrids

Geogrids are plastics formed into an open, grid-like configuration. The apertures (i.e., openings between the adjacent sets of longitudinal and transverse ribs) of geogrids are large enough, typically between 10 and 100 mm (0.4 and 4.0 in.), to allow soil particles to strike-through from one side of a geogrid to the other. Because of the large apertures, geogrids have been used almost exclusively for reinforcement applications.

Two major types of geogrids are available. One type is *flexible geogrids*, or overlapping geogrids. They are textile-like and formed by overlapping perpendicular polymer strands using a weaving or knitting process, bonding the strands at their junctions, then encasing them with a polymer-based plasticized coating. Flexible geogrids may be of woven or knitted geogrids, depending on the process by which they are formed. The other type is *stiff geogrids*, or punched-and-drawn geogrids. They are manufactured by drawing a perforated polymer sheet.

Most flexible geogrids are made from PVC-coated or polypropylene-coated polyester, while most stiff geogrids are made from high density polyethylene (HDPE) or polypropylene. Flexible geogrids are made about the same strength in both directions or stronger in one direction, as shown in Figure 4.4. Rigid geogrids are available in three forms: uniaxial (stronger in the drawn direction), biaxial (about the same strength in the two drawn directions), or triaxial (with isosceles triangular apertures, about the same strength in the three directions), as shown in Figure 4.5. For cost considerations, uniaxial geogrids are more widely used than other geogrids in reinforced soil walls.

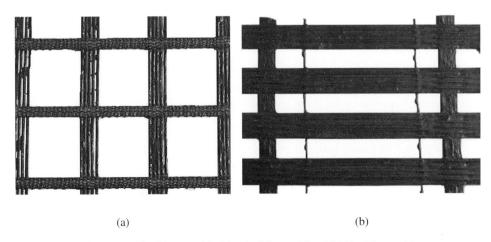

(a)   (b)

Figure 4.4 Flexible geogrids: (a) uniaxial geogrid and (b) biaxial geogrid

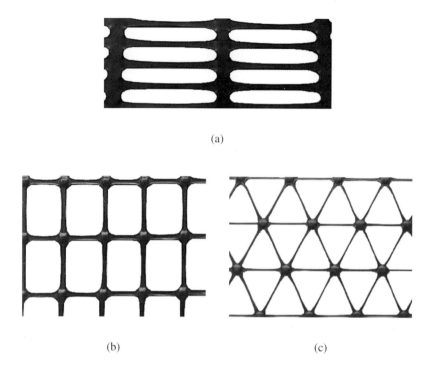

(a)

(b)   (c)

Figure 4.5 Rigid geogrids: (a) uniaxial geogrid, (b) biaxial geogrid, and (c) triaxial geogrid

4.1.3   Geocells

Geocells, as shown in Figure 4.6(a), are a three-dimensional geosynthetic honey-combed cellular confinement system developed primarily for slope protection or earth retention applications. Geocells are typically formed by extruding from polymeric materials into strips welded together ultrasonically in series. In field installation, the strips are expanded in the opposite direction to form the side-walls of cells (which is often textured or perforated) to accommodate soil. Fill is placed inside the cells in an expanded state and is compacted after placement. The cellular confinement is said to reduce the lateral movement of soil particles and form a stiffened mat that can distribute vertical loads over a wider area.

There is a thicker version of geocells (up to 1 m or 3 ft thick, typical), known as a *geogrid mattress* or *geocell mattress*. A geogrid/geocell mattress is a continuous cellular mat fabricated on site by securing vertically positioned geogrids at their intersection with polymer rods or metallic bodkins (see Figure 4.6(b)).

As opposed to planar geosynthetic reinforcement (in the form of sheets) of which lateral confinement is induced by soil–reinforcement interface friction, geocells use their sidewalls to provide lateral confinement. For fill materials that cannot develop sufficiently high interface friction with other fill particles (hence are difficult to compact) or with geosynthetic sheets, geocells may prove to be better reinforcing geosynthetics. An example of such a fill material is uniform-size (i.e., poorly graded) aggregates.

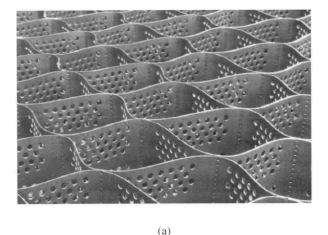

(a)

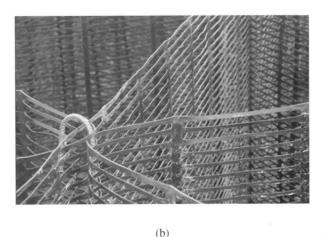

(b)

Figure 4.6 Geocells: (a) a geocell made with ultrasonically-welded high-density polyethylene (HDPE) (courtesy of Strata Systems, Inc.) and (b) a geocell mattress secured at intersections by metallic bodkins

4.1.4 Geocomposites

As noted earlier in Section 4.1, a number of different types of geosynthetic products have been manufactured to fulfill different individual functions, with each type of geosynthetic product having some distinct advantages over others in a given application. Geocomposites are formed by combining two or more types of geosynthetic products in the manufacturing process to obtain the combined advantages of the different geosynthetic products. A long array of geocomposites is now available (see Figure 4.7(a)). For instance, a needled geotextile/geogrid composite combines the drainage function of a needle-punched geotextile (when deployed in low permeability backfill) and the reinforcing function of a geogrid. Geocomposite drains can be prefabricated strip drains formed by encasing a nonwoven geotextile (functioning as a filter) over a fluted or dimpled polymeric unit (functioning as a conduit for fluid flow), as shown in Figure 4.7(b).

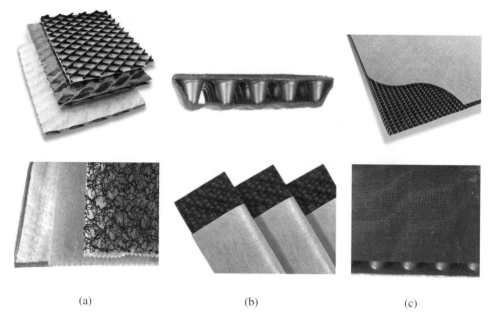

(a)   (b)   (c)

Figure 4.7 Examples of (a) geocomposite, (b) geocomposite strip drain, and (c) geocomposite mat drain

Geocomposite drains can also be drainage mats formed by sandwiching a polymeric drain unit between nonwoven geotextiles, as shown in Figure 4.7(c). Geocomposite drains have been installed along excavation limits (behind the retained earth zone) to minimize subsurface water entering the reinforced zone of reinforced soil walls.

For obvious reasons, there has been considerable advancement and innovation of geocomposites in recent years. This trend will likely continue for many years to come as new applications of geosynthetics are realized, and new and improved geosynthetics products are developed.

4.1.5   Description of Geosynthetics

It has been suggested that the description of a geosynthetic be given by following a certain order (Holtz et al., 1997):

- polymer type (polypropylene, polyester, high density polyethylene, etc.)
- fiber/yarn type, if applicable (monofilament, multifilament, slit film, etc.)
- manufacturing process, if applicable (woven, nonwoven, needle-punched, etc.)
- primary type of geosynthetic (geotextile, geogrid, geocomposite, etc.)
- mass per unit area and/or thickness, when appropriate
- additional information related to specific applications.

Examples: a polyester multifilament woven geotextile, 240 g/m^2 (7 oz/yd^2); a polypropylene staple filament needle-punched nonwoven geotextile, 350 g/m^2 (10.5 oz/yd^2); a polyethylene extruded biaxial geogrid, with 25 mm × 25 mm (1.0 in. × 1.0 in.) openings.

4.1.6   Costs

The costs of geotextiles, geogrids, and geocomposites depend largely on polymer type, manufacturing process, weight, and locality. In terms of polymer type, the costs in descending order are polyester > polyethylene > polypropylene. In terms of manufacturing process, weaving usually costs more than the nonwoven forming process. Within nonwoven geotextiles, resin-bonding generally costs more than heat-bonding, with needle-punching being the least expensive process.

As of 2018, the costs of nonwoven geotextiles per square meter were typically between \$0.15 and \$1.5 for light-weight geotextiles (weight 100–170 g/m^2 or 3–5 oz/yd^2), between \$0.3 and \$3 for medium-weight geotextiles (weight 200–270 g/m^2 or 6–8 oz/yd^2), and between \$0.7 and \$4 for heavy-weight geotextiles (weight 300–550 g/m^2 or 9–16 oz/yd^2). Woven geotextiles usually cost between \$0.25 and \$5 per square meter. The cost of geogrids per square meter typically ranges from \$0.50 to \$2.50, but can be as high as \$5 to \$8 (Note: 1 $m^2 \approx 1.2$ yd^2).

4.2   Mechanical and Hydraulic Properties of Geosynthetics

The mechanical properties of geosynthetics, including tensile stiffness/strength properties, creep properties, soil–geosynthetic interface properties, tear strength, puncture strength, fatigue strength, and seam strength, have been addressed in a number of books on geosynthetics (e.g., Koerner, 1994, 2005; Van Santvoort, 1994; Holtz et al., 1997), and hundreds of journal articles. In this section, we shall discuss the first three aforementioned mechanical properties as related to the analysis and design of GRS structures. Hydraulic properties related to GRS are briefly discussed at the end of this section.

4.2.1   Load–Deformation Properties of Geosynthetics

The constitutive behavior of civil engineering materials is typically expressed in terms of stress versus strain relationships. The constitutive relationships of geosynthetics are commonly expressed in terms of load/width versus deformation (Note: The term *deformation* can be regarded as being interchangeable with *strain*). The main reason that load/width is used, as opposed to stress (load/area), is because the thickness of some geosynthetics is pressure-dependent. It is important to note that the slope of the load/width versus strain curve is not *Young's modulus* (E), rather it is the product of Young's modulus (E) and thickness (t). The product of $E \cdot t$ has been used to denote the *tensile stiffness* of a geosynthetic product.

The load–deformation behavior of geosynthetics, commonly characterized by the behavior a uniaxial loading condition, can be divided into two groups: behavior in an unconfined condition (also referred to as *in-air* or *in-isolation* behavior) and behavior in a confined condition. The former refers to a condition where a geosynthetic specimen is tested in isolation, while the latter refers to a condition where a geosynthetic specimen is tested under pressure or soil confinement to better simulate the operative condition of a geosynthetic product.

(A)   Unconfined Load–Deformation Tests

A number of different types of unconfined load–deformation test device with different specimen sizes, clamping methods, and clamp sizes have been developed to measure the in-air or in-isolation load–deformation behavior of geotextiles and geogrids. Figure 4.8 shows schematic diagrams of four test methods for measuring the load–deformation and strength properties of geosynthetic products in an unconfined condition, including (i) the grab tensile test (ASTM D4632 for geotextiles), (ii) the wide-width tensile test (ASTM D4595 for geotextiles, with ASTM D6637 as a parallel test for geogrids), (iii) the plane-strain test, and (iv) the biaxial test. The results of the first two test methods have routinely been reported by geosynthetics manufacturers and used in design (especially the wide-width tensile test) and specifications. The plane-strain test and the biaxial test, although arguably give better simulation of the typical loading conditions of geosynthetics in reinforced soil structures, are not as widely available; their use has been limited to research purposes.

The grab tensile test for geotextiles (ASTM D4632), Figure 4.8(a), uses a test specimen of $B = 100$ mm (4.0 in.), $L = 150$ mm (6.0 in.), and $W = 25$ mm (1.0 in.) gripped by a

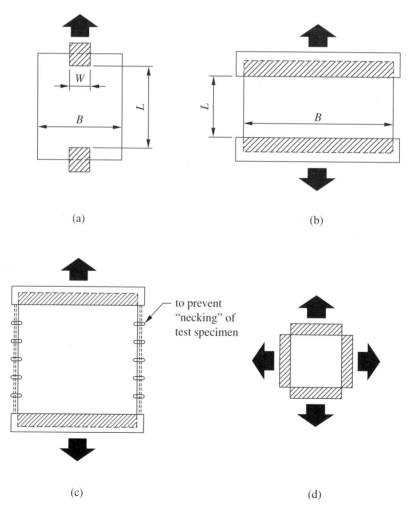

(a)

(b)

(c)

(d)

Figure 4.8 Test methods for measuring the load–deformation relationship or strength of geotextiles: (a) grab tensile test, (b) wide-width tensile test, (c) plane-strain tensile test, and (d) biaxial tensile test (modified after Myles, 1987)

pair of jaws along the centerline at opposing ends. The value of the grab tensile strength has been reported by the manufacturer for many geotextiles. For GRS walls, however, grab tensile strength can only be used as an index property, and could be used only as a reference for establishing comparative strengths of different geotextiles.

The wide-width tensile test for geotextiles (ASTM D4595), see Figure 4.8(b), uses a geotextile specimen of $B = 200$ mm (8.0 in.) and $L = 100$ mm (4.0 in.). The specimen is loaded uniaxially at a constant strain rate of $10 \pm 3\%$ per minute, under a temperature of $21 \pm 2°C$, and by a pair of fixed clamps or wedge clamps. Unlike the grab tensile tests, the grips of wide-width tensile tests extend over the entire width of the test specimen. Wide-width ultimate tensile strength and stiffness at 5% strains have been commonly reported by the manufacturers of geotextiles that are intended for reinforcing applications. The value of the wide-width strength (and sometimes 5% stiffness as well) has often been used in design and analysis of GRS and GMSE walls.

Serrated wedges such as the one shown in Figure 4.9(a) have been used to grip geotextiles in wide-width tensile tests of geotextiles. For higher strength geotextiles (say, strength greater than about 50 kN/m), slippage or damage may occur in the grip area of the test specimen before ultimate strength is reached, therefore roller grips (in combination with extensometers), see Figure 4.9(b), or special clamps may be needed. One special clamping method that has had good success is to treat the grip area of the geotextile specimen with a high-strength epoxy, and, if needed, reinforce the grip area with metal plates to prevent deformation of the specimen in the gripping area (Ling et al., 1992; Wu and Su, 1987), as shown in Figure 4.9(c).

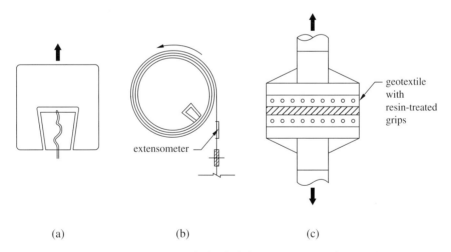

(a) (b) (c)

Figure 4.9 Methods to grip test specimens for load–deformation testing of geotextiles: (a) serrated wedge grips, (b) roller grips, and (c) rigid grips

For measuring the tensile properties of geogrids, test methods for single rib and multiple ribs (ASTM D6637) have been used. To perform a multiple ribs test, the width of the geogrid specimen needs to be at least 200 mm (8 in.) and contain a minimum of five ribs in the cross-direction. At the start of a test, the distance between the clamps or the distance from centerline to centerline of the rollers should be adjusted to the greater distance of three junctions or 200 ± 3 mm (8.0 ± 0.1 in.), such that at least one transverse rib is contained centrally within the gauge length. Also, at least one clamp must be supported by a free swivel or universal joint to allow the clamp to rotate in the plane of the geogrid specimen. The test method also recommends a strain rate of $10 \pm 3\%$ per minute.

It should be noted that with a specimen aspect ratio (width/gauge length) of 2, as stipulated in the wide-width tensile test for geotextiles (ASTM D 4595), a significant *Poisson effect* (also known as *necking*) can occur for most nonwoven geotextiles and geogrids. The Poisson effect is negligible for woven geotextiles of which the deformation of transverse yarns is hardly affected by the deformation of longitudinal yarns (Wu and Tatsuoka, 1992). The Poisson effect generally results in a weaker load–extension response. Figure 4.10 shows the influence of the specimen aspect ratio on the load–deformation relationships of a nonwoven geotextile. It is interesting that the figure also includes the load–deformation curve of a specimen with an infinite aspect ratio, achieved by sewing the geosynthetic specimen into a cylindrical shape end-to-end, as shown in Figure 4.11.

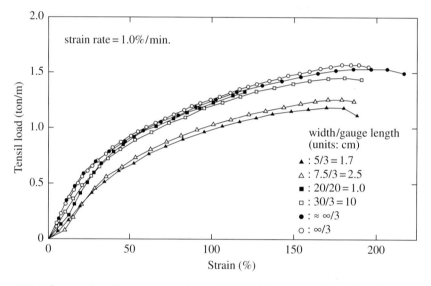

Figure 4.10 Influence of specimen aspect ratio on the load–deformation relationship of a nonwoven geotextile in an unconfined condition (modified after Yamauchi, 1987)

The plane-strain tensile test shown in Figure 4.8(c) is designed to minimize the Poisson effect in a test specimen. To obtain a more accurate load–deformation relationship of woven geotextiles and geogrids that are deployed in a *continuous* manner in field installation (as opposed to overlapping one sheet over the other), a larger aspect ratio (say, no less than 6) has been recommended for geosynthetic products sensitive to the specimen aspect ratio (Ling et al., 1992), i.e. for most nonwoven geotextiles and geogrids.

The load–deformation curves of geosynthetics can vary significantly depending on polymer type and manufacturing method. Representative uniaxial load–deformation curves are depicted in Figure 4.12. The curves shown in this figure are for general references only; actual curves may vary considerably.

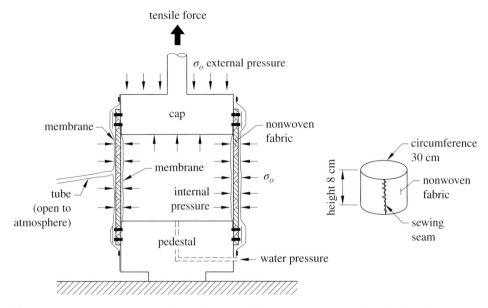

Figure 4.11 Setup of a geosynthetic specimen of an infinite aspect ratio in the uniaxial load–deformation testing of geotextiles (modified after Yamauchi, 1987)

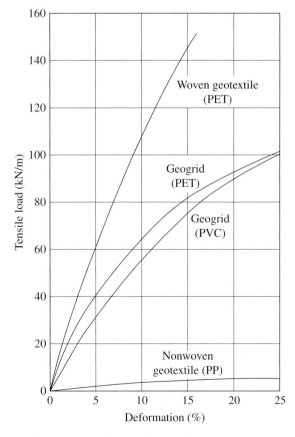

Figure 4.12 Uniaxial load–deformation relationships for different representative geosynthetics of comparable unit weights

The load–deformation behavior of geosynthetics measured in the laboratory is also affected by the rate of extension (i.e., strain rate) and temperature. Generally speaking, the tensile stiffness and strength of geosynthetics tend to increase with increasing strain rate. This may be an important factor to consider in design. The strain rates in actual field installation are typically much smaller than those used in laboratory testing. The effect, however, is different for different polymer types (Van Santvoort, 1994). The influence of strain rate is usually rather significant for polyethylene geosynthetics (e.g., see Figure 4.13(a)) and negligible for polyester geosynthetics, unless the strain is very large (greater than about 7–9% strain for the polyester geotextile shown in Figure 4.13(b)).

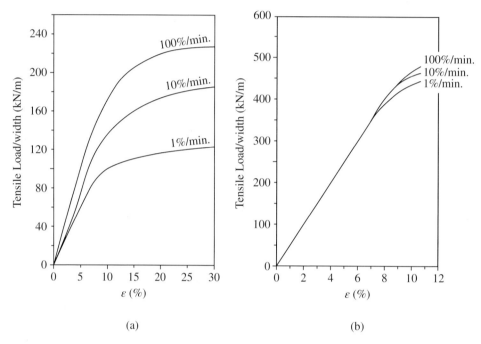

(a)

(b)

Figure 4.13 Illustrative uniaxial load–deformation relationships of (a) a polyethylene geotextile and (b) a polyester geotextile, as influenced by strain rate

Temperature is also known to affect the tensile stiffness and strength of geosynthetics. Tensile stiffness and strength of geosynthetics tends to increase with decreasing temperature (e.g., see Figure 4.14). The load–deformation properties of geosynthetics have typically been reported at an ambient temperature of $21 \pm 1°C$ ($70 \pm 2°F$). In situations where geosynthetic reinforcement is expected to be in an environment where in-ground temperatures may go well above $21°C$ ($70°F$), substantial reduction in strength and stiffness should be used in the design. On the other hand, in situations where the in-ground temperatures may be much below $21°C$ ($70°F$), the geosynthetic reinforcement will likely be less ductile, i.e. having a lower elongation to break (Bonaparte et al., 1987). For most applications in the U.S., the in-ground temperature is typically lower than $21°C$ ($70°F$), hence reported load–deformation properties at $21 \pm 1°C$ ($70 \pm 2°F$) in those applications are slightly on the conservative side in terms of stiffness and strength, and unconservative in terms of ductility.

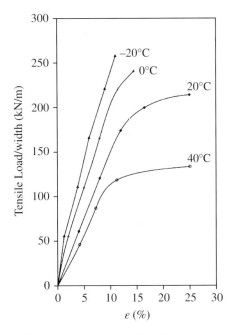

Figure 4.14 Illustrative uniaxial load–deformation relationships of a polyethylene geotextile, as influenced by temperature

(B) *Confined Load–Deformation Tests*

Since geosynthetics reinforcements are subjected to soil confinement in actual installation, many *confined tensile tests* have been proposed (e.g., McGown, et al., 1982; Christopher et al., 1986; Siel et al., 1987; Leshchinsky and Field, 1987; Kokkalis and Papacharisis, 1989). A common objective of these confined tensile tests is to determine the load–deformation properties of geosynthetics under typical operative conditions in field applications. Wu (1991) has given an overview of various confined tests and pointed out that confined tests are only needed for pressure-sensitive geosynthetics (i.e., needled geotextiles). The shortcomings of each confined tests reported in the literature were discussed.

A common *flaw* with many confined tensile tests for investigating the load–deformation behavior of geosynthetics is that the measured load–deformation behavior inadvertently includes soil–geosynthetic interface friction as part of the resistance to deformation. Take the confined test method shown in Figure 4.15 as an example, as the soil is held "stationary", in order for the geosynthetic specimen to deform under applied tensile loads, it must also overcome the frictional resistance induced at the soil–geosynthetic interface. Because the soil in a GRS structure will likely deform *with* the geosynthetic reinforcement until the reinforced soil structure is approaching failure, any test method that includes soil–geosynthetic interface shear resistance for determination of load–deformation behavior is unsafe. In other words, such a confined test will give a poor representation of typical operating conditions and is unsafe. For a confined test to

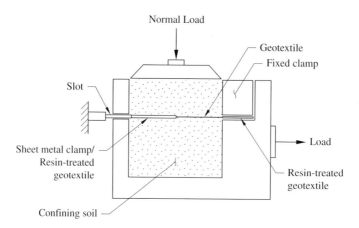

Figure 4.15 Schematic of a confined load–deformation test of which the measured loads include soil–geosynthetic interface frictional resistance, i.e., interface slippage is forced to occur

give representative (and conservative) load–deformation behavior, the soil should only provide pressure confinement, not frictional resistance at the interface (Wu, 1991).

Soil-confined tests, such as the uniaxial tension test proposed by Ling et al. (1992) and a modified triaxial test proposed by Wu and Arabian (1990), in which tensile loads are applied to a soil-confined geosynthetic specimen *and* the confining soil without inducing soil–geosynthetic interface friction, are shown in Figure 4.16. Because the only function of the soil in these tests is to transmit confining pressure, the soil confinement in these tests can be replaced by a flexible membrane to exert pressure confinement (Ling et al., 1992).

A confined tensile test that is conducted by enclosing a geosynthetic specimen by flexible membrane and using a vacuum to supply confining pressure (see Figure 4.17) is referred to as an *intrinsic confined tensile test*. The load–deformation behavior obtained from an intrinsic confined test is essentially the same as that obtained from a confined in-soil test where soil–geosynthetic interface slippage is not forced to occur (see Figure 4.18). Among geosynthetic products used as reinforcement for reinforced soil walls, needle-punched geotextiles have been found to be the only type of geosynthetics of which load–deformation behavior is affected appreciably by pressure confinement within the practical range of confining pressure in actual wall installation. For other types of geosynthetics (i.e., woven geotextiles and geogrids), the load–deformation behavior is essentially the same in unconfined and confined conditions. In summary, other than nonwoven needle-punched geotextiles, an unconfined test should be used for measuring load–deformation behavior. For nonwoven needle-punched geotextiles, an intrinsic confined test is recommended.

Uniaxial load–deformation relationships of geosynthetics are generally nonlinear after the strain exceeds a certain value. For design purposes, the load–deformation behavior of a given geosynthetic product can be characterized by an ultimate strength and a stiffness value. The latter can be the *secant* modulus over the anticipated range of

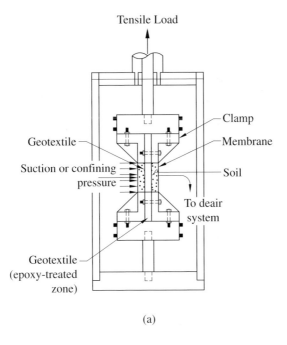

Tensile Load

Geotextile

Clamp

Membrane

Suction or confining
pressure

Soil

To deair
system

Geotextile
(epoxy-treated
zone)

(a)

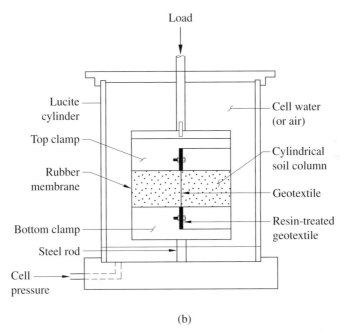

Load

Lucite
cylinder

Cell water
(or air)

Top clamp

Cylindrical
soil column

Rubber
membrane

Geotextile

Bottom clamp

Resin-treated
geotextile

Steel rod

Cell
pressure

(b)

Figure 4.16 Schematic of soil-confined load–deformation tests without including the soil–
geosynthetic interface frictional resistance in the measured loads: (a) a uniaxial tension
test (Ling et al., 1992) and (b) a modified triaxial test

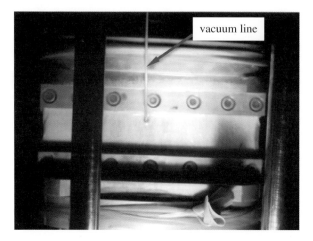

Figure 4.17 Test specimen in the intrinsic confined load–deformation test; the specimen is enclosed by an air-tight rubber membrane with a vacuum line attached

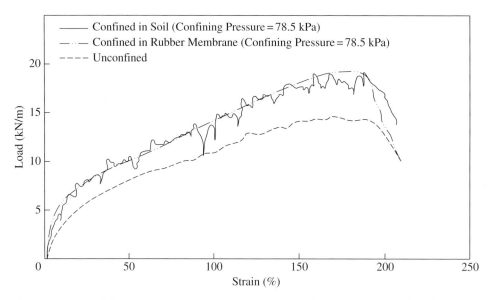

Figure 4.18 Load–deformation relationships of a nonwoven needle-punched geotextile under different confinement conditions (Ling et al., 1992)

strain, which is the slope of a straight line drawn between the point of largest anticipated strain and the origin, as shown in Figure 4.19. Note that the secant modulus gives the average value of the tangent moduli over the range of strains under consideration.

For most woven geotextiles and geogrids in reinforcing applications, a conservative value for the largest strain under service loads can be taken as 1.0–3.0%. Product data published annually by the Industrial Fabrics Association International (IFAI) (known as the *Geosynthetics Specifier's Guide*) have reported stiffness at 5% strain and ultimate

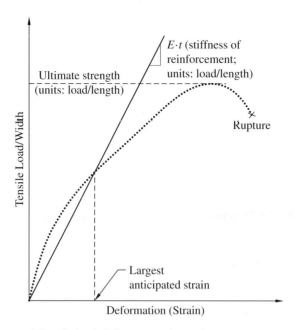

Figure 4.19 Secant modulus of a load–deformation relationship over an anticipated range of strain

strength obtained from ASTM D4595 (for geotextiles) and D6637 tests (for geogrids). These values have been referred to in routine designs. The reported stiffness values at 5% strain, although conservative, are too low for more sophisticated analysis. Note again that the slope of the uniaxial load–deformation curve is the product of E (Young's modulus) and t (thickness) of the geosynthetic product. To carry out a plane-strain finite element analysis, the area (A) and Young's modulus (E) of geosynthetic reinforcement are usually required as input parameters. Since the axial stiffness of geosynthetic reinforcement is the product of E and A, as long as the correct product ($E \cdot A$) is used, the actual values of E or A do not matter in the analysis. If the area (A) is set equal to the nominal thickness (t), the value of Young's modulus E is then the secant slope of the uniaxial load–deformation curve divided by the nominal thickness (t).

4.2.2 Creep of Geosynthetics and Soil–Geosynthetic Composites

Creep is a term refers to time-dependent deformation of a material under a constant static stress. Geosynthetics, being manufactured with polymers, are considered creep susceptible. Most civil engineers are comfortable with steel, concrete, and timbers in part because they are generally not creep susceptible. Creep has been a major concern for geosynthetic reinforced soil structures.

Stress level, polymer type, manufacturing method, and temperature are known to affect the creep potential of geosynthetics. Figure 4.20 shows the creep behavior of different polymers at stress levels of 20% and 60%. It is seen that polypropylene (PP) and polyethylene (PE) generally exhibit much higher creep deformation than polyester (PET) and polyamide (PA). For geotextiles that are to be loaded over a long period of

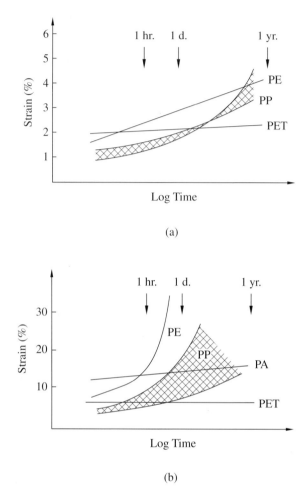

Figure 4.20 Creep behavior of different polymers at two load levels: (a) at 20% of the ultimate level and (b) at 60% of the ultimate level (den Hoedt, 1986)

time (up to 75–100 years), permissible loads on the order of 40–50% of the ultimate tensile strength have been recommended for polyester and polyamide. For polypropylene and polyethylene, recommended permissible loads have been much lower, on the order of 20–25% of the ultimate strength.

The allowable tensile load of geosynthetic reinforcement used in the prevailing design guides is typically determined by applying a safety factor and a combination of reduction factors to the limiting strengths of geosynthetics determined by short- or long-term laboratory tests. These tests are conducted by applying uniaxial tensile loads *directly* to a geosynthetic specimen (in a confined or unconfined condition) without any regard to soil–geosynthetic interaction. It is important to note that conducting uniaxial creep tests for a geosynthetic product in the confinement of soil does not mean the soil–geosynthetic interaction is accounted for.

   McGown et al. (1982) developed a fairly sophisticated uniaxial tension test device to measure the creep behavior of geotextiles under soil confinement (see Figure 4.21). They conducted a number of tests on different geotextiles and concluded that soil confinement can influence significantly the creep behavior of some geosynthetics. McGown et al. presented the results of their creep tests in the form of a diagram shown in Figure 4.22. The load–deformation relationship that brings a geosynthetic specimen to

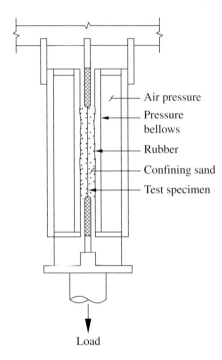

Figure 4.21 Schematic of a uniaxial tension test device for investigating the creep behavior of geotextiles under soil confinement (modified after McGown et al., 1982)

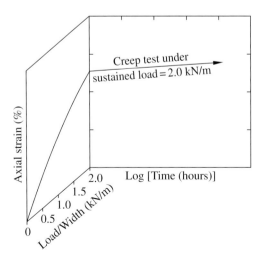

Figure 4.22 A diagram for presenting the creep test results of a geosynthetic specimen

a prescribed load level is plotted on the left side of the diagram, with time-dependent creep deformation behavior shown on the right side of the diagram. The sustained load as well as the strain at the onset of the creep test can be seen clearly from the diagram.

Previous discussions on confined load–deformation tests (see subsection (B) of Section 4.2.1) are equally applicable to confined creep tests. That is, since soil and geosynthetic in the operational condition of GRS walls will likely deform in a compatible manner until failure is approached, confined creep tests that include soil–geosynthetic interface resistance in the measured loads are unconservative (i.e., will significantly underestimate creep potential and creep deformation). Note that the test device shown in Figure 4.21 is conceptually sound for studying the creep behavior of pressure-sensitive geosynthetics because the confining soil and geosynthetic specimen in this test device will deform in a compatible manner. However, the intrinsic confined test described in subsection (B) of Section 4.2.1 is a much simpler alternative. An intrinsic creep test can be conducted by simply applying stained axial loads to a pressure-confined test specimen under a prescribed vacuum pressure. It is important to emphasize that the intrinsic test is necessary only for pressure-sensitive geosynthetic products (i.e., needle-punched geotextiles). For non-pressure-sensitive geosynthetics (such as nonwoven geotextiles and geogrids), an *unconfined* creep test would be sufficient for investigating the long-term creep behavior.

Since temperature is known to influence the rate of creep of geosynthetics, creep tests should be conducted over the range of temperatures anticipated under the in-service condition of the reinforced soil structure. This does, however, require extensive testing at different temperatures over long periods of time. In the absence of such information, time-shifting techniques may be utilized judiciously to account for the temperature effect. Time shifting is especially needed if the anticipated in-service temperatures are higher than the ambient temperatures under which the creep data are obtained. If these data are used directly in design computations without temperature correction, creep deformation will be underestimated.

Whether or not creep is a design issue for a geosynthetic reinforced soil structure must be evaluated by considering the deformation characteristics of the reinforced soil mass as a whole, i.e., the soil–geosynthetic interaction must be properly accounted for. It can be misleading to evaluate the long-term creep potential of a reinforced soil structure based on the results of *element* creep tests performed on geosynthetic specimens alone, as has (inappropriately) been stipulated by most prevailing design guides. Wu and Helwany (1996) developed a simple plane-strain long-term soil–geosynthetic interaction test in which the soil and geosynthetic are allowed to deform in an interactive manner. They reported results for two long-term creep tests: one with a clay and the other with a sand. For the sand-backfill test, creep deformation essentially ceased within 100 minutes after initiation of the test; whereas the clay-backfill test experienced continuing creep deformation over the entire test period (18 days); shear failure occurred in the soil on the 18th day. Element tests on the geosynthetic alone would have underestimated the maximum strain by 250% in the clay-backfill test, and overestimated the maximum strain by 400% in the sand-backfill test.

To examine soil–geosynthetic interactive behavior in GRS mass by a laboratory test, a simple device was developed by Ketchart and Wu (2001, 2002) by streamlining the test apparatus used in Wu and Helwany's study (1996). The simplified test is referred to as the *soil–geosynthetic interactive performance* (SGIP) test and is shown in Figure 4.23. The procedure of the SGIP test can be described as follows:

Step 1   Treat the interior side walls of a SGIP test bin by applying a thin layer of silicon grease to the side walls and covering them with a sheet of latex membrane to minimize soil–wall friction/adhesion (for side wall lubrication, see Section 1.4.3).

Step 2   Lay a sheet of geosynthetic covering the base of the test bin; place and compact the fill material in lifts inside the cuboidal space formed by the side walls and movable supporting plates until the fill reaches the mid-height of the test bin.

Step 3   Lay a second sheet of geosynthetic specimen over the surface of fill at the mid-height; continue with placing and compacting the fill material until it reaches the full height of the test bin.

Step 4   Lay a third sheet of the geosynthetic over the top of the compacted fill; position a loading plate over the third geosynthetic sheet.

Step 5   Apply increasing vertical loads to the top surface of the soil–geosynthetic composite until a prescribed sustained load is reached; release the movable supporting plates and measure vertical and lateral time-dependent deformation while maintaining a constant sustained load; continue monitoring the deformation over a planned test period.

The fill material and geosynthetic specimen used in the test should be the same as those to be used in actual construction, the fill material should be compacted to the anticipated dry density and water content, and the sustained load should correspond to the largest vertical stress anticipated in the reinforced soil structure. The above procedure can only be used for fill material that has sufficient cohesion to allow the test to be conducted in an unconfined condition. For fill material with little or no cohesion, a confining pressure can be applied to the soil–geosynthetic composite by enclosing the entire test specimen in a rubber membrane in an air-tight condition with vacuum pressure applied.

Ketchart and Wu (2001, 2002) have reported results of SGIP tests on a number of specimens of soil–geosynthetic composites, with a few tests conducted under elevated temperatures to accelerate creep of geosynthetics. A polypropylene nonwoven geotextile, of which the creep rate under 52°C (125°F) has been determined to be between 150 and 250 times faster than that under ambient temperature, was selected for the tests. The fill material was a road base (with 20% of fines, prepared at 95% relative compaction and 2% wet-of-optimum moisture) and the soil–geosynthetic composite was subjected to a sustained surcharge of 100 kPa (15 psi) in a 52°C (125°F) environment (see Figure 4.24(a)). Creep deformation of the soil–geotextile composite under the elevated temperature was found to be very small and decreased rapidly with time, as seen in Figure 4.24(b). Creep deformation ceased completely after approximately 12 days, which suggests that creep deformation of the soil–geosynthetic composite would have become negligible after

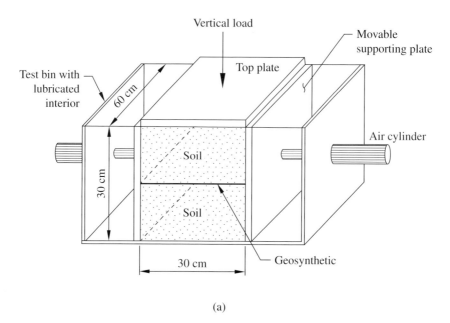

(a)

(b)

Figure 4.23 The soil–geosynthetic interactive performance (SGIP) test: (a) schematic diagram and (b) a SGIP test loaded by a universal tester (Ketchart and Wu, 2001)

(a)

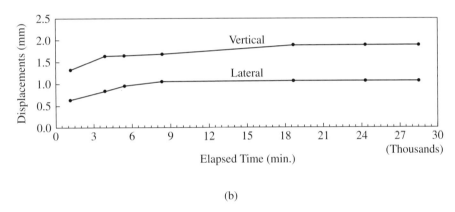

(b)

Figure 4.24 A SGIP test conducted under elevated temperature: (a) test being conducted in a 52°C (125°F) environment inside a temperature incubator and (b) a set of test results (Ketchart and Wu, 2001)

about 6.5 years under ambient temperature. The study concludes that creep of soil–geosynthetic composites is strongly influenced by the creep characteristics of the confining soil. With a well-compacted granular backfill (such as a road base material), creep of soil–geosynthetic composite is negligible under a sustained surcharge of 100 kPa, much higher than typical design loads of GRS walls in non-load-bearing applications.

A modified SGIP test was subsequently developed to accommodate larger/taller soil–geosynthetic composite specimens, either for greater reinforcement spacing or for soils of larger grain sizes. The modified SGIP test is shown in Figure 4.25. Note that a confining pressure in the photo is being applied to the soil–geosynthetic composite via a vacuum through a membrane enclosing the entire test specimen (the vacuum is needed for fill materials of little cohesion).

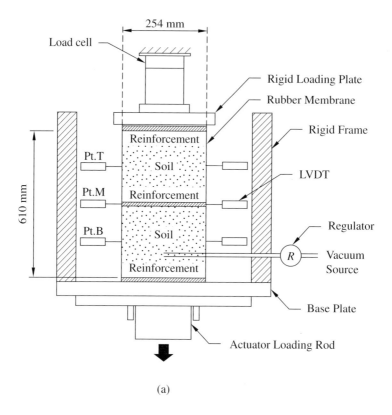

(a)

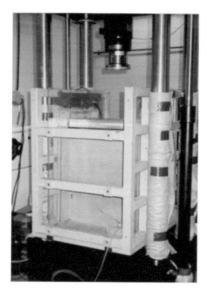

(b)

Figure 4.25 A modified SGIP test for taller specimens: (a) schematic diagram of test setup and (b) loaded by a universal tester (modified after Ketchart and Wu, 2001)

There are two unique features in the different generations of the SGIP test. First, a sustained load is applied to the soil, which in turn transfers the loads to the geosynthetic, replicating the typical load transfer mechanism in GRS structures (i.e., loads in geosynthetic reinforcement are transferred from soil through soil–geosynthetic interface bonding to the geosynthetic reinforcement, rather than loads being applied directly to geosynthetic reinforcement as in most creep tests). Second, both the soil and geosynthetic are allowed to deform in an interactive manner under a plane-strain condition — a very important feature in any properly conducted creep tests of a GRS mass.

Crouse and Wu (2003) summarized the measured behavior of seven full-scale reinforced soil walls that had been monitored for an extended period of time to examine the long-term performance. The reinforced soil walls, representing a variety of wall types with granular backfill, including the Glenwood Canyon wall (Bell et al., 1983), the Tanque Verde–Wrightown–Pantano Roads project wall (Berg et al., 1986; Collin et al., 1994), the Norwegian Geotechnical Institute (NGI) project, known as the NGI wall (Fannin and Hermann, 1992), the Japan Railway Test Embankment project, known as the JR wall (Tatsuoka et al., 1992), the Highbury Avenue, the London Ontario project, known as the Highbury wall (Bathurst, 1992), the FHWA Algonquin wall (Simac et al., 1990; Christopher et al., 1994), and the Seattle Preload Fill project, known as the Seattle wall (Allen et al., 1992). The maximum creep strains in the geosynthetic reinforcement were less than 1.5% in all seven walls. Also, the creep strain rate in all these cases decreases with time, and there exists an approximately linear relationship between log (creep rate) and log (time) for all the walls.

Allen and Bathurst (2003) analyzed the long-term creep data of 10 full-scale in-service geosynthetic walls. Post-construction, long-term wall face deformation data show that the geosynthetic wall face deformation for the properly designed walls is generally less than 25–30 mm (1.0–1.2 in.) during the first year of service, and less than 35 mm (1.4 in.) over the design lifetime for walls that are lower than 13 m (40 ft) in height. Allen (1991) and Bathurst et al. (1988) also studied geosynthetic reinforcement strains measured in geosynthetic reinforced soil walls and slopes. The maximum strains were found to be on the order of 1–2% or less.

Helwany and Wu (1995) reported some interesting results of finite element analysis on long-term behavior of two 3.0-m high geosynthetic reinforced soil walls. The walls were identical in every respect except that one had a clayey backfill and the other a granular backfill. Figure 4.26 shows time-dependent displacement fields of the two walls. The difference in the displacement fields is evident. In the clay-backfill wall, the maximum strain in the geosynthetic reinforcement increases by 3.5% from the end of construction to 15 years afterwards. In the granular-backfill wall, on the other hand, the increase in maximum strain over the same time period is negligible to very small. Note that the very large differences in the displacements and strains occur despite the levels of loads in the two walls being comparable. The analysis suggests clearly that the time-dependent behavior of the backfill plays a crucial role in the creep deformation of a soil–geosynthetic composite. Similar findings have been reported in studies

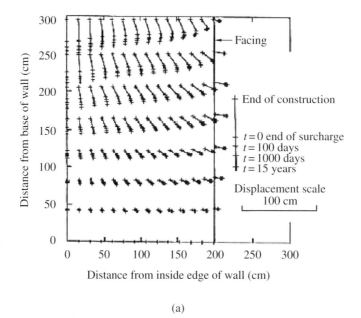

(a)

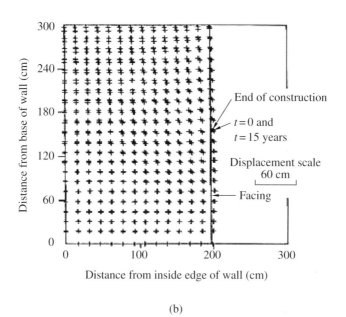

(b)

Figure 4.26 Time-dependent displacement fields of two GRS walls: (a) a clayey backfill wall and (b) a granular backfill wall, as obtained from finite element analysis (Helwany and Wu, 1995)

conducted by Li and Rowe (2008), Skinner and Rowe (2005), Rowe and Taechakumthorn (2008), Bergado and Teerawattanasuk (2008), Liu and Won (2009), Liu et al. (2009), and Li et al. (2011).

A design protocol for predicting the long-term creep deformation of a GRS mass has been proposed by Crouse and Wu (2003). The protocol involves conducting a SGIP test by employing the soil and geosynthetic to be used in a design. Because GRS structures with well-compacted granular soil have been known to have little creep deformation, the protocol is only needed when a "questionable" on-site soil (in terms of long-term deformation) is being considered as backfill for a GRS structure or when the load level is unusually high.

4.2.3   Stress Relaxation of Geosynthetics

Stress relaxation is a term used to describe the phenomenon that stress in a material is relieved with progression of time when the material is kept under a constant strain condition. A material which would deform with time under a constant stress is said to be susceptible to creep. Such a material tends to experience stress relaxation if the deformation is held constant. Geosynthetics are known to be susceptible to both creep and stress relaxation. In general, polypropylene and polyethylene are much more susceptible to creep and stress relaxation than polyester. Since deformation of well-compacted granular soils tends to become quite small soon after load applications and soil–reinforcement tends to deform in a compatible manner (unless when a soil–geosynthetic composite is approaching failure), stress relaxation in geosynthetic reinforcement is a prevalent phenomenon in geosynthetic reinforced soil structures when a well-compacted granular fill is employed, especially if the geosynthetic reinforcement is polypropylene or polyethylene.

Let us begin the discussion of stress relaxation by reviewing the relationships among stress, strain, and time. Load–deformation–time relationships of polymeric geosynthetics can be expressed in terms of three different sets of planar curves: (i) the *isostress curve* (commonly referred to as the *creep curve*), see Figure 4.27(a), (ii) the *isochronous* curve, see Figure 4.27(b), and (iii) the *isostrain* curve (also referred to as the *relaxation curve*), see Figure 4.27(c). For some creep-susceptible materials, the relationships among *time*, *stress*, and *strain* can be expressed by a curved surface in the stress–strain–time domain. The three curves shown in Figures 4.27(a), (b), and (c) are then the projections of this curved surface onto three perpendicular planes: the isochronous plane, the isostress plane, and the isostrain plane. If a set of projected curves on one of the three planes is known, the projected curves on the other two planes can be obtained as long as the material has a *unique* stress–strain–time relationship. Limited creep and relaxation tests data have suggested that a unique curved surface exists only for polyester geosynthetics. Example 4.1 illustrates how a set of isostrain (relaxation) curves can be obtained from a set of isostress (creep) curves. This implies that a material susceptible to creep will also be susceptible to stress relaxation.

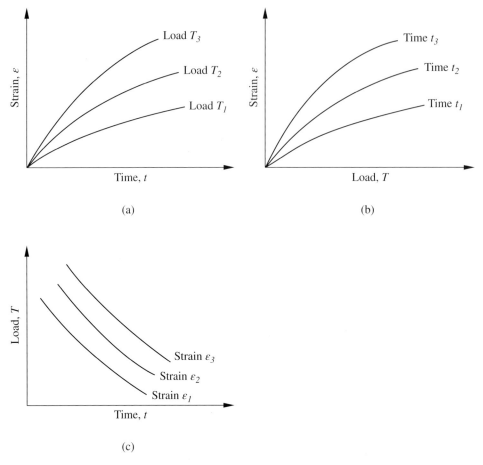

Figure 4.27 Two-dimensional representation of load–strain–time relationships: (a) isostress curves (or creep curves), (b) isochronous curves, and (c) isostrain curves (or relaxation curves)

Example 4.1   Creep tests were performed on three specimens of a polyester geosynthetic product with each specimen being subject to three different sustained loads: T_1, T_2 and T_3 ($T_1 < T_2 < T_3$). For each test, a sustained load (T) is applied to one end of the specimen, time-dependent elongations are measured, and corresponding strains are calculated. The results of the creep tests, i.e., creep curves, are shown in Figure Ex. 4.1(a). Determine the relaxation curves of the geosynthetic from the creep curves, assuming there is a unique time–stress–strain relationship for the geosynthetic product.

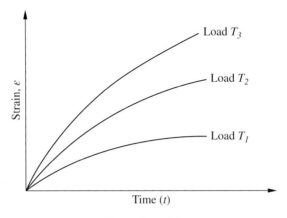

Figure Ex. 4.1(a)

Solution:

i)   Select a strain, ε_a, draw a horizontal line from ε_a, and locate points a, b, and c on the creep curves corresponding to T_3, T_2, and T_1, respectively, as seen in Figure Ex. 4.1(b).

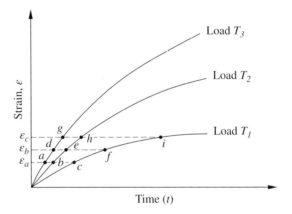

Figure Ex. 4.1(b)

ii) Use the coordinates of points a, b, and c to construct a relaxation curve corresponding to ε_a, see the lowest curve in Figure Ex. 4.1(c).

iii) Select a different strain, ε_b ($\varepsilon_b > \varepsilon_a$), draw a horizontal line from ε_b and locate points d, e, and f on the creep curves corresponding to T_3, T_2, and T_1, respectively, as seen in Figure Ex. 4.1(b).

iv) Use the coordinates of points d, e, and f to construct a relaxation curve, see the middle curve in Figure Ex. 4.1(c).

v)   Select a third strain, ε_c ($\varepsilon_c > \varepsilon_b$) and repeat the same procedure as before to construct a third relaxation curve by using the coordinates of points g, h, and i, as seen in Figure Ex. 4.1(c).

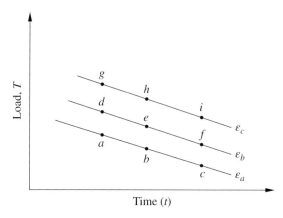

Figure Ex. 4.1(c)

Data obtained from isostrain (relaxation) tests of geosynthetics have revealed that reduction of loads in geosynthetic reinforcement would occur early and rapidly, and decrease with time at a decreasing rate. Kaliakin et al. (2000) have reported results of relaxation tests for 11 geogrids produced by three different manufacturers (see Figure 4.28 for 3 sets of test results). The reductions in loads in the first 100 minutes of the geogrids are seen greater than those in the ensuing month (43,000 minutes).

Unlike creep deformation, stress relaxation is not *visible*. Moreover, because *loads in geosynthetics cannot be measured* (other than at the extremities), stress relaxation is not as well understood as creep. We can, however, conceptualize stress relaxation of a geo-synthetic product by visualizing a simple experiment. Let us picture a creep-susceptible strip of geosynthetic with one end affixed to the ceiling and the other end attached to a heavy dead weight. Due to the weight, the geosynthetic will experience creep deforma-tion with time. If we are to "stop" the deformation from occurring after the geosynthetic reaches a certain prescribed strain and maintain the strain to be unchanged from that point on, the weight will have to be reduced; to wit, stress relaxation will have to occur. A similar phenomenon will occur when a creep-susceptible geosynthetic reinforcement is embedded in a well-compacted granular fill. When the geosynthetic is subject to loads, time-dependent creep deformation will occur in the geosynthetic reinforcement. Since deformation of the granular fill will cease after a certain time has elapsed, relaxation in the geosynthetic product will then occur after the granular fill ceases to deform. Studies have suggested that relaxation may initiate when the rate of deformation becomes very small, before deformation comes to a completely stop (e.g., Ketchart and Wu, 2001).

Full-scale experiments of reinforced soil walls have also suggested that significant stress relaxation may occur within a short time after construction provided that well-compacted granular soil is used in construction. An interesting example is a reinforced soil wall experiment conducted by the Public Works Research Institute (PWRI) of Japan, referred to as the PWRI test wall. The wall was 6.0-m high with concrete block facing, and the fill was a well-graded silty sand. The reinforcement was a HDPE polymer grid with a tensile strength of 56.4 KN/m on vertical spacing of 0.5 m. Two different reinforcement lengths were used: 3.8 m (primary) and 1.3 m (secondary). In less than three days after the wall was constructed, the grid reinforcement was *cut* sequentially at

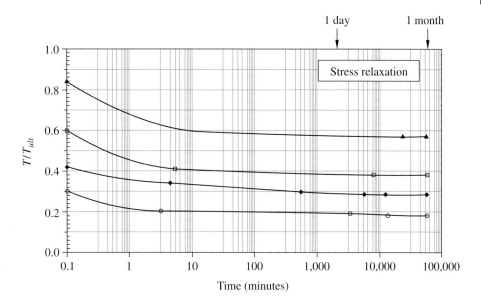

(a)

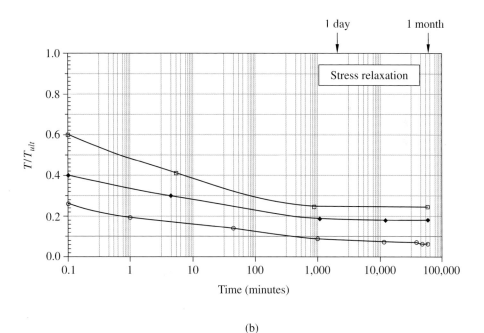

(b)

Figure 4.28 Normalized isostrain curves obtained from stress relaxation tests for different geogrids: (a) a flexible polyester woven geogrid, (b) a rigid HDPE punched-and-drawn geogrid, and (c) a rigid polypropylene punched-and-drawn geogrid (Kaliakin et al., 2000)

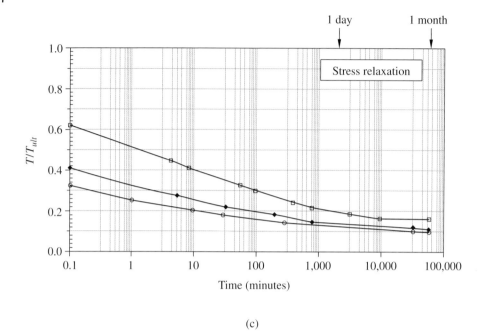

(c)

Figure 4.28 (Continued)

pre-selected locations as shown in Figure 4.29(a) (Note: The numbers noted on the reinforcement layers indicate the sequence of cutting). Figure 4.29(b) shows the maximum horizontal wall displacement in response to the cutting. It is seen that there was negligible displacement (on the order of 0.1 mm) from Cut No. 1 to Cut No. 50. This suggests that the loads in the geosynthetic reinforcement up to Cut No. 50 have likely reduced to negligible values at the time of cutting. In an attempt to validate the observation in the PWRI test wall, a similar experiment was conducted by Ketchart and Wu on a 3.0-m high reinforced soil wall built with a compacted well-graded crushed granular fill and a 70 kN/m woven polypropylene geotextile on 0.3 m spacing (Wu, 2001). The experiment confirmed the observed behavior of the PWRI test wall, in that severing geosynthetic reinforcement within a short time after construction causes negligible deformation (less than 0.1 mm) of the reinforced soil mass.

In GRS walls, geosynthetic reinforcement is typically subject to the highest load at the end of construction (for non-load-bearing applications) or right after placement of heavy loads (in load-bearing applications, such as girders in bridge abutments). The safety margin against overstress of geosynthetic reinforcement is therefore the lowest at those points in time. Soon after this, stress relaxation will likely occur, and the safety margin with respect to overstress of geosynthetic reinforcement will likely increase with time.

In the literature, loads in geosynthetic reinforcement are sometimes reported, and those loads are typically obtained from (i) load–deformation relationships obtained from the wide-width tensile test (ASTM D4595 for geotextiles, ASTM D6637 for

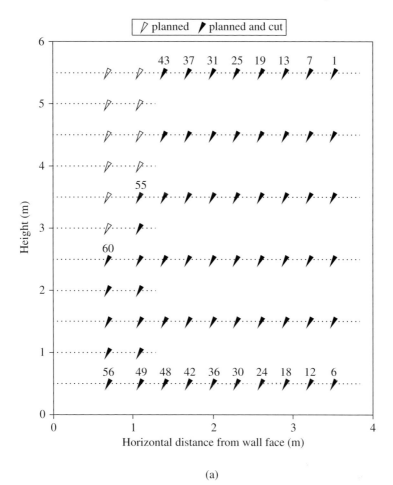

(a)

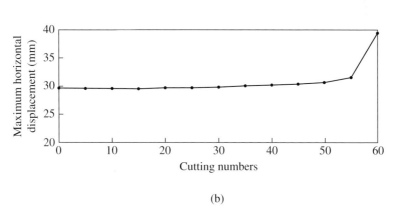

(b)

Figure 4.29 The PWRI test wall: (a) sequence of cutting geosynthetic reinforcement as indicated by numbers above reinforcement layers (cutting stopped at Cut No. 60 when failure occurred) and (b) maximum horizontal wall displacement in response to cutting of geosynthetic reinforcement

geogrids) or (ii) isochronouous curves (usually obtained from results of creep tests using a procedure similar to that given in Example 4.1). The former is vastly unreliable as the load–deformation relationship is affected by such factors as strain rate, temperature, specimen aspect ratio, and confinement (see subsection (A) of Section 4.2.1). For the latter, other than polyester geosynthetics, a unique time–stress–strain relationship does not exist. Therefore, reinforcement loads obtained from this approach are deemed unreliable (other than for polyester geosynthetics).

Researchers have developed methods that consider geosynthetics as elasto-visco-plastic material to determine reinforcement loads (e.g., Peng et al., 2010; Kongkitkul et al., 2010). The viscous property is that of isotach which states that current load in geosynthetic reinforcement is a function of the current strain and the current strain rate, not time. Based on isotach viscosity, creep can be described as a process of increasing strain at decreasing strain rate under a sustained load; load relaxation can be described as a process of decreasing load when the change in the elastic strain (negative) is balanced by the change in the inelastic strain (positive). The approach arguably offers the most rational prediction of reinforcement loads. However, since no method is available to measure loads in geosynthetic reinforcement (other than at the extremities), there has not been direct verification of loads obtained from this approach.

4.2.4   Soil–Geosynthetic Interface Properties

Most design guides of reinforced soil structures require that the tensile reinforcement has sufficient interface bond strength with the fill material. Interface bonding is needed to allow the geosynthetic reinforcement to effectively carry tensile loads and to prevent slippage of reinforcement that may lead to instability.

The interface bond strength between the soil and geosynthetic reinforcement has been evaluated by two test methods: the *direct shear interface test* and the *pullout test*, as shown in Figure 4.30. The former applies shear forces to the soil–geosynthetic interface and the latter applies tensile forces to the geosynthetic. In addition to the difference in force application mechanism, the two test methods also differ in geometric configuration, loading/stress path, and boundary conditions. Published data have indicated that interface friction angles determined by the two tests are usually different (Collios et al., 1980; Ingold, 1983; Richards and Scott, 1985; Rowe et al., 1985; Koerner, 1986; Juran et al., 1988). For geosynthetic reinforcement embedded in a dense sand under low overburden pressure, the pullout test tends to give a higher interface bond strength than the direct shear interface test. For geosynthetic reinforcement embedded in loose sand under high overburden pressure, the interface bond strengths obtained from the two tests have been found to be similar. Depending on the application, one of the tests is likely to be more appropriate than the other. For example, the pullout test is more appropriate for evaluating the potential pullout of reinforcement from the anchored zone (see Figure 4.31(a)). The direct shear interface test, on the other hand, is more appropriate for evaluating the slide-out failure between a soil wedge and the geosynthetic reinforcement beneath it (see Figure 4.31(b)).

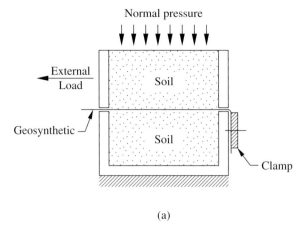

(a)

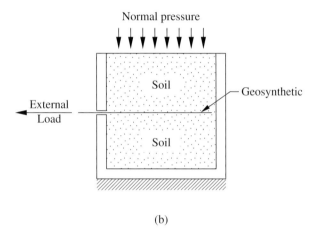

(b)

Figure 4.30 Test methods for evaluating the interface bond strength between soil and geosynthetic: (a) direct shear interface test and (b) pullout test

Direct shear interface tests come in different degrees of constraint on deformation of the geosynthetic specimen and soil–geosynthetic contact area. Figure 4.32 shows three direct shear test devices with different constraints to deformation of the geosynthetic specimen: free geosynthetic, fixed geosynthetic, and fixed geosynthetic with an enlarged base. The test with an enlarged base allows the area of interface shear to remain constant during testing. To perform a fixed geosynthetic direct shear interface test, the geosynthetic (usually geotextiles) is affixed to a wooden block, either by simply covering tightly over the block with clamps at the two ends or affixing the entire surface area of geosynthetic specimen to the block by adhesives. The choice of which type of direct shear interface test to perform should be determined based on the anticipated constraint conditions of the geosynthetic reinforcement in actual applications. The fixed geosynthetic test is usually the test of choice for finite element analysis where the interface element may be simulated as a pair of *contacting points*. Regardless of the type of direct shear interface test, interpretation of test results for soil–geosynthetic bonding

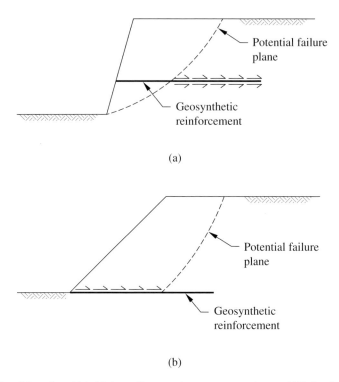

Figure 4.31 Conditions for which (a) the pullout test is more appropriate and (b) the direct shear interface test is more appropriate

strength is performed by the same procedure as for direct shear tests performed on soils (see Section 1.4.1). The same principles regarding selection of specimen deformation constraint also apply to pullout tests.

Pullout tests vary widely in the dimensions of test bin, length and width of reinforcement, and constraint on deformation of the geosynthetic specimen. The variations have caused serious confusion, especially with interpretation of test results. The *interface pullout formula* developed by Sobhi and Wu (1996) is recommended for interpretation of the results measured by pullout tests. The formula was derived based on postulates deduced from findings of large-scale laboratory pullout tests and finite element analysis. The interface pullout formula is capable of describing with very good accuracy the relationship between tensile force and displacement for any given length of geosynthetic test specimen used in a laboratory pullout test.

The *interface pullout formula* (Eqn. (4-1)) gives the relationship between tensile force per unit width (T) at any given section and the location of the section (denoted by coordinate x) along the length of the test specimen:

$$T = (F + E \cdot t) e^{\left(\frac{2\sigma_n f}{E \cdot t} x\right)} - E \cdot t \tag{4-1}$$

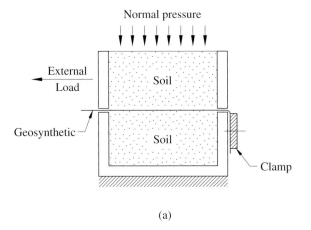

(a)

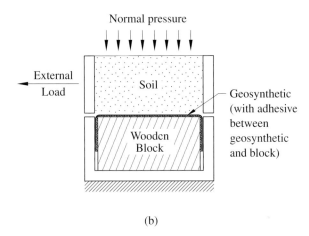

(b)

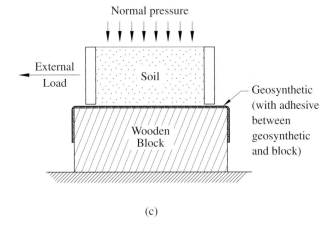

(c)

Figure 4.32 Three types of direct shear interface tests for evaluation of soil–geosynthetic interface properties with different deformation constraints of geosynthetic specimen: (a) free geosynthetic, (b) fixed geosynthetic, and (c) fixed geosynthetic with an enlarged base

where F = applied pullout force per unit width at the load-application end of test specimen, E = Young's modulus of geosynthetic reinforcement (obtained from an uniaxial load–deformation test), t = nominal thickness of geosynthetic reinforcement (Note: The product of E and t is the slope of load–deformation curve when the load is expressed as tensile load/width, see Figure 4.19), σ_n = vertical pressure applied on the planar surface of the geosynthetic specimen, and f = coefficient of friction of the soil–geosynthetic interface.

The interface pullout formula can be used to predict and interpret pullout test results. The following are some example applications, including: (i) to determine the coefficient of friction at the soil–geosynthetic interface, (ii) to predict the applied pullout force needed to bring about complete pullout failure for a given length of geosynthetic specimen, and (iii) to predict displacement at any point for a given applied pullout force at or before pullout failure. Some details for the example applications of the interface pullout formula are given below.

Case A: Use the interface pullout formula to determine the coefficient of friction f at the soil–geosynthetic interface.

Solution: Given the applied force per unit width at failure (F_f) as determined in a pullout test and the total length of the geosynthetic specimen (L), the friction coefficient (f) can be determined by:

$$f = \frac{E \cdot t \cdot \ln\left(\dfrac{F_f}{E \cdot t} + 1\right)}{2\,\sigma_n\,L} \tag{4-2}$$

Some pullout tests do not reach a failure state when terminated. In which case, if the *active length* (the length of test specimen over which movement occurred under a given applied pullout force) and the corresponding applied pullout force (prior to failure) are known, these values can be used in lieu of total length (L) and pullout force at failure (F_f) in Eqn. (4-2).

Case B: Use the interface pullout formula to predict the applied pullout force needed to bring about complete pullout failure for a given length of geosynthetic.

Solution: The applied pullout force F (per unit width) needed to bring about complete pullout failure of a specimen for total geosynthetic length L is:

$$F = E \cdot t \left[e^{\left(\frac{2\sigma_n f}{E \cdot t}\right) L}[-1] \right] \tag{4-3}$$

Case C: Use the interface pullout formula to predict the displacement at any point along the length of a geosynthetic specimen when subject to a given pullout force.

Solution: Given a pullout force F (before or at failure), the displacement u at a point along the geosynthetic reinforcement with a coordinate x_b can be determined by:

$$u = \frac{1}{E \cdot t} \int_{x_a}^{x_b} \left[-E \cdot t + (F + E \cdot t) e^{\frac{2\sigma_n f}{E \cdot t} x} \right] dx \tag{4-4}$$

where x_a is the coordinate of the active length (the length over which deformation occurred in a pullout test), which can be determined as:

$$x_a = \frac{E \cdot t}{2\sigma_n f} \ln \left(\frac{E \cdot t}{F + E \cdot t} \right) \tag{4-5}$$

Note that $x = 0$ corresponds to the section where pullout forces are applied to the geosynthetic specimen.

4.2.5 Hydraulic Properties of Geosynthetics

Filtration and *drainage* are water flow-related functions of geosynthetics that may be involved in design and analysis of reinforced soil walls. The term *filtration* refers to how well a geosynthetic product may retain soil particles as water flows *across* the plane of geosynthetic product, while the term *drainage* refers to how easy water may travel *within* the plane of a geosynthetic product. The cross-plane permeability is commonly expressed in terms of permittivity (ψ) and in-plane permeability in terms of transmissivity (θ), which are defined as:

permeability, $\psi = \dfrac{k_n}{t}$

transmissivity, $\theta = k_h t$

where k_n = Darcy's coefficient of permeability (or hydraulic conductivity) normal to the plane of geosynthetic product

k_h = Darcy's coefficient of permeability (or hydraulic conductivity) within the plane of geosynthetic product

t = thickness of geosynthetic product

and the flow rate (q) can be calculated as:

$$q_{(cross\text{-}plane)} = \psi (\Delta h) A \tag{4-6}$$

$$q_{(in\text{-}plane)} = \theta \left(\frac{L}{\Delta h} \right) W \tag{4-7}$$

where Δh = loss of total head

A = total area of geosynthetic product involved in cross-plane flow

L = total length of geosynthetic product involved in in-plane flow

W = width of geosynthetic product involved in in-plane flow.

When permittivity (ψ) and transmissivity (θ) are used to calculate cross-plane and in-plane flow rates, the thickness of the geosynthetics is not part of the calculations. In fact, this is a reason to use permittivity and transmissivity in lieu of Darcy's coefficients of permeability because ψ and θ allow a designer to make a direct comparison of the flow rates for different geosynthetics.

The value of permittivity (ψ) of a geotextile (with common units of sec^{-1}) can be determined by ASTM Method D4991 "Water Permeability of Geotextiles by Permittivity" by either a constant head test or a falling head test. For determination of the value of transmissivity (θ) of a geotextile (with common units of m^2/sec or cm^2/sec), a number of test devices are available, including ASTM D4716. Descriptions of these test methods have been given by Koerner (2005).

To provide sufficient in-plane flow capability to fulfill intended drainage function, a geosynthetic product needs to be of adequate thickness and/or have an adequately high permeability within its plane. Woven or heat-bounded nonwoven geotextiles have very low transmissivity and cannot be used as drains. The only type of geotextile that can be used effectively as a drain is needle-punched nonwoven geotextile, for which typical values of transmissivity are in the range of 10^{-5} to 10^{-7} m^2/sec.

The flow capability of geotextiles tends to decrease with time because of clogging in the geotextile structures. A number of test methods have been proposed to assess the potential of excessive clogging, including the long-term flow test (Koerner and Ko, 1982), the U.S. Army Corps of Engineers' gradient ratio test (CW-02215), and the hydraulic conductivity ratio test (Williams and Abouzakham, 1989). Geotextile filter design and drainage design have been given by Holtz et al. (1997).

4.3 Advantages and Disadvantages of Geosynthetics as Reinforcement

Use of geosynthetics (geotextiles, geogrids, geocomposites, and geocells) as reinforcement for GRS walls can have the following advantages over other reinforcement materials, especially metals:

- In addition to serving the function of reinforcing, geosynthetic reinforcement also serve one or more other functions, e.g., facilitate drainage of soil, retain soil grains and maintain soil integrity, and separate different soils or aggregates (especially when subject to repeated external loads) during construction and service life.

- Reinforced soil walls with geosynthetics as reinforcement can be more adaptable to low-quality backfills.

- Geosynthetics are more durable and have stronger resistance to corrosion and bacterial action than metallic reinforcement.

- Walls using geosynthetics as reinforcement are more flexible than those using metallic reinforcement, hence can tolerate greater foundation settlement. Figure 4.33 shows a wrapped-face GRS wall of which the crest remained an even surface despite a large cavity (approximately 2.0 m wide and 1.3 m high) formed near its base due to erosion. Note that, other than the area immediately above the cavity, the rest of the wall was unaffected by the cavity. It was several years from the time the cavity was discovered before any repair work was undertaken.

- Geosynthetics are generally lower in cost compared to metallic reinforcement.
- Geosynthetics usually come in rolls hence are easier to transport to remote construction sites and deploy, and are readily available in many varieties.
- Geosynthetics are easier to install than metallic reinforcement.
- Geosynthetics have higher sustainability and lower CO_2 footprint than metallic reinforcement.

Figure 4.33 A wrapped-face GRS wall with a large cavity developed near the base due to erosion

On the other hand, there are a number of drawbacks of using geosynthetics inclusion as reinforcement:

- Geosynthetics may be susceptible to chemical and biological degradation. Some geosynthetics may deteriorate when exposed directly to ultraviolet light over a prolonged period of time.
- Geosynthetics can be susceptible to creep when the fill material is not well compacted or have a stronger tendency to deform with time than the geosynthetic reinforcement.
- Construction operation, especially fill compaction, may cause damage to geosynthetics during installation. Care may need to be exercised to prevent potential problems.

References

Allen, T.M. (1991). Determination of Long-Term Strength of Geosynthetics: a State-of-the-Art Review. *Proceedings, Geosynthetics'91, IFAI, Volume 1, Atlanta, Georgia, USA, February 1991*, pp. 351–379.

Allen, T.M. and Bathurst, R.J. (2003). *Prediction of Reinforcement Loads in Reinforced Soil Walls*. Final Research Report, Washington State Department of Transportation and Federal Highway Administration, 290 pp.

Allen, T.M., Christopher, B.R., and Holtz, R.D. (1992). Performance of a 12.6 m High Geotextile Wall in Seattle, Washington. In *Geosynthetic-Reinforced Soil Retaining Walls*, Wu (ed.). A.A. Balkema, Rotterdam, pp. 81–100.

Bathurst, R.J. (1992). Case Study of a Monitored Propped Panel Wall. In *Geosynthetic-Reinforced Soil Retaining Walls*, Wu (ed.). A.A. Balkema, Rotterdam, pp. 159–166.

Bathurst, R.J., Benjamin, D.J. and Jarrett, P.M. (1988). Laboratory Study of Geogrid Reinforced Soil Walls. In *Geosynthetics for Soil Improvement*, Holtz (ed.). ASCE Geotechnical Special Publication No. 18, Nashville, Tennessee, May 1988, pp. 178–192.

Bell, J.R., Barrett, R.K. and Ruckman, A.C. (1983). *Geotextile Earth-Reinforced Retaining Wall Test: Glenwood Canyon, Colorado*. Transportation Research Record, No. 916, Washington, D.C., pp. 59–69.

Berg, R.R., Bonaparte, R., Anderson, R.P, and Chouery, V.E. (1986). Design, Construction and Performance of Two Geogrid Reinforced Soil Retaining Walls. *Proceedings, 3rd International Conference on Geotextiles, Vienna, Austria*, pp. 401–406.

Bergado, D.T. and Tearawattanasuk, C. (2008). 2D and 3D Numerical Simulations of Reinforced Embankments on Soft Ground. *Geotextiles and Geomembranes*, 26(1), 39–55.

Bonaparte, R., Holtz, R.D., and Giroud, J.P. (1987). Soil Reinforcement Design Using Geotextiles and Geogrids. In *Geotextile Testing and the Design Engineer*, ASTM STP 952, Fluet (ed.). American Society of Testing and Materials, Philadelphia, pp. 69–116.

Christopher, B.R., Holtz, R.D., and Bell, W.D. (1986). New Tests for Determining the In-Soil Stress-Strain Properties of Geotextiles. *Proceedings, 3rd International Conference on Geotextiles, Vienna, Austria*, pp. 683–688.

Christopher, B.R., Bonczkiewicz, C., and Holtz, R.D. (1994). Design, Construction and Monitoring of Full-Scale Test Soil Walls and Slopes. In *Recent Case Histories of Permanent Geosynthetic Reinforced Soil Retaining Walls*, Tatsuoka and Leshchinsky (eds.). A.A. Balkema, Rotterdam, pp. 45–60.

Collin, J.G., Bright, D.G., and Berg, R.R. (1994). Performance Summary of the Tanque Verde Project – Geogrid Reinforced Soil Retaining Walls. *Proceedings, Earth Retaining Session, ASCE Convention, Atlanta, Georgia*.

Collios, A., Delmas, P., Gourc, J.P., and Giroud, J.P. (1980). Experiments on Soil Reinforcement with Geotextiles. *The Use of Geotextiles for Soil Improvement, ASCE National Convention, Portland, Oregon*, pp. 53–73.

Cook, D.I. (2003). Geosynthetics. *Rapra Review Reports*, 14(2), 120.

Crouse, P. and Wu, J.T.H. (2003). Geosynthetic-Reinforced Soil (GRS) Walls. *Journal of Transportation Research Board*, 1849, 53–58.

den Hoedt, G. (1986). Creep and Relaxation of Geotextile Fabrics. *Geotextiles and Geomembranes*, 4(2), 83–92.

Fannin, R.J. and Hermann, S. (1992). Geosynthetic Strength – Ultimate and Serviceability Limit State Design. *Proceedings, ASCE Specialty Conference on Stability and Performance of Slopes & Embankments II, University of California, Berkeley, California,* pp. 1411–1426.

Helwany, S. and Wu, J.T.H. (1995). A Numerical Model for Analyzing Long-Term Performance of Geosynthetic-Reinforced Soil Structures. *Geosynthetics International,* 2(2), 429–453.

Holtz, R.D., Christopher, B.R., and Berg, R.R. (1997). *Geosynthetic Engineering.* Bitech Publisher, 452 pp.

INDA, Association of Nonwoven Fabrics Industry, 10 East 40th Street, New York, NY 10016, USA.

Ingold, T.S. (1983). Laboratory Pull-Out Testing of Grid Reinforcements in Sand. *ASTM Geotechnical Testing Journal,* 6(3), 101–111.

Ingold, T.S. and Miller, K.S. (1988). *Geotextiles Handbook,* Thomas Telford Publishing, 152 pp.

Jones, C.J.F.P. (1985). *Earth Reinforcement and Soil Structures.* Butterworths and Company, London, 183 pp.

Juran, I., Knochenmus, G., Acar, Y.B., and Arman, A. (1988). Pull-Out Response of Geotextiles and Geogrids (Synthesis of Available Experimental Data). In *Geosynthetics for Soil Improvement,* Holtz (ed.). Geotechnical Special Publication No. 18, ASCE, Nashville, Tennessee, pp. 92–111.

Kaliakin, V.N., Dechasakulsom, M., and Leshchinsky, D. (2000). Investigation of the Isochrone Concept for Predicting Relaxation of Geogrids. *Geosynthetics International,* 7(2), 79–99.

Ketchart, K. and Wu, J.T.H. (2001). *Performance Test for Geosynthetic-Reinforced Soil Including Effects of Preloading.* FHWA-RD-01-018, Turner-Fairbank Highway Research Center, Federal Highway Administration, McLean, Virginia, 282 pp.

Ketchart, K. and Wu, J.T.H. (2002). A Modified Soil–Geosynthetic Interactive Performance Test for Evaluating Deformation Behavior of GRS Structures. *ASTM Geotechnical Testing Journal,* 25(4), 405–413.

Koerner, R.M. (1986). *Direct Shear/Pull-Out Tests on Geogrids.* Report No. 1, Department of Civil Engineering, Drexel University, Philadelphia.

Koerner, R.M. (1994). *Designing with Geosynthetics,* 3rd edition. Prentice Hall Publisher, Upper Saddle River, New Jersey, 783 pp.

Koerner, R.M. (2005). *Designing with Geosynthetics,* 5th edition. Prentice Hall Publisher, Upper Saddle River, New Jersey, 796 pp.

Koerner, R.M. and Ko, F.K. (1982). Laboratory Studies of Long-Term Drainage Capability of Geotextiles. *Proceedings, 2nd International Conference on Geotextiles, Las Vegas, Nevada, Volume 1,* pp. 91–95.

Kokkalis, A. and Papacharisis, N. (1989). A Simple Laboratory Method to Estimate the In-Soil Behavior of Geotextiles. *Geotextiles and Geomembranes,* 8, 147–157.

Kongkitkul, W., Tatsuoka, F., Hirakawa, D., Sugimoto, T., Kawahata, S., and Ito, M. (2010). Time Histories of Tensile Force in Geogrid Arranged in Two Full-Scale High Walls. *Geosynthetics International,* 17(1), 12–33.

Leshchinsky, D. and Field, D.A. (1987). In-Soil Load Elongation, Tensile Strength and Interface Friction of Nonwoven Geotextiles. *Proceedings, Geosynthetics'87, New Orleans, Louisiana,* pp. 238–249.

Li, A.L. and Rowe, R.K. (2008). Effects of Viscous Behaviour of Geosynthetic Reinforcement and Foundation Soils on Embankment Performance. *Geotextiles and Geomembranes*, 26(4), 317–334.

Li, F.L., Peng, F.L., Tan, Y., Kongkitkul, W., and Siddiquee, M.S.A. (2011). FE Simulation of Viscous Behavior of Geogrid-Reinforced Sand under Laboratory Scale Plane-Strain-Compression Testing. *Geotextiles and Geomembranes*, doi:10.1016/j.geotexmem.2011.09.005.

Ling, H., Wu, J.T.H., and Tatsuoka, F. (1992). Short-Term Strength and Deformation Characteristics of Geotextiles under Typical Operational Conditions. *Geotextiles and Geomembranes*, 11(2), 185–219.

Liu, H. and Won, M.S. (2009). Long Term Reinforcement Load of Geosynthetic-Reinforced Soil Retaining Walls. *Journal of Geotechnical and Geoenvironmental Engineering, ASCE*, 135(7), 875–889.

Liu, H., Wang, X., and Song, E. (2009). Long-Term Behavior of GRS Retaining Walls with Marginal Backfill Soils. *Geotextiles and Geomembranes*, 27, 295–307.

McGown, A., Andrawes, K.Z., and Kabir, M.H. (1982). Load-Extension Testing of Geotextiles Confined In Soil. *Proceedings, 2nd International Conference on Geotextiles, IFAI, Volume 3, Las Vegas, Nevada, USA*, pp. 793–798.

Mills, C.W. (2005). *Geosynthetic Bunker Liners: A Proposed Design Methodology for Golf Course Improvement*. GSP 142, ASCE. Waste Containment and Remediation, GeoFrontiers, pp. 1–10.

Myles, B. (1987). A Review of Existing Geotextile Tension Testing Methods. In *Geotextile Testing and the Design Engineer*, Fluet (ed.). ASTM STP 952, pp. 57–68.

Peng, F.L., Li, Y.T., and Kongkitkul, W. (2010). Effects of Loading Rate on Viscoplastic Properties of Polymer Geosynthetics and its Constitutive Modeling. *Polymer Engineering & Science*, 50(3), 550–560.

Richards, E.A. and Scott, J.D. (1985). Soil Geotextile Frictional Properties. *Proceedings, 2nd Canadian Symposium on Geotextiles and Geomembranes, Edmonton, Alberta, Canada*, pp. 13–24.

Rowe, R.K. and Taechakumthorn, C. (2008). Combined Effect of PVDs and Reinforcement on Embankments over Rate-Sensitive Soils. *Geotextiles and Geomembranes*, 26(3), 239–249.

Rowe, R.K., Ho, S.K., and Fisher, D.G. (1985). Determination of Soil–Geotextile Interface Strength Properties. *Proceedings, 2nd Canadian Symposium on Geotextiles and Geomembranes, Edmonton, Alberta, Canada, September 1985*, pp. 25–34.

Siel, B.D., Wu, J.T.H., and Chou, N.S. (1987). In-Soil Stress-Strain Behavior of Geotextiles. *Proceedings, Geosynthetics'87, New Orleans, Louisiana*, pp. 260–265.

Simac, M.R., Christopher, B.R., and Bonczkiewicz, C. (1990). Instrumented Field Performance of a 6 m Geogrid Soil Wall. In *Geotextiles, Geomembranes and Related Products*, Hoedt (ed.). A.A. Balkema, Rotterdam, pp. 53–59.

Skinner, G.D. and Rowe, R.K. (2005). Design and Behaviour of a Geosynthetic Reinforced Retaining Wall and Bridge Abutment on a Yielding Foundation. *Geotextiles and Geomembranes*, 23(3), 235–260.

Sobhi, S. and Wu, J.T.H. (1996). An Interface Pullout Formula for Extensible Sheet Reinforcement. *Geosynthetics International*, 3(5), 565–582.

Tatsuoka, F., Murata, O., and Tateyama, M. (1992). Permanent Geosynthetic-Reinforced Soil Retaining Walls Used for Railway Embankments. In *Geosynthetic-Reinforced Soil Retaining Walls*, Wu (ed.). A.A. Balkema, Rotterdam, pp. 101–130.

Van Santvoort, G.P.T.M. (1994). *Geotextiles and Geomembranes in Civil Engineering.* A.A. Balkema, Rotterdam, 578 pp.

Williams, N.D. and Abouzakham, M.A. (1989). Evaluation of Geotextile/Soil Filtration Characteristics Using the Hydraulic Conductivity Ratio Analysis. *Geotextiles and Geomembranes*, 8(1), 1–26.

Wu, J.T.H. (1991). Measuring Inherent Load-Extension Properties of Geotextiles for Design of Reinforced Structures. *ASTM Geotechnical Testing Journal*, 14(2), 157–165.

Wu, J.T.H. (2001). *Revising the AASHTO Guidelines for Design and Construction of GRS Walls.* Colorado Department of Transportation, Report No. CDOT-DTD-R-2001-16, 148 pp.

Wu, J.T.H. and Arabian, V. (1990). Cubical and Cylindrical Tests for Measuring In-soil Load-Extension Properties of Geotextiles. *Proceedings, 4th International Conference of Geotextiles, Geomembranes and Related Products, The Hague, Volume 2*, pp. 785.

Wu, J.T.H. and Helwany, S. (1996). A Performance Test for Assessment of Long-Term Creep Behavior of Soil–Geosynthetic Composites. *Geosynthetics International*, 3(1), 107–124.

Wu, J.T.H. and Su, C. (1987). Soil–Geotextile Interaction Mechanism in Pullout Test. *Geosynthetics '87 Conference, Industrial Fabrics Association International, New Orleans, Louisiana, Volume 1*, pp. 250–259.

Wu, J.T.H. and Tatsuoka, F. (1992). Discussion of "Laboratory Model Study on Geosynthetic Reinforced Soil Retaining Walls." by I. Juran and B. Christopher. *Journal of Geotechnical Engineering, ASCE*, 118(3), 496–498.

Yamauchi, H. (1987). Use of Nonwoven geotextiles for Reinforcing Clayey Embankment. (in Japanese). Doctoral Dissertation, University of Tokyo, Japan, 507 pp.

5

Design of Geosynthetic Reinforced Soil (GRS) Walls

In this chapter we shall address the design of geosynthetic reinforced soil (GRS) walls, including design concepts, design methods, and related issues for both mechanically stabilized earth (MSE) walls with geosynthetics as reinforcement (referred to as GMSE walls) and geosynthetic reinforced soil (GRS) walls. The differences between GMSE and GRS have been explained in Chapter 3. Among current design methods for reinforced soil walls, most are applicable to GMSE walls with a handful of them being applicable specifically to GRS walls. Design methods for MSE walls with metallic reinforcement have been presented elsewhere, e.g., Schlosser and Segrestin (1979), Smith and Pole (1980), and Mitchell and Villet (1987), and will not be addressed here.

This chapter begins with a description of basic design concepts, followed by an overview of design methods. In addition, six recent advances related to the design of GRS walls are discussed in some detail, including:

- required reinforcement stiffness and strengths
- evaluation of pullout stability
- lateral movement of wall face
- required long-term strength of geosynthetic reinforcement
- connection stability of concrete block facing
- required reinforcement length.

Step-by-step design procedures for four design methods are then given, including:

- the U.S. Forest Service (USFS) design method
- the AASHTO ASD design guidance for GMSE walls
- the NCHRP design method for GRS bridge abutment walls
- the GRS-NLB design method for non-load-bearing GRS walls.

Geosynthetic Reinforced Soil (GRS) Walls, First Edition. Jonathan T.H. Wu.
© 2019 John Wiley & Sons Ltd. Published 2019 by John Wiley & Sons Ltd.

A few popular design methods, including the National Concrete Masonry Association (NCMA) method (2010), the FHWA-NHI design method (Berg et al., 2009), the AASHTO LRFD design guidelines (2014), and the FHWA GRS-IBS design method (Adams et al., 2011a), are not included in this book as they have been presented in a comprehensive manner in respective design guides.

It is noteworthy that, although not addressed specifically in any of the design methods, drainage design for surface and subsurface water, including seasonal change of drainage conditions, must be a component of design. Inadequate drainage has been identified as one of the leading causes of poor performance (especially excessive deformation) of reinforced soil walls.

5.1 Fundamental Design Concepts

Reinforced soil walls typically consist of two major components: a reinforced soil mass and facing (see Figure 3.23). Design of reinforced soil walls involves satisfying internal stability and external stability. Internal stability refers to stability *within* the reinforced soil wall, including the reinforced soil mass and the facing, if facing is present. External stability, on the other hand, refers to the stability of the reinforced soil mass and facing as a whole in relation to the soil beneath and behind them.

The internal stability of a reinforced soil wall is checked to ensure that the reinforced soil mass and facing in themselves are sufficiently stable so that they will behave as a coherent and stable unit. To ensure stability of the reinforced soil mass, stability analysis needs to be carried out to make sure that reinforcement in the reinforced soil mass is not overstressed or ruptured (referred to as *tensile rupture failure*, see Figure 5.1(a)), and that the bonding between soil and reinforcement is maintained (referred to as *pullout failure*, see Figure 5.1(b)). For GMSE, the internal stability check is performed by treating reinforcement layers as *quasi-tiebacks* to stabilize potential failure wedges. Therefore, such design methods are collectively referred to as *quasi-tieback wedge methods*. Generally speaking, the required minimum *strength* of reinforcement for a GMSE wall is governed by rupture failure, and the required minimum *length* of reinforcement for a GMSE wall is governed by pullout failure. Note that the

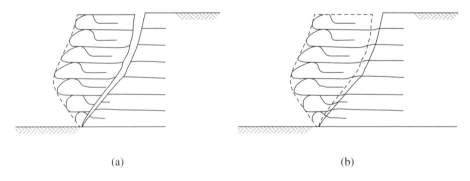

(a)   (b)

Figure 5.1 Potential failure modes for checking the internal stability of a GMSE wall: (a) tensile rupture failure of reinforcement and (b) pullout failure of reinforcement

required minimum reinforcement strength and length are determined in a rather differ-ent manner for GRS; this will be discussed toward the end of this section.

Facing failure may occur within the facing or at the connection (if present) between the facing and the reinforced soil mass. Potential failure of discrete block facing may occur in three modes:

1) rupture failure of reinforcement at the connection between the facing and the reinforced soil mass behind the facing (see Figure 5.2(a))

2) slip-out failure of the reinforcement between facing blocks (see Figure 5.2(b))

3) interface shear failure of the block facing, either between block and block or between block and reinforcement (see Figure 5.2(c)).

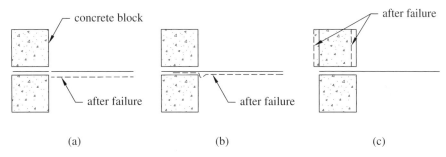

(a)   (b)   (c)

Figure 5.2 Potential failure modes of discrete block facing: (a) rupture failure of reinforcement at connection, (b) slip out failure of reinforcement between facing units, and (c) interface shear failure between adjacent facing blocks

External stability of a reinforced soil wall, as noted earlier, is evaluated to ensure that the reinforced soil wall as a whole is stable. It is carried out by considering the reinforced soil mass along with the facing as a *rigid soil wall* with earth pressures acting behind and beneath the reinforced soil mass. The stability of the rigid soil wall is checked by methods similar to those for stability analysis of rigid earth retaining structures, namely, to check the stability against four potential failure modes (see Figure 5.3): lat-eral sliding failure, foundation bearing failure, eccentricity (lift-off) failure, and overall rotational slide-out failure. External stability has been found to be a major cause for most reinforced soil walls that experience either failure or poor performance.

In addition to the external and internal failure modes shown in Figures 5.1–5.3, failure involving a combination of both types of failure mode, within and around a rein-forced soil mass (see Figure 5.4), may also occur. This mode of failure, referred to as *compound failure*, should also be evaluated as part of the external stability check.

It is important to point out that when a reinforced soil wall with closely spaced geosynthetic reinforcement is designed by considering the reinforced soil mass as a soil–geosynthetic composite (as opposed to a soil mass being stabilized by geosynthetic quasi-tieback members), evaluation of the internal stability of a reinforced soil mass is very different from that described above for GMSE walls. For GRS walls designed with

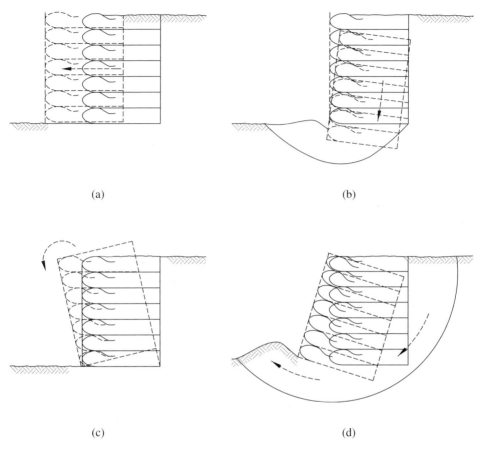

Figure 5.3 Potential failure modes for checking the external stability of a reinforced soil wall: (a) lateral sliding failure, (b) bearing failure, (c) eccentricity (lift-off) failure, and (d) overall rotational slide-out failure

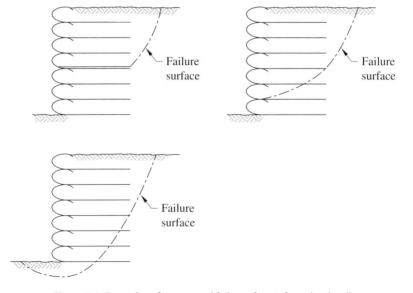

Figure 5.4 Examples of compound failure of a reinforced soil wall

the concept of soil–geosynthetic composites, the required tensile stiffness and strength of geosynthetic reinforcement is typically governed by the load-carrying capacity of the composite resulting from soil–reinforcement interaction (see Section 5.3.1), or by pre-scribed limiting lateral wall movement (see Section 5.3.3) whichever is greater. The required length of geosynthetic reinforcement, on the other hand, is typically governed by external stability checks for various modes of external failure and generally *not* by pullout of reinforcement (cf. Section 5.3.2).

5.2   Overview of Design Methods

All prevailing design methods for reinforced soil walls with geosynthetics as reinforcement are based on the limiting equilibrium method of analysis. The design methods require that both internal and external stability be checked. For evaluation of internal stability, most methods adopt an earth pressure approach, while a handful of others adopt a slope stability approach. Examples of design methods based on the earth pressure approach include the U.S. Forest Service (USFS) method (Steward et al., 1977, revised 1983), the Broms method (1978), the Bonaparte et al. method (1987), the GeoServices method (Giroud, 1989), the Christopher et al. method (1990), the FHWA-NHI method (Berg et al., 2009), the NCMA method (2010), the AASHTO guidance (2002, 2014), and the K-stiffness method (Allen et al., 2004; Bathurst et al., 2008). Examples of design methods based on the slope stability approach are the Schmertmann et al. method (1987) and the Leshchinsky–Perry method (1987).

The slope stability approach adopts the same procedure as that used in conventional slope stability analysis, but with modifications to account for the presence of tensile reinforcement. Different methods of the slope stability approach differ in the assumed shape of potential failure surfaces and orientation of reinforcement forces along a potential failure surface. For example, the Leshchinsky–Perry method assumes rotational (log-spiral) and translational (planar) failure surfaces. The Schmertmann et al. method assumes straight-line and bilinear failure surfaces. The simplified Bishop method and the Spencer method, commonly used for analysis of unreinforced soil slopes, have been modified for analysis of reinforced soil slopes. Because of the effort and time required for design computations, design methods based on the slope stability approach have to rely on either design software or design charts.

For design methods that are based on the earth pressure approach, a potential failure plane (which occurs through the reinforced soil mass) and an earth pressure diagram (which acts on the vertical plane along the back of the reinforced soil mass) are assumed. Two sets of safety factors (or variations) are determined for each layer of reinforcement, one against rupture failure and the other against pullout failure. The safety factors for every layer need to meet certain minimum criteria to ensure sufficient safety margins. The safety factor against rupture failure for a given reinforcement layer is evaluated as the ratio of reinforcement strength and the resultant force in the tributary zone of the layer. On the other hand, the safety factor against pullout failure for a given reinforcement layer is evaluated as the ratio of resisting (or stabilizing) force to driving (or destabilizing) force. The resisting force is the resultant of soil–reinforcement interface frictional resistance along the length of reinforcement extending beyond the

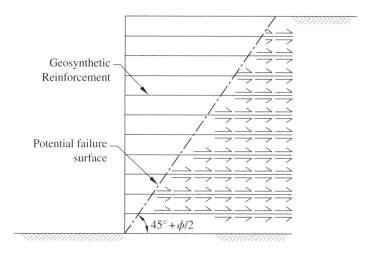

Figure 5.5 Commonly assumed failure plane in the earth pressure design methods, and frictional resistance against pullout failure in the anchored zone

assumed failure plane of the layer (see Figure 5.5); on the other hand, the driving force is the resultant of the assumed lateral earth pressure in the tributary zone of the layer. This method of stability analysis is very similar to the way a tieback system is designed. As noted earlier in Section 5.1, design methods that considered geosynthetic reinforcement as quasi-tieback tension members are collectively referred to as quasi-tieback wedge methods.

Most earth pressure design methods assume that the failure plane of reinforced soil walls follows the Rankine active failure plane for an unreinforced soil mass under uniform vertical surcharge. The failure plane will initiate at the heel and slope upwards at an angle of $45° + \phi/2$ from the horizontal, where ϕ is the angle of internal friction of the soil (in terms of effective stress, ϕ', for free-draining backfill), as shown in Figure 5.5. The earth pressure design method proposed by the NCMA, a trade organization for masonry blocks, is the sole exception to the use of the Rankine active failure plane. It assumes the lateral earth thrust and failure plane follow the Coulomb theory (see Section 2.3), which accounts for soil–facing interface friction, and is less conservative than the Rankine theory (hence gives a smaller earth thrust on a wall).

Claybourn and Wu (1992) compared designs obtained from seven design methods (two slope stability methods and five earth pressure methods) with measured results of a field-scale loading experiment for a timber-facing GRS wall conducted under well-controlled conditions (see Figure 5.6). The GRS wall in the experiment was constructed with granular backfill and a very weak nonwoven geotextile, and loaded vertically in increments until failure occurred in a plane-strain condition (Wu, 1992). Table 5.1 lists the predictive failure surcharge obtained based on the seven design methods. All seven methods give much lower failure loads (ranging from 0 to 7.3 psi) than the measured failure load of 29 psi (200 kPa). The observed failure mode was excessive bulging of the facing as a result of failing nails in the forming elements of the wall face (cf. Section 6.1.4). It is important to point out that the predicted failure pressures shown in Table 5.1 are determined with all safety factors being set equal to 1.0,

(a)   (b)

Figure 5.6 A field-scale timber facing GRS test wall in plane-strain condition: (a) test bin and (b) excessive bulging of facing at failure, under a vertical surcharge load of 29 psi (200 kPa)

Table 5.1 Failure surcharge pressures predicted by various design methods with a safety factor of 1.0

Design Method (with $F_s = 1.0$)	Surcharge Load at Failure (measured value = 29 psi)
U.S. Forest Service (Steward et al., 1977, revised 1983)	0.7 psi
Broms (1978)	6.2 psi
Collin (1986)	7.3 psi
Bonaparte et al. (1987)	0.9 psi
Leshchinsky and Perry (1987)	5.2 psi
Schmertmann et al. (1987)	6.0 psi
Giroud (1989)	0 psi (cannot be built)

even though all the design methods stipulate safety factors greater than 1.0 should have been used in intermediate computation steps. Had the safety factors stipulated by the design methods been used, the failure load would have been much smaller than those given in Table 5.1. In other words, the differences between the measured and predicted failure loads as seen in Table 5.1 are a direct indication of the deficiencies of the design methods. They are not a reflection of any intended safety margins. The comparison indicates that all of those design methods are very conservative (with the closest predictive failure load being off by a factor of 4, and the lowest predictive value saying the wall could not be built).

Claybourn (1990) and Claybourn and Wu (1993) also reported a study to compare designs obtained from six earth pressure methods by incorporating recommended safety factors stipulated by the respective methods. The designed reinforcement layouts for a 12-ft (3.6-m) high wall are shown in Figure 5.7. The differences among the

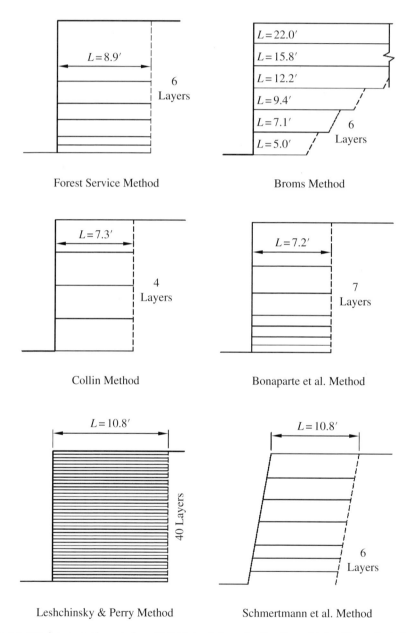

Figure 5.7 Reinforcement layouts for a 12-ft high reinforced soil wall, as obtained from six design methods, by incorporating recommended safety factors stipulated in respective design methods (Claybourn, 1990)

different methods are quite dramatic. Upon further investigation, it was found that in addition to differences in assumed lateral earth pressure diagrams for the different methods, a major source of the differences came from the safety factors. Figure 5.8 shows the design reinforcement layouts with all safety factors set to 1.0 for a 20-ft (6.1-m) high wall. It is seen that the differences for the six design methods are much

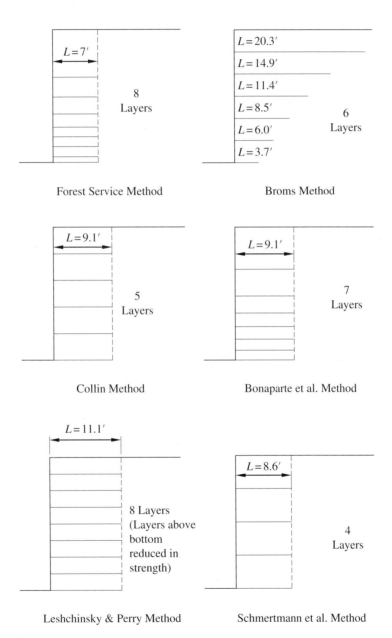

Figure 5.8 Reinforcement layouts for a 20-ft high wall, as obtained from six design methods, by setting all safety factors = 1.0 (Claybourn, 1990)

smaller compared to those seen in Figure 5.7. The fact is that the values of the safety factors stipulated in all the design methods are somewhat arbitrary; they are not established based on sufficient empiricism or sound analysis of a reliable and sufficiently large database. Because of uncertainties associated with design parameters and computation procedure as well as a general desire to obtain a conservative design, different values of safety factors have been assigned. Some safety factors are applied to soil strength parameters, and others to calculated forces, moments, etc. Individual safety factors, ranging from 1.2 to 1.5, appears rather "innocent" and reasonable. However, they can be very different as a whole. The issue of the somewhat arbitrary safety factors remains an unresolved problem in all design methods of reinforced soil walls. Use of the load and resistance factor design (LRFD) approach attempts to resolve this issue in an analytical manner; however, a lack of reliable and sufficient database has made it a still unresolved issue.

Load-bearing GRS walls (such as bridge abutment walls) face a unique challenge in that there are heavy bridge loads applied near the wall face. The National Cooperative Highway Research Program initiated a study on the subject, and a design method for GRS abutment walls has been developed (Wu et al., 2006). The design method was based on full-scale experiments and extensive finite element analysis. Two performance criteria were established to determine allowable bearing pressure for GRS bridge sills: a sill settlement limit criterion and a critical shear strain distribution criterion in the reinforced soil mass. The allowable bearing pressures of bridge sills are a function of the friction angle of the fill, reinforcement vertical spacing, sill width, and type of sill (isolated sill or integrated sill). The default value for reinforcement vertical spacing was set as 0.2 m (8 in.), which was shown to offer significant benefits over the larger reinforcement spacing commonly used in GMSE. To ensure satisfactory performance and acquire an adequate margin of safety, reinforcement spacing greater than 0.4 m is not recommended for GRS abutments under any circumstances.

Realizing the significant benefits of closely spaced geosynthetic reinforcement, the U.S. Federal Highway Administration (FHWA) subsequently developed a comprehensive design and construction guidance for a GRS bridge abutment system, known as the geosynthetic reinforced soil – integrated bridge system (GRS-IBS) (Adams et al., 2011a,b). The GRS-IBS stipulates that geosynthetic reinforcement spacing should not exceed 0.3 m (12 in.). To date, hundreds of GRS-IBS bridge abutments have been built. These bridge abutments have seen excellent performance.

Most reinforced soil walls with geosynthetics as reinforcement built in North America have been GMSE walls with concrete block facing. Approximately 60,000 GMSE walls have been constructed on U.S. highways. Most of these walls have been designed by following the AASHTO guides (2002, 2014), the FHWA-NHI method (Berg et al., 2009), or the NCMA method (2010). Typical reinforcement spacing of GMSE walls is in the range of 0.3–1.0 m (or 12–40 in.). GMSE walls, however, have been found to experience an unusually high failure rate. Berg (2010) and Valentine (2013) have estimated a failure rate as high as 5%. The NCMA's estimate of the failure rate is 2–8%. There appears to be no consensus about the causes for the very high failure rate, yet more and more walls are being constructed on a daily basis. In any case, the high rate of failure of GMSE, a few folds higher than conventional earth retaining walls, is unacceptable in civil engineering practice.

Holtz (2010) indicates that bulging of wall face is an issue for GMSE walls. In a recent survey, many state Departments of Transportation (DOTs) listed excessive bulging and deformation as a top concern for GMSE walls (Gerber, 2012). Koerner and Koerner (2013) reported a database of failed GMSE walls. Of the 171 walls reported, 44 were excessive deformation and 127 were at least partial collapse. Some main statistics for the 171 walls are as follows:

- 96% were private (as opposed to public) financed walls
- 78% were located in North America
- 71% were masonry block faced SRW walls
- 65% were 4–12 m high
- 91% were geogrid reinforced; the other 9% were geotextile reinforced
- 86% failed within 4 years after construction
- 61% used silt and/or clay backfill in the reinforced soil zone
- 72% had poor-to-moderate compaction
- 98% were caused by improper design or construction (none were caused by geosynthetic manufacturing failures)
- 60% were caused by internal or external water (the remaining 40% were caused by internal or external soil-related issues).

GRS walls with closely spaced reinforcement have been proven to be a viable alternative to GMSE. The fundamental concept of GRS has been addressed in Section 3.1. Other than reinforcement spacing, which spells an obvious difference between GRS and GMSE, the two reinforced soil systems have different reinforcing mechanisms. As opposed to the quasi-tieback stabilization mechanism for GMSE, nine different reinforcing mechanisms have been identified for GRS (see Section 3.3). GRS has seen a very low failure rate (zero, according to a prominent GRS wall builder of over two decades).

In recent years, there have been a number of major advances related to the design of closely spaced GRS. Six new advances are described in some detail in Section 5.3.

5.3   Recent Advances in the Design of GRS Walls

This section describes the six major advances related to the design of GRS walls:
- required reinforcement stiffness and strength
- evaluation of pullout stability
- lateral movement of wall face
- required long-term strength of geosynthetic reinforcement
- connection stability of concrete block facing
- required reinforcement length.

5.3.1 Required Reinforcement Stiffness and Strength

In most prevailing design methods of reinforced soil walls and bridge abutments, the required minimum reinforcement strength, $T_{required}$, at depth z (measured from the top of the wall) has been determined as:

$$T_{required} = \sigma_h \cdot s_v \cdot F_s \qquad (5\text{-}1)$$

where $T_{required}$ = the required minimum strength for reinforcement at depth z, σ_h = average horizontal stress at depth z, s_v = vertical spacing of reinforcement at depth z, and F_s = safety factor.

In the prevailing design methods, σ_h has been assumed to be the same as that in an unreinforced soil (i.e., by following the Rankine active earth pressure theory). It is interesting to note that this assumption in fact goes against the mechanism of apparent confining pressure (Section 3.3.1) and the mechanism of restraining lateral deformation of soil (in Section 3.3.3) where geosynthetic inclusion is said to reinforce a soil through soil–reinforcement interface friction and results in increased confining pressure and lateral stress in the soil.

Eqn. (5-1) states that the reinforcement in a reinforced soil mass needs to have a minimum strength which is equal to the lateral stress in the soil mass multiplied by reinforcement spacing and a safety factor. The equation appears to be quite reasonable (after all, that's what *force equilibrium* is all about). However, if the values of σ_h and F_s are to remain unchanged for any reinforcement spacing, as has typically been assumed in the prevailing design methods of reinforced soil structures, $T_{required}$ is then linearly proportional to s_v (i.e., the ratio $T_{required}/s_v$ = a constant). This implies that reinforcement spacing and reinforcement strength will have exactly the same influence on the performance of a reinforced soil wall. In other words, larger reinforcement spacing can be fully compensated by using reinforcement of proportionally higher strength (e.g., if the reinforcement spacing is doubled, the same performance can be achieved simply by using a reinforcement of twice the strength). This fundamental design equation (Eqn. (5-1)) has encouraged designers to use large reinforcement spacing along with high strength reinforcement because larger reinforcement spacing usually means shorter construction time.

Field-scale experiments conducted to examine the effects of reinforcement spacing versus reinforcement strength have demonstrated convincingly that reinforcement spacing has a much stronger influence on the performance of a geosynthetic reinforced soil mass than reinforcement strength. Examples include the mini pier experiments by Adams et al. (2002, 2007), unconfined compression tests by Elton and Patawaran (2005), and the GSC tests by Pham (2009) and Wu et al. (2013). These experiments have been described in Section 3.2.

Wu et al. (2010) developed an analytical equation for calculating the ultimate load-carrying capacity, q_{ult}, of a GRS mass with closely spaced reinforcement (or a soil–geosynthetic composite). With a cohesionless backfill, the ultimate load-carrying capacity of a soil–geosynthetic composite can be expressed as:

$$q_{ult} = \left[\sigma_c + W \left(\frac{T_{ult}}{s_v} \right) \right] K_P = \left[\sigma_c + 0.7^{\left(\frac{s_v}{6d_{max}} \right)} \left(\frac{T_{ult}}{s_v} \right) \right] K_P \qquad (5\text{-}2)$$

in which q_{ult} = ultimate load-carrying capacity of a GRS mass, σ_c = lateral confining pressure, W = a factor signifying the comparative roles of reinforcement spacing and maximum particle size of fill material in a soil–reinforcement composite, s_v = reinforcement spacing, d_{max} = maximum grain size of fill material, T_{ult} = ultimate strength of reinforcement, and K_P = coefficient of Rankine passive earth pressure.

To determine whether a given reinforcement meets the load–deformation requirements in a design, it will be prudent to check if reinforcement meets the following two criteria:

1) Stiffness criterion: the stiffness of reinforcement must be sufficiently large to ensure satisfactory performance under service loads.

2) Strength criterion: the ultimate strength of the reinforcement must be sufficiently large to provide a sufficient safety margin under the limiting condition over the lifetime of the earth structure.

To ensure adequate ductility, the ultimate strength needs to correspond to a sufficiently large strain to prevent sudden collapse (the smallest strain to ensure adequate ductility is herein referred to as "ductility strain", $\varepsilon_{ductility}$). To wit, two parameters are needed for prescribing reinforcement in design: reinforcement stiffness ($T_{@\ service\ \varepsilon}$) and reinforcement strength (T_{ult}), with the latter being associated with a minimum value of ductility strain ($\varepsilon_{ductility}$). For example, if $T_{@\ service\ \varepsilon}$ and T_{ult} shown in Figure 5.9

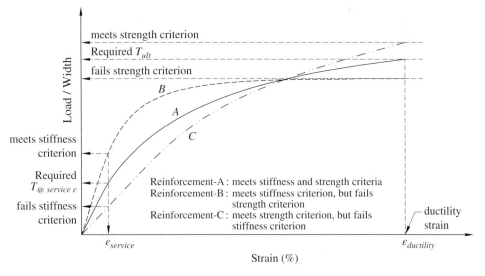

Figure 5.9 Three reinforcements which meet or fail a set of reinforcement design criteria: required reinforcement strength T_{ult} and reinforcement stiffness $T_{@\ service\ strain}$

represent the minimum required values of stiffness and strength determined from design computations (after applying safety factors or reduction factors), and $\varepsilon_{ductility}$ is a prescribed value of ductility strain, then Reinforcement-A would just meet both the stiffness and strength criteria, Reinforcement-B would meet the stiffness criterion but fail the strength criterion, and Reinforcement-C would meet the strength criterion but fail the stiffness criterion. In this example, only Reinforcement-A meets both stiffness and strength criteria; neither Reinforcement-B nor Reinforcement-C can be selected.

The tensile stiffness is defined as the tensile resistance at a prescribed service strain level. It is recommended that a tensile strain of 2.0% be taken as the prescribed service strain for specification of the minimum required reinforcement stiffness of GRS bridge abutments (Wu et al., 2011; Adams et al., 2011b). For non-load-bearing applications, the recommended strain level is relaxed from 2.0% to 3.0% as the vertical and lateral displacements are not as critical to the intended function of the walls.

For conservatism in design, the required minimum reinforcement stiffness (in the direction perpendicular to wall face), T_{ult}, calculated by Eqn. (5-2), can be set equal to $T_{@\,\varepsilon=2\,or\,3\%}$. In other words, the required minimum reinforcement stiffness, $T_{@\varepsilon=2\,or\,3\%}$, can be determined as:

$$T_{@\,\varepsilon=2\,or\,3\%} = \left[\frac{\sigma_h - \sigma_c}{0.7^{\left(\frac{s_v}{6d_{max}}\right)}} \right] s_v \qquad (5\text{-}3)$$

in which $T_{@\,\varepsilon=2\,or\,3\%}$ = required minimum tensile stiffness at $\varepsilon = 2$ or 3% (2% for load-bearing applications; 3% for non-load-bearing applications), σ_h = horizontal stress at a depth where required stiffness is being sought, σ_c = lateral confining pressure of the GRS mass, s_v = reinforcement spacing, and d_{max} = maximum grain size of the backfill.

For a GRS wall or bridge abutment wall with dry-stacked concrete block facing without any connection enhancement elements (such as pins, lips, or keys), the value of σ_c can be estimated as $\sigma_c = \gamma_b \cdot D \cdot \tan \delta_b$, where γ_b is the *bulk* unit weight of the facing block (bulk unit weight = total weight of a unit block divided by the *exterior* volume of the block), D is the depth of a unit block (in the direction perpendicular to the wall face), and δ_b is the friction angle between adjacent facing blocks (or between geosynthetic reinforcement and the facing block, if the geosynthetic is sandwiched between adjacent blocks). The value of σ_c is usually very small for dry-stacked light-weight block facing and can conservatively be assumed to be zero. The largest value of σ_h typically corresponds to the base of a wall and can be estimated by assuming that the Rankine active condition applies. Namely, $\sigma_h = (\gamma H + q) K_A = (\gamma H + q) \tan^2(45° - \phi/2)$, where γ is the unit weight of the compacted backfill, H is the total height of the wall, q is the uniform vertical surcharge pressure over the wall crest, and K_A is the coefficient of Rankine active earth pressure.

To ensure sufficient ductility and satisfactory long-term performance, the required minimum value of reinforcement strength (in the direction perpendicular to the wall face), T_{ult}, can be determined by imposing a *ductility and long-term factor*, F_{dl}, on $T_{@\,\varepsilon=2\,or\,3.0\%}$ determined from Eqn. (5-3), i.e.,

$$T_{ult} = F_{dl} \cdot T_{@\,\varepsilon=2\,or\,3.0\%} \tag{5-4}$$

The tensile strain corresponding to T_{ult} on the load–deformation curve of the geosynthetic should be at least 5–7% to ensure sufficient ductility. The $\varepsilon_{ductility} = 5$–7% is a suggested value from limited experience. The issue of long-term design, including recommended values of F_{dl} (the ductility and long-term factor) for different soils and geosynthetics, is further addressed in Section 5.3.4.

Validation of the analytical models for minimum required reinforcement strength (for $F_s = 1.0$ and $F_{dl} = 1.0$) has been carried out by comparing calculated reinforcement loads at failure with those obtained from field-scale experiments, including the soil–geosynthetic composite (SGC) tests (see Section 3.2.3) and Elton and Patawaran's tests (see Section 3.2.2). Note that these tests were selected for the validation because they use specimen sizes that are large enough to provide an adequate representation of soil–reinforcement composite behavior in actual field installation and were loaded to failure. Loads in reinforcement cannot be measured, however the reinforcement loads are known to equal the ultimate strengths when a reinforcement is at rupture failure.

Table 5.2 shows a comparison of results obtained from Eqns. (5-1) and (5-4) with "measured" results of the SGC tests. The largest difference in reinforcement loads at failure for Eqn. (5-4) is 16%, whereas the difference for Eqn. (5-1) is as high as 47%.

Table 5.2 Comparisons of reinforcement loads at failure between Eqns. (5-1) and (5-4) and results of the SGC tests

Parameter	Test Designation			
	Test 2 $(T_f,\,s_v)$	Test 3 $(2T_f,\,2s_v)$	Test 4 $(T_f,\,2s_v)$	Test 5 $(T_f,\,s_v)$
Reinforcement Strength, T_f (kN/m)	70	140	70	70
Reinforcement Spacing, s_v (m)	0.2	0.4	0.4	0.2
Vertical Pressure at Failure in Test (kPa)	2,700	1,750	1,300	1,900
Lateral Confining Pressure, σ_3 (kPa)	34	34	34	0
Maximum Reinforcement Load (kN/m) at Failure, by Eqn. (5-1) with $F_s = 1.0$	62.4	74.4	50.5	41.2
Difference between Eqn. (5-1) and Test Results*	−11%	−47%	−28%	−41%
Maximum Reinforcement Load (kN/m) at Failure, by Eqn. (5-4) with $F_{dl} = 1.0$	79.4	124.1	75.4	58.8
Difference between Eqn. (5-4) and Test Results*	+13%	−11%	+8%	−16%

Note: Soil parameters: internal friction angle, $\phi = 50°$; cohesion, $c = 70$ kPa; unit weight, $\gamma_{backfill} = 24$ kN/m³; maximum grain size, $d_{max} = 33$ mm.
*negative value: smaller than measured; positive value: larger than measured.

A similar comparison for the Elton and Patawaran tests is shown in Table 5.3. The largest difference in reinforcement loads for Eqn. (5-4) is 13%, whereas it is as high as 74% for Eqn. (5-1). Note again that because loads in reinforcement cannot be measured (except at the extremities), the comparisons can only be made in an *at-failure* state when the reinforcement loads are known to be the respective strength values. The analytical model (Eqn. (5-4)) is clearly an improved tool over Eqn. (5-1) for determination of required reinforcement strength.

Table 5.3 Comparisons of reinforcement loads at failure between Eqns. (5-1) and (5-4) and results of Elton and Patawaran's unconfined compression tests

Parameter	Reinforcement Designation						
	TG 500	TG 500	TG 600	TG 700	TG 800	TG 1000	TG 028
Reinforcement Strength, T_f (kN/m)	9	9	14	15	19	20	25
Reinforcement Spacing, s_v (m)	0.15	0.30	0.15	0.15	0.15	0.15	0.15
Vertical Pressure at Failure in Test (kPa)	230	129	306	292	402	397	459
Maximum Reinforcement Load (kN/m) at Failure, by Eqn. (5-1) with $F_s = 1.0$	4.47	2.35	6.95	6.49	10.08	9.91	11.94
Difference between Eqn. (5-1) and Test Results*	−50%	−74%	−50%	−57%	−47%	−50%	−52%
Maximum Reinforcement Load (kN/m) at Failure, by Eqn. (5-4) with $F_{dl} = 1.0$	9.02	9.56	14.02	13.10	20.34	20.01	24.09
Difference between Eqn. (5-4) and Test Results*	0%	+6%	0%	−13%	+7%	0%	−4%

Note: Soil parameters: internal friction angle, $\phi = 40°$; cohesion, $c = 27.6$ kPa; unit weight, $\gamma_{backfill} = 18.8$ kN/m³; maximum grain diameter, $d_{max} = 13$ mm.
*negative value: smaller than measured; positive value: larger than measured.

The *load-carrying capacity equation* (Eqn. (5-4)) for calculation of T_{ult} of GRS composites has also been validated by available "measured" data from field-scale experiments as of 2011 (Adams et al., 2011b). Figure 5.10 shows a comparison of the load-carrying capacity equation and measured data compiled by the Federal Highway Administration. A good agreement between all measured test data and calculated values is seen.

5.3.2 Evaluation of Pullout Stability

As noted at the beginning of this chapter, the design of GRS walls involves checking external stability and internal stability. External stability refers to the stability of the reinforced soil mass in relation to the surrounding soil. Internal stability, on the other hand, refers to stability within the body of the reinforced soil wall. To carry out an internal stability check, two modes of failure have commonly been examined: tensile rupture of reinforcement and pullout of reinforcement from the reinforced soil mass. A pullout stability check has routinely been performed in the design of GMSE walls.

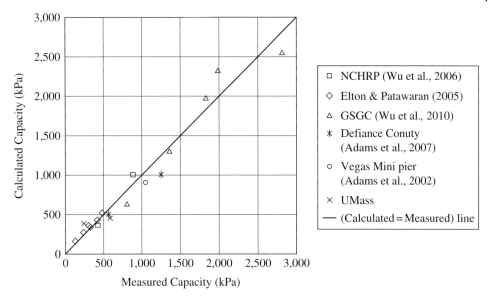

Figure 5.10 Comparison between measured and calculated load-carrying capacities of soil–geosynthetic composites, with the calculated values determined from Eqn. (5-4) for $F_{dl} = 1.0$ (Adams et al., 2011b)

The check of reinforcement pullout has traditionally been performed by requiring the reinforcement to be long enough that the resisting force (due to soil–reinforcement interface friction or interface bonding in general) behind an assumed failure plane of a reinforcement layer is sufficiently greater than the driving force (due to earth thrust in the tributary zone) of the reinforcement layer for *each and every* reinforcement layer in a reinforced soil mass. A minimum safety factor of 1.5 and a minimum anchored length of 0.9 m (3 ft) have commonly been specified to ensure pullout stability.

The safety factor against pullout failure for reinforcement *layer-i* (at depth z_i) has been evaluated as:

$$Fs_i = \frac{2\gamma z_i \tan\delta \left[L_t - (H - z_i)\tan\left(45° - \frac{\phi}{2}\right) \right]}{\sigma_{hi\,(max)} \cdot S_v} = \left(\frac{Pr_i}{Pd_i}\right) \tag{5-5}$$

In Eqn. (5-5), Pr_i is the resisting pullout force of *layer-i*, Pd_i is the driving pullout force of *layer-i*, L_t is the total reinforcement length of *layer-i* (as measured from the backface of the wall), $\sigma_{hi(max)}$ is the maximum horizontal stress in the tributary zone of *layer-i*, and δ is the angle of friction at the soil–reinforcement interface.

Let us consider a question: when performing a pullout stability check for a reinforced soil wall with closely spaced reinforcement, is it rational to require that each and every layer of reinforcement be safe against pullout failure? To address this question, we have to ask ourselves another question: can a single reinforcement layer in a closely spaced reinforced soil system experience pullout *by itself* while pullout of adjacent reinforcement layers does not occur? The answer has to be a resounding "no." In other words, requiring each and every layer to be safe against pullout failure is overly conservative for the pullout stability check of a closely spaced reinforced soil mass. This may help explain why there has not been any pullout failure reported in actual GRS walls. If pullout failure is to occur, it will have to occur as a group of adjacent layers.

A *group pullout safety factor* is defined for more realistic evaluation of the pullout stability of walls with tight reinforcement spacing. To wit, the stabilizing and destabilizing forces in a pullout stability check should be evaluated for a group of reinforcement layers, rather than for any single layer. The next question that needs to be addressed is: how many layers would constitute a realistic group? To be conservative in design, the number of layers in a group should be kept to a minimum. The smallest group involving multiple layers of reinforcement is theoretically a group of two. However, for a group of two layers to experience pullout failure, both will be right next to an immobilized layer immediately above or below. This will still be overly conservative for walls with closely spaced reinforcement. Therefore, the smallest group that is reasonably conservative would be "a group of three".

It is recommended that a *group pullout safety factor* against pullout failure be used to check pullout failure by following the *rule of three*, i.e., a group pullout safety factor for *any three consecutive* reinforcement layers must not be less than a certain limiting value to ensure pullout stability (see Figure 5.11). The group pullout safety factor, $Fs_{(group)}$, with depth of the mid-layer z_i, is defined as:

$$Fs_{i(group)} = \frac{Pr_{i-1} + Pr_i + Pr_{i+1}}{Pd_{i-1} + Pd_i + Pd_{i+1}} \tag{5-6}$$

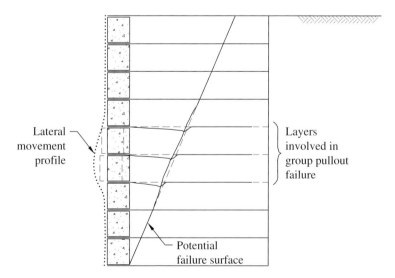

Figure 5.11 Schematic diagram of group pullout failure evaluated by "the rule of three"

in which subscripts $i - 1$, i, and $i + 1$ denote three consecutive reinforcement layers at depths z_{i-1}, z_i, and z_{i+1}, respectively.

Assuming the critical slip surface is to follow the Rankine active failure theory, the pullout resisting forces (Pr_i) and driving forces (Pd_i) for a GRS wall with a horizontal crest and a "vertical backslope" (i.e., with a vertical face in the back of the facing) while subjected to uniform vertical surcharge q, are (all other parameters are as defined for Eqn. (5-5)):

$$Pr_i = 2(\gamma z_i + q)\tan\delta\left[L_t - (H - z_i)\tan\left(45° - \frac{\phi}{2} \right) \right] \tag{5-7}$$

$$Pd_i = \left(\frac{1}{W} \right) K_A (\gamma z_i + q) s_v \tag{5-8}$$

The parameter $W = 0.7^{\left(\frac{s_v}{6d_{max}}\right)}$ in Eqn. (5-8) was previously introduced in Eqn. (5-2) to account for soil–reinforcement interaction (Wu et al., 2010). Due to frictional resistance at the soil–reinforcement interface, the lateral stress within a reinforced soil mass has been found to be higher than that in an unreinforced soil mass (described by the Rankine earth pressure theory as being equal to $K_A\sigma_v$).

As noted above, the group pullout safety factor against reinforcement pullout, as expressed by Eqn. (5-6), is the sum of the resisting pullout forces for *any* three consecutive layers divided by the sum of the driving pullout forces for the same three layers. Since there is no reinforcement layer above the very top reinforcement layer or beneath the bottom layer of a wall, the values of the "non-existent" reinforcement layers (i.e., Pr_{i-1} and Pd_{i-1} for the top layer, and Pr_{i+1} and Pd_{i+1} for the bottom layer) should be taken as zero. A minimum group pullout safety factor of 1.5 is recommended. If the any of the $F_{s(group)}$ in a GRS wall is less than 1.5, the reinforcement length should be increased until all values of $F_{s(group)}$ of the wall are at least 1.5.

5.3.3 Lateral Movement of Wall Face

A GRS wall with concrete block facing is inherently a "flexible" wall system. A sound design method for GRS walls therefore should include determination of lateral movement to ensure that it does not deform excessively under service loads. The AASHTO (2002, 2014) and FHWA-NHI (Berg et al., 2009) design guides adopt a simple chart proposed by Christopher et al. (1990) to give a rough estimate of the largest lateral wall movement. The chart is easy to use; however, it does not include the influence of some key factors such as the stress–strain–strength properties of the fill material, the load–deformation properties of reinforcement, and facing rigidity, all of which have been shown to influence significantly lateral deformation of GRS walls (e.g., Tatsuoka, 1992; Rowe and Ho, 1993; Helwany et al., 1999; Bathurst et al., 2006). Therefore, the chart is not useful, even for preliminary determination of the batter of a wall only.

A number of methods have been developed to estimate the lateral movement of reinforced soil walls, including those proposed by Giroud (1989), Wu (1994), Jewell and

Milligan (1989), and Chew and Mitchell (1994). Among them, the Jewell–Milligan method has consistently been shown to provide the closest agreement with results of finite element methods of analysis (Macklin, 1994). The Jewell–Milligan method, however, ignores the effect of facing rigidity, hence is applicable only to reinforced soil walls with negligible facing rigidity (such as wrapped-faced GRS walls). Wu and Pham (2010) modified the Jewell–Milligan method to include the influence of facing rigidity. The Wu–Pham method can be used to predict the lateral displacement profile of a reinforced soil wall with block facing during construction and under design loads. This method can also be used to determine the required reinforcement stiffness and strength for a specified allowable maximum lateral wall movement in design.

In the Wu–Pham method, the value of Δ_i, the lateral displacement of a GRS wall at depth z_i, is determined by the following equation:

$$
\Delta_i = 0.5 \left[\frac{K_h \left(\gamma z_i + q \right) s_v - \gamma_b \, D s_v \tan \delta_b \left(1 + \tan \delta_b \tan \beta \right)}{K_{reinf}} \right] \cdot
$$

$$
\left(H - z_i \right) \left[\tan \left(45° - \frac{\psi}{2} \right) + \tan \left(90° - \phi_{ds} \right) \right] \tag{5-9}
$$

where K_h = lateral stress coefficient of reinforced fill (for simplicity, K_h can be taken as being equal to K_A), K_{reinf} = stiffness of reinforcement ($K_{reinf} = E \cdot t$, see Figure 4.19), H = wall height, γ = unit weight of reinforced fill, γ_b = bulk unit weight of facing block, ϕ_{ds} = effective direct shear friction angle of soil, ψ = angle of dilation of soil, β = friction angle between the soil and backface of the wall, and δ_b = friction angle between vertically adjacent blocks (or between reinforcement and block, if the reinforcement is sandwiched between blocks). If frictional resistance between backface of the wall and reinforced fill is ignored (i.e., set $\beta = 0$ for conservatism, Eqn. (5-9) reduces to:

$$
\Delta_i = 0.5 \left[\frac{K_h \left(\gamma z_i + q \right) s_v - \gamma_b \, D s_v \tan \delta_b}{K_{reinf}} \right] \left(H - z_i \right) \left[\tan \left(45° - \frac{\psi}{2} \right) + \tan \left(90° - \phi_{ds} \right) \right]
$$

$$
\tag{5-10}
$$

Wu and Pham (2010) compared (Eqn. 5-10) with the Jewell–Milligan method. As to be expected, when $\gamma_b = 0$ (no facing resistance), the lateral displacement profiles determined by the Wu–Pham method are identical to those obtained by the Jewell–Milligan method. As γ_b increases, the differences between the two methods become increasingly pronounced.

Measured lateral displacements data of a field-scale experiment (Hatami and Bathurst, 2005, 2006) has been used to validate the Wu–Pham method. The wall, as shown in Figure 5.12, is a modular block facing reinforced soil wall. Figure 5.13 shows comparisons of lateral displacements of the wall under surcharge pressures of 50 kN/m and 70 kN/m, including measured displacements (Hatami and Bathurst, 2006), displacements calculated by the Jewell–Milligan method (1989), and displacements calculated by the Wu–Pham method (2010). It is seen that the lateral movement under both surcharge pressures calculated by the Wu–Pham method is in good

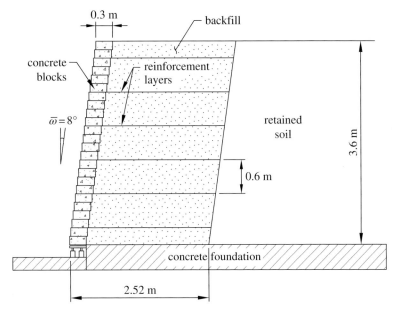

Figure 5.12 Schematic diagram of a field-scale modular block facing reinforced soil wall experiment (modified after Hatami and Bathurst, 2006)

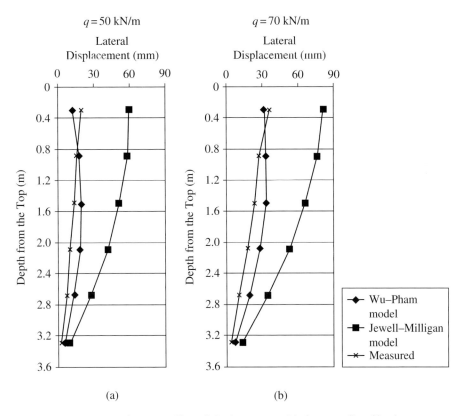

Figure 5.13 Comparisons of measured lateral displacements with those predicted by the Jewell–Milligan method and the Wu–Pham method, under two surcharge pressures: (a) $q = 50$ kN/m and (b) $q = 70$ kN/m (Wu and Pham, 2010)

agreement with the measured values, and the agreement is better than that of the Jewell–Milligan method. Note that the measured data indicate the wall has a higher bending stiffness than that indicated by both analytical methods.

If the calculated maximum lateral displacement of a wall is found to be greater than a certain prescribed tolerable value, the lateral movement can be reduced by choosing a reinforcement of higher stiffness or by employing smaller reinforcement spacing.

5.3.4 Required Long-Term Strength of Geosynthetic Reinforcement

In the prevailing design methods of reinforced soil walls with geosynthetics as reinforcement (e.g., AASHTO guidelines, 2002, 2014; FHWA-NHI manual; Berg et al., 2009; NCMA manual, 2010), the required strength of geosynthetic reinforcement is determined by applying three reduction factors to account for installation damage, creep, durability, together with a safety factor to the ultimate strength of geosynthetic reinforcement. In the form of an equation, the allowable reinforcement strength, T_a, is

$$T_a = \frac{T_{ult}}{FS \cdot RF} = \frac{T_{ult}}{FS \cdot RF_{ID} \cdot RF_{CR} \cdot RF_D} = k \cdot T_{ult} \tag{5-11}$$

in which T_{ult} is the ultimate wide-width strip tensile strength of the geosynthetic, as per ASTM D4595 for geotextiles and D6637 for geogrids, based on the *minimum average roll value* (MARV) of the product (Note: MARV is the strength that is two standard deviations below the mean tensile strength); FS is the overall safety factor to account for various uncertainties; RF is a combined reduction factor to account for possible loss of strength during design life, which is equal to $RF_{ID} \cdot RF_{CR} \cdot RF_D$, where RF_{ID} is a reduction factor for installation damage; RF_{CR} is a reduction factor for creep; and RF_D is a reduction factor for chemical and biological degradation.

For the FHWA-NHI method, RF_{ID} = 1.1–3.0, RF_{CR} = 1.6–5.0, RF_D = 1.15–2.0, and FS = 1.0 are recommended, resulting in k = 3–50%. Different states' Departments of Transportation have adopted variations for the reduction factors and safety factor, hence different k-values. Take the Colorado State Department of Transportation, for example, the following values have been employed: for preapproved geosynthetic products, FS = 1.78, RF_{ID} = 1.1–1.5, RF_{CR} = 2.2–2.7, and RF_D = 1.1, resulting in k = 13.6–17.0%; for non-preapproved geosynthetic products, k = 2.8–6.6%. In other words, the allowable design strength is on the order of 15% of MARV for preapproved products and 5% of MARV for non-preapproved products. These very low k-values have excluded nearly all geotextiles in design of reinforced soil walls. Many privately financed GRS walls have ignored the recommended k-values and have used geotextiles in design and construction. Most of these walls, many constructed over two decades ago, have performed satisfactorily.

In the following, we shall begin with a discussion of studies and findings of the three reduction factors in Eqn. (5-11), namely, long-term degradation reduction factor, installation damage reduction factor, and creep reduction factor. A discussion on long-term design strength along with some recommendations then follows.

(A)   *Long-Term Degradation*

Studies conducted by Elias (2000) and Allen and Bathurst (2003) have indicated that long-term degradation of geosynthetics in the in-soil environment in typical reinforcement applications is very small. Of particular interest is a field-scale experiment described in Section 4.2.3 and shown in Figure 4.29. The experiment conducted by the Public Research Institute (PWRI) of Japan is enlightening especially in regard to the influence of long-term degradation of geosynthetic reinforcement. The experiment was initially designed to examine the failure surface of a 6.0-m high GRS wall. After the wall was constructed, layers of geogrid reinforcement embedded in the wall were severed at preselected sections. The horizontal movement of the wall due to cutting of the reinforcements was essentially zero until the cutting approached approximately $0.2H$ from the wall face (H = wall height), at which time lateral movement increased by about 10 mm. The PWRI experiment suggests that stress relaxation in geosynthetic reinforcement might have completed at the time of cutting (within three days after construction). If stress relaxation is to reduce reinforcement loads to a very small value or zero, the geosynthetic reinforcement would no longer be serving any reinforcing function shortly after construction, hence long-term degradation of the reinforcement will no longer be an issue. Cutting of reinforcements in the experiment can be viewed as an extreme form of degradation, in that the reinforcement is degraded into pieces.

(B)   *Installation Damage*

Studies performed to investigate installation/construction damage of geosynthetics (e.g., Hufenus et al., 2005) have indicated that most installation damage does not severely affect the load–deformation behavior of woven geotextiles and geogrids, if at all, until damage levels become quite high. In applications where tensile resistance of reinforcement at low elongations is relevant (typical in GRS walls), the effects of moderate installation damage has been found to be very small. It has been suggested that RF_{ID} in Eqn. (5-11) be designated as being close to unity, on the order of 1.0 to 1.1.

Allen and Bathurst (1996) have provided convincing evidence that installation damage would have little, if any, effect on creep strains and creep rates for typical levels of installation damage in full-scale geosynthetic reinforced retaining walls. They concluded that in many cases, installation damage will have a negligible effect on the long-term strength at working stress levels (i.e., the geosynthetic behaves as if it is not damaged).

(C)   *Creep*

Geosynthetics, being polymeric materials, are generally considered susceptible to creep under sustained loads. Stress level, polymer type, manufacturing method, and temperature have been known to affect the creep potential of geosynthetics. When evaluating the creep potential of geosynthetic reinforcement in reinforced soil structures, however, it can be misleading to evaluate the creep potential based on laboratory tests performed by applying a sustained load directly to a geosynthetic specimen, as has been stipulated in the prevailing design guides. This is because the interaction between geosynthetic reinforcement and the time-dependent deformation behavior of the confining soil must be taken into account when evaluating the long-term creep behavior of geosynthetic reinforced soil structures. If the confining soil has a tendency to deform faster than the geosynthetic reinforcement (in isolation or subject to a confining pressure for pressure-sensitive geosynthetics), the geosynthetic reinforcement will impose a restraining effect on the time-dependent deformation of the soil through interface bonding forces. Conversely, if the confining soil tends to deform slower than the geosynthetic reinforcement (in isolation or subject to a confining pressure for pressure-sensitive geosynthetics), the confining soil will then restrain the creep deformation of the reinforcement. This restraining effect is a direct result of soil–reinforcement interaction wherein redistribution of stresses in the confining soil and changes in tensile loads in the reinforcement will occur over time in an interactive manner (Wu and Helwany, 1996).

Creep of geosynthetics and geosynthetic reinforced soil mass has been discussed in some detail in Section 4.2.2. Field measured data have strongly suggested that creep deformation of geosynthetic reinforcement will not be a design issue when well-compacted granular fill is used as backfill in the reinforced soil zone (Crouse and Wu, 2003; Allen and Bathurst, 2003).

(D)   *Discussion on Long-Term Design Strength and Recommendations*

To account for installation damage, creep, and durability of geosynthetic reinforcement, the prevailing design guides have suggested that the design strength be obtained by applying a combination of reduction factors to the short-term strength (Eqn. (5-11)). The value of the combined reduction factor typically depends on whether a geosynthetic is on a preapproved list; this value is typically about 5% of short-term strength for non-preapproved geosynthetics, and 15% for preapproved geosynthetics. This is hardly a sound approach, for three reasons.

- Soil–reinforcement interaction is known to be critical to creep of geosynthetic reinforcement, yet the approach is based on tests where loads are applied directly to geosynthetics specimens without any regard to soil–geosynthetic interaction (see Sections 4.2.1(B) and 4.2.2).

- There is no evidence that a compounding effect of installation damage and creep would occur. In fact there is evidence to the contrary (e.g., Allen and Bathurst, 1996; Greenwood, 2002), hence multiplication of individual reduction factors in Eqn. (5-11) is not justified.

- Creep is a deformation problem resulting from soil–geosynthetic interaction, yet the approach employs simplistic reduction factors, of which the values are not based on a reliable and sufficiently large database, to account for influences of these factors on complicated long-term creep deformation.

It is an indisputable fact that creep of geosynthetic reinforcement in a GRS is strongly influenced by time-dependent deformation behavior of the soil surrounding the geosynthetic reinforcement. A rational assessment of long-term design strength, therefore, must take the confining soil into account. If the time-dependent behavior of a given backfill is in question, a laboratory test such as the SGIP test (Wu and Helwany, 1996; Ketchart and Wu, 2002), shown in Figures 4.23 and 4.25, is recommended for evaluation of the potential creep deformation of a GRS structure.

When a GRS wall is constructed with well-compacted granular backfill, long-term deformation and stability are generally not major design concerns. In this case, the use of a single combined factor to account for long-term effects, uncertainty, and ductility is recommended. The value of this combined factor needs to be a function of soil properties and, to a lesser extent, geosynthetic type. Moreover, as noted in Section 5.3.1, when designing a GRS wall, the required minimum values of reinforcement stiffness, reinforcement strength and ductility strain still need to be specified.

For GRS walls with well-compacted granular fill, the required minimum tensile stiffness in the direction perpendicular to the wall face ($T_{@\,\varepsilon=2\,or\,3.0\%}$) can be determined by Eqn. (5-3) (Note: Use 2% for bridge abutments and 3% for non-load-carrying walls are recommended). On the other hand, the required minimum reinforcement strength (T_{ult}) can be obtained by imposing a *ductility and long-term factor* F_{dl} on the required minimum tensile stiffness to ensure sufficient ductility and satisfactory long-term performance, i.e., $T_{ult} = F_{dl} \cdot T_{@\,\varepsilon=2\,or\,3.0\%}$ (Eqn. (5-4)). In addition, the tensile strain corresponding to T_{ult} on the load–deformation curve of the geosynthetic needs to be greater than the ductility strain (5–7% for most applications, based on a limited database) to ensure sufficient ductility.

The following values for *ductility and long-term factor* F_{dl} in Eqn. (5-4) are recommended for design of GRS walls constructed with well-compacted granular fill materials:

- Plasticity index (*PI*) is 3 or less:

    F_{dl} = 3.5 for all geosynthetics

- Plasticity index (*PI*) is between 4 and 6:

    F_{dl} = 5.5 for polypropylene geosynthetics

    F_{dl} = 5.0 for polyethylene geosynthetics

    F_{dl} = 4.0 for polyester geosynthetics

5.3.5 Connection Stability of Concrete Block Facing

Tatsuoka (1992) has provided strong evidence that facing rigidity has a significant effect on wall performance and the distribution of tensile loads in the reinforcement. For a reinforcement layer that is securely connected to rigid facing, the largest tensile load in the reinforcement layer will occur at the wall face. On the other hand, if the facing rigidity is low, the largest tensile load will occur away from the wall face.

This section focuses on concrete block facing of a GRS wall, particularly the requirements for connection stability. The facing connection requirements are influenced by lateral earth pressure exerted on the facing, which is reflected by tensile loads in the geosynthetic reinforcement at the wall face. For a concrete block GRS wall, if lightweight dry-stacked blocks (e.g., concrete masonry units, CMU) are employed, the global bending rigidity is rather low, hence the reinforcement load at the facing will be rather small and connection requirements will be little to none. On the other hand, if heavy modular blocks or strong facing connection enhancement elements (e.g., pins, lips, or keys) are employed, the lateral earth pressure on the wall face will be significantly higher.

Speaking of lateral earth pressure, an explanation of the difference between lateral pressure and lateral stress is in order. According to Webster's dictionary, the term *pressure* is used to describe the intensity of forces exerted against *an opposing body*, whereas the term *stress* is used to describe the intensity of an applied force or system of forces that tends to strain or deform a body. For a gravity wall or a cantilever wall, the lateral stress in the soil mass is considered equal to the lateral soil pressure on the wall. However, lateral stress within a reinforced soil mass can be quite different from lateral earth pressure exerted by a reinforced soil mass on the wall face. The difference between pressure and stress is especially relevant for closely spaced GRS walls. Due to the constraints to lateral deformation in a reinforced soil mass (resulting from frictional resistance at the soil–reinforcement interface), the lateral earth *pressure* against the wall face can be much smaller than the lateral *stress* within a reinforced soil mass (Wu, 2001, 2007). This point is elaborated below.

For a hypothetical reinforced soil wall of which the reinforced soil mass experiences no lateral deformation, the lateral earth pressure on the facing at all depths will be zero. The lateral soil stress within the reinforced soil mass however is not zero; it will typically increase with depth, much like in an unreinforced earth retaining wall. We shall begin the discussion on lateral earth pressure and lateral stress in a GRS wall by examining some analysis results. Chou and Wu (1993) conducted a study to investigate the behavior of GRS walls on 0.3 m reinforcement spacing by the finite element method of analysis. The lateral pressures/stresses along three sections of a GRS wall were examined: lateral earth pressure against the wall face, lateral earth pressure behind a reinforced soil mass, and lateral stresses within the reinforced soil mass along the plane of maximum reinforcement tensile loads, as shown in Figure 5.14. It is seen that the pressures/

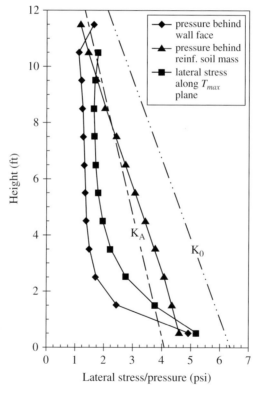

Figure 5.14 Lateral stress/pressure profiles along three different sections of a GRS wall: lateral pressure on wall face, lateral pressure on the back of reinforced soil mass, and lateral stress along the plane of maximum tensile loads (Chou and Wu, 1993)

stresses in those three sections are quite different. The lateral earth pressure against the wall face, the smallest of the three, is nearly constant with depth except near the base of the wall where there is greater constraint to deformation. The lateral stress along the plane of maximum reinforcement tensile loads is roughly parallel to the lateral earth pressure profile on the facing, but greater in magnitude. The earth pressure behind the reinforced soil mass, which is commonly used to evaluate the external stability of reinforced soil walls, is seen to be rather close to the Rankine active earth pressure.

In an idealized situation where every reinforcement layer in a reinforced soil wall is able to restrain perfectly lateral deformation of the soil, the lateral pressure on the wall face will be zero at the exact depth of the reinforcement layer. Between any two adjacent reinforcement layers, the lateral earth pressure will vary from zero at one layer, increase somewhat with depth due to gravity, and decrease back to zero at the other layer. We can therefore picture the condition between "any" two adjacent reinforcement layers as that of a *bin*. Since reinforcement in an actual wall is not perfectly rigid, a *bin pressure diagram* (see Figure 5.15) has been recommended for design by Wu (2001, 2007). The lateral earth pressure within any bin would increase approximately linearly with depth to a limiting value of $\sigma_h = K_A \cdot \gamma \cdot s_v$ before decreasing to a small value at the next reinforcement layer. The bin pressure diagram shown in Figure 5.15 is a *reasoned* pressure diagram and is applicable when the following four conditions are met:

1) The facing offers insignificant restraint to lateral movement of the reinforced soil mass.

2) The fill material behind the facing is free-draining so that there is little or no hydraulic pressure involved.

3) The soil–reinforcement interface bonding is maintained under the design load.

4) The reinforcement is sufficiently stiff.

Note that the bin pressure is independent of wall height, and is only a function of reinforcement spacing and the strength parameters of the soil. The resultant lateral thrust of a bin (T_{bin}) increases with the square of reinforcement spacing, i.e., $T_{bin} \propto s_v^2$ (s_v = reinforcement spacing), and is independent of depth. The bin pressure "postulate" has been found to collaborate well with field-scale experiments of GRS where the lateral earth pressure was measured by fatback pressure cells (Iwamoto et al., 2015) and field measurement data of GRS–IBS abutments in Louisiana (Saghebfar et al., 2017).

For concrete block GRS walls in non-load-bearing applications, the bin pressure diagram is applicable to walls with light-weight block facing. For GRS bridge abutments, the bin pressure diagram is applicable only up to a certain load level (available data have suggested the limiting surcharge pressure is about 200 kPa). A safety factor of 2.0 is recommended for checking connection requirements when the bin pressure diagram is employed. For concrete block GRS walls with tight reinforcement spacing (say, spacing ≤ 0.3 m) and reinforcement is placed at every course of facing blocks, the lateral earth thrust has been found to be *always* smaller than the frictional resistance between facing blocks, hence no special design provisions for facing connection stability are needed for walls in these conditions.

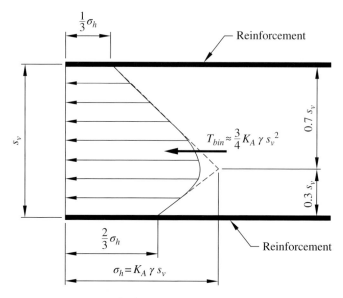

Figure 5.15 The bin pressure diagram (Wu, 2001)

To be on the conservative side, Wu and Payeur (2015) ignored the bin effect and assumed that the lateral earth pressure against the facing follows the Rankine active earth pressure (i.e., the pressure increases with depth along the wall height). Wu and Payeur derived a set of *connection force equations* based on force equilibrium to evaluate the stability of block facing. The driving (de-stabilizing) force D_i and the resisting (stabilizing) force R_i at the geosynthetic–block interface at depth z_i are:

Geosynthetic–block interface:

$$D_i = 2nh\,K_A\left(\gamma z_i + q\right) - 2nh\left[\frac{B\,\gamma_b\left(\tan\delta_{gb}\right)}{1-\left(\tan\delta_{gb}\right)\left(\tan\delta_{sb}\right)}\right] \qquad (5\text{-}12)$$

$$R_i = 2B\,\gamma_b\,z_i\left(\tan\delta_{gb}\right) + 2\,nh B\,\gamma_b\left[\frac{\left(\tan\delta_{sb}\right)\left(\tan^2\delta_{gb}\right)}{1-\left(\tan\delta_{gb}\right)\left(\tan\delta_{sb}\right)}\right] \qquad (5\text{-}13)$$

If the number of facing blocks between adjacent reinforcement layers, denoted by n, is 2 or more, connection failure at the block–block interface should also be checked. The driving force D_i and resisting forced R_i at the block–block interface at depth z_i become (applicable only to $n \geq 2$):

Block–block interface:

$$D_i = 2\,(n-1)\,h\,K_A\left(\gamma z_i + q\right) - 2\,(n-1)\,h\left[\frac{B\gamma_b\left(\tan\delta_{bb}\right)}{1-\left(\tan\delta_{bb}\right)\left(\tan\delta_{sb}\right)}\right] \qquad (5\text{-}14)$$

$$R_i = 2\,B\,\gamma_b\,z_i\left(\tan\delta_{bb}\right) + 2\left(n-1\right)h\,B\,\gamma_b\left[\dfrac{\left(\tan\delta_{sb}\right)\left(\tan^2\delta_{bb}\right)}{1-\left(\tan\delta_{bb}\right)\left(\tan\delta_{sb}\right)}\right] \tag{5-15}$$

in which n = number of courses of facing blocks between adjacent reinforcement layers, h = height of each facing block (use the largest h-value if more than one block height is employed), γ = unit weight of soil, ϕ = friction angle of soil, K_A = coefficient of Rankine active earth pressure ($K_A = \tan^2(45° - \phi/2)$), B = *equivalent depth* of facing block unit (see explanation below), γ_b = *bulk unit weight* of facing block (see also explanation below), δ_{bb} = block–block friction angle, δ_{gb} = geosynthetic–block friction angle, δ_{sb} = soil–block friction angle, and q = uniform vertical surcharge on the wall crest. The equivalent depth of a facing block is equal to the total horizontal surface area of the facing block divided by the width of the facing block. For a block containing hollow cells on its surface, the equivalent depth will be smaller than the exterior depth of the block. Similarly, the *bulk* unit weight of a facing block is the total weight of the block divided by the exterior volume of the block. For a block containing hollow cells, the bulk unit weight will be smaller than the unit weight of the block material.

Since geosynthetic reinforcement can only resist tensile force, when driving force D_i determined from Eqn. (5-12) or (5-14) turns out to be a negative value (i.e., compressive), it should be set equal to zero. Connection failure is said to occur whenever $D_i \geq R_i$.

Using the *connection force equations* (i.e., Eqns. (5-12) to (5-15)), Wu and Payeur (2015) examined driving and resisting forces for block walls under some common conditions. The following are highlights of their findings:

- For reinforced soil walls on 0.2 m and 0.4 m reinforcement spacing (i.e., $n =$ 1 and $n = 2$), the *net connection forces* (Note: Net connection force = resisting force – driving force) are found to be always positive (i.e., stable) for reinforced soil walls with light-weight facing blocks and without any connection enhancement elements. Figure 5.16 shows the profiles of net connection forces at the geosynthetic–block interface, which is typically more critical than at the block–block interface for reinforcement spacing of 0.2 m, 0.4 m, and 0.6 m (i.e., $n = 1$, $n = 2$, and $n = 3$). The net connection forces in the top 0.5 m of GRS walls are, however, rather small for all values of reinforcement spacing. This suggests that the common practice of adding concrete mix in the cells of the top 2–3 courses of light-weight blocks to tie these blocks together (see Section 6.1.1 and Figure 6.1) is a sensible measure.

- For reinforced soil walls on 0.6 m reinforcement spacing (i.e., $n = 3$), the net connection force is quite sensitive to the frictional angle at the geosynthetic–block interface; unless the geosynthetic–block interface friction is sufficiently high (35° or higher in most cases), facing connection failure will likely occur.

- Tight reinforcement spacing (say, spacing ≤ 0.3 m) is very beneficial for improving the safety margin of concrete block GRS walls against facing failure.

- For reinforced soil walls with friction-connected light-weight concrete block facing on reinforcement spacing of 0.2 m and 0.4 m (i.e., $n = 1$ and $n = 2$), the net connection forces would increase with depth (see Figure 5.16) and there is no limit to wall height in terms of connection failure. The satisfactory performance of a 16.7-m high wall (with $s_v = 0.2$ m) as shown in Figure 5.17 (cf. Figure 3.28(b), photo taken in late spring) is therefore of no surprise. Note that this is not the case for reinforcement spacing of 0.6 m (i.e., $n = 3$), see Figure 5.16. The common perception that a higher wall is more susceptible to connection failure (e.g., Dallaire, 2001) is true only for walls on larger reinforcement spacing.

- As can be expected, the stability of facing connections can be improved appreciably by increasing the bulk unit weight of facing blocks, the equivalent depth of facing blocks, and/or the friction angle of the backfill (e.g., with better fill compaction).

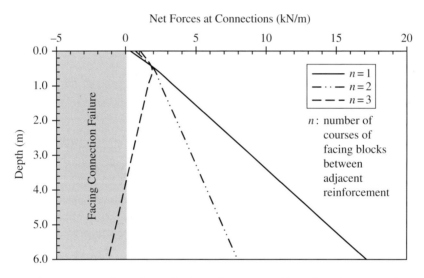

Figure 5.16 Net connection forces (sum of resisting forces minus driving forces) at geosynthetic–block interface with 1, 2 and 3 blocks between successive reinforcement layers under a surcharge load of 10 kPa (Wu and Payeur, 2015)

Many builders of concrete modular block geosynthetic walls have routinely opted for using pins, blocks with lips or keys, and/or heavy-weight blocks to presumably improve facing stability. These measures, while increasing the resisting forces of the facing, also *attract* more loads to the wall face (and to the connection between blocks), hence increase the earth thrust exerted on the wall face (i.e., larger than Rankine thrust). As a result, the resisting and driving forces at the facing connections are both increased. The use of facing connection enhancement measures may not necessarily improve facing stability. The net effect can be evaluated by the connection force equations, Eqns. (5-12) to (5-15). Note again that the driving forces in these equations were derived conservatively based on Rankine earth pressure theory; the bin effect was not accounted for in the derivation.

Figure 5.17 A 16.7-m (55-ft) high GRS wall with light-weight frictionally connected concrete block facing, seen is an ice cliff that builds and recedes all winter long (courtesy of Bob Barrett)

5.3.6 Required Reinforcement Length

A uniform reinforcement length of $0.6H$ to $0.7H$ (where H = total wall height at face), with a minimum length of 2.4 m (8 ft), is common for reinforced soil walls and abutments. This is a result of three major design guidelines: the AASHTO Bridge Design Specifications, later AASHTO LRFD Bridge Design Specifications (AASHTO, 2002, 2014), the FHWA NHI manual (Berg et al., 2009), and the NCMA manual (NCMA, 2010). The NCMA design manual, widely used in the private sector, requires a minimum reinforcement length of $0.6H$. A minimum length of $0.7H$, on the other hand, has been specified in the FHWA and AASHTO guides that are used routinely in the public sector. The standard of practice in Europe and Asia uses a similar criterion for minimum reinforcement length: $0.7H$ for routine applications and $0.6H$ for low lateral load applications with a minimum length of 3 m (BSI, 1995; GEO, 2002). Brazilian guides, perhaps the most conservative of all, require a minimum reinforcement length of $0.8H$.

Despite the popularity of the 0.6–$0.7H$ rule, uniform reinforcement lengths as small as 0.3–$0.4H$ have been employed with good success for construction of GRS walls. Shorter reinforcement length is considered a worthy subject because it can result in significant cost savings, especially in situations where construction involves excavation of existing ground. Another important design concern is with short reinforcement lengths in the lower part of a GRS wall (referred to as *truncated-base walls*). In situations where excavation into rock or stiff deposits to allow for a uniform reinforcement length would involve significantly higher costs, use of shorter reinforcement lengths in the lower part of a wall usually proves to be very cost-beneficial. There are, however, design concerns with such a measure.

The following begins with a review of research and case histories on short reinforcement lengths and concludes with a discussion of related issues. Both uniform reinforcement length walls and truncated base walls are addressed. GRS walls involving a *constrained fill* zone (i.e., walls where rock, heavily overconsolidated soil, or a nailed wall is present within a short distance from the wall face) are also discussed.

(A)   Research and Case Histories

i)   Reinforced Soil Walls with Uniform Reinforcement Length

When it comes to short reinforcement lengths, the research by Fumio Tatsuoka and the Japan Railway (Tatsuoka et al., 1992, 1997; Tatsuoka, 2008) was among the first. Beginning in the mid-1980s, Tatsuoka and his associates at the Japan Railway Technical Research Institute developed a GRS wall system with full-height rigid facing, referred to as the CIP-FHF (cast-in-place full-height facing) system (Section 3.4.3). Due to its superior performance in resisting strong earthquakes and heavy rainfalls over conventional cantilever walls and metallic reinforced earth walls, the CIP-FHF system became the "default" wall type for retaining walls constructed by the Japan Railway. To date, over 120 km of CIP-FHF GRS walls have been built all over Japan with great success.

As noted in Section 3.4.3, the CIP-FHF system involves a two-stage construction procedure. In the first stage, a GRS wall with geosynthetic reinforcement of 0.3–0.4H in length and typical spacing of 0.3 m is constructed using stacked gravel gabions as facing. The wall is allowed to deform under its self-weight before starting the second stage of construction, which involves installing full-height rigid concrete facing (cast-in-place) over the gabion face GRS wall. The rigid concrete facing is attached to the reinforced soil mass via protruded steel bars that are embedded in the GRS mass during the first stage. Despite the short reinforcement lengths of 0.3–0.4H, none of the walls have ever experienced failure. Both granular and cohesive backfills have been used in the construction of CIP-FHF wall systems. Tatsuoka and his associates suggested that overturning may be the most critical mode of failure for reinforced soil walls with short reinforcement lengths of 0.3–0.4H, whereas lateral sliding is typically the governing failure mode for walls with reinforcement lengths of 0.6–0.7H.

In a rebuttal to Tatsuoka's comments regarding CIP-FHF GRS walls vs. reinforced earth walls (with steel strips as reinforcement), Segrestin (1994), of Terre Armée International, stated that a number of reinforced earth walls with reinforcement length as low as 0.45H had been constructed successfully. Segrestin also stated that a very tall (10.5-m high) reinforced earth wall with reinforcement length of 0.48H had demonstrated satisfactory performance. Segrestin maintained that the basic mechanism and behavior of reinforced soil structures with reinforcement lengths between 0.4H and 0.7H were identical, and this had been validated by a finite element parametric study conducted by Terre Armée.

Bastick (1990) conducted finite element analysis of MSE walls undergoing changes due to reduced reinforcement length. It was found, similar to Terre Armée's finding, that the performance of an MSE wall would remain practically the same as long as the reinforcement lengths were kept above 0.4–0.5H and the reinforcement spacing stayed the same. The finding was confirmed by a full-scale experiment with reinforcement length of 0.48H loaded to an average surcharge of 840 kPa.

There are two distinct situations when very short reinforcement lengths find their way into reinforced soil walls: (a) walls with a *constrained fill* zone, where a rock or heavily over-consolidated soil outcrop or an existing nailed wall is present in the back of the reinforced soil wall, in which the space constraint makes the commonly used reinforcement length of 0.6–0.7*H* impractical (Lawson and Yee, 2005), and (b) walls with the end of reinforcement being anchored by metal plates or geosynthetic loops (Fukuoka et al., 1986; Brandl, 1998). Situation (a) is in line with the scope of this subject and will be elaborated further.

Lawson and Yee (2005) proposed a design-and-analysis method for reinforced soil retaining walls involving a constrained fill zone. Within the constrained reinforced fill zone, the full active failure wedge is unable to develop because of the close proximity of the rigid zone behind the reinforced fill. In this method, the magnitude of horizontal thrust acting on the wall face, P_h, is evaluated as $P_h = 0.5 K \gamma H^2$. Lawson and Yee concluded that for walls with reinforcement lengths greater than 0.5*H*, the Rankine active wedge can fully develop within the granular fill zone, hence the lateral earth pressure coefficient, K, is equal to K_A. However, if reinforcement length is less than 0.5*H*, the full active wedge cannot develop fully and the magnitude of the lateral earth pressure coefficient will decrease with decreasing reinforcement lengths.

A glaring example of reduced lateral thrust for walls with a constrained fill zone is a very tall wall constructed by Lin et al. (1997) in Taiwan, a country where very heavy seasonal rainfall occurs annually. The wall was 39.5 m high, constructed in six tiers (three 8-m tall tiers and three shorter tiers). The geogrid reinforcement length was 1.5 m, or 0.19*H*. All tiers, despite having to carry the soil weight from upper tiers, performed satisfactorily even with the very short reinforcement length. It is interesting to note that the reinforcement length of 0.19*H* happens to coincide with the finding of a numerical study (Vulova, 2000; Vulova and Leshchinsky, 2003). The numerical study also agreed well with Tatsuoka's assertion that overturning will be the control failure mode for a MSE wall with short reinforcement lengths.

Morrison et al. (2006) performed centrifuge tests on *shored MSE walls* where a constrained fill zone is present due to the shoring and reported the following:

- Reinforcement lengths in the range of 0.25–0.6*H* generally produce stable wall systems.
- Reinforcement lengths of 0.25*H* or smaller generally produce outward deformation followed by an overturning collapse of the MSE mass under increasing gravitational levels.
- For reinforcement lengths less than 0.6*H*, deformation produced a "trench" at the shoring interface, interpreted to be the result of tension; the trench was not observed with reinforcement lengths of 0.6*H* or greater.
- A conventional MSE wall with retained fill and a reinforcement length of 0.3*H* was observed to be stable up to an acceleration level of 80*g*, which represents a prototype height of approximately 27 meters.

It has been suggested that a shorter reinforcement length may result in larger lateral displacements and likely larger settlement as well. A finite element study conducted by Chew et al. (1991) showed that reducing reinforcement length from 0.7*H* to 0.5*H* caused about a 50% increase in lateral deformation. Ling and Leshchinsky (2003) reported that, with a reinforcement length of $0.5H$, a reinforced soil wall would produce satisfactory performance considering the maximum displacement mobilized in the reinforcement layers. A study by Liu (2012) suggested that the larger lateral displacement when shorter reinforcement is used may be a result of larger lateral deformation of the soil behind the reinforced zone.

It is of interest to note that a reinforced soil mass with closely spaced reinforcement and well-compacted fill tends to behave as a coherent mass. This behavior was observed in two loading experiments of full-scale concrete block facing GRS bridge abutment walls, referred to as the NCHRP GRS test abutments (Wu et al., 2006, 2008). Two different woven geotextiles of different values of stiffness and strength were used as reinforcement, each 3.15 m long and on 0.2 m spacing. The backfill was a non-plastic silty sand, and the abutment was loaded by applying increasing vertical loads onto a strip footing near the front face. A tension crack was observed on the wall crest in both experiments. The tension crack was located exactly at the extent of the reinforced zone (i.e., 3.15 m from the backface of the facing blocks) under an applied pressure between 150 and 200 kPa. The location of the tension cracks suggests that the reinforced soil mass behaves as a coherent mass for GRS walls with closely spaced reinforcement. This also explains why eccentricity (lift-off) has been the most critical failure mode with short reinforcement lengths of $0.3–0.4H$, especially for walls with closely spaced reinforcement. The only exception is when a constrained fill zone is present.

ii)   Reinforced Soil Walls with a Truncated Base

A reinforced soil wall with shortened reinforcement lengths in the lower part of the wall, known as a *truncated-base wall*, is often used when excavation to allow for uniform reinforcement lengths is cost prohibitive. The reduction in reinforcement lengths commonly takes one of two forms: stepped truncation (reducing lengths in groups of 2 to 4 reinforcement layers) and trapezoidal truncation (reducing lengths approximately linearly with depth).

A truncated base for MSE walls is allowed in the FHWA NHI manual. The manual provides general guides for a truncated base, and states that this provision should only be considered if the base of the MSE wall is founded on rock or competent soil; competent soil is a soil which exhibits minimal post-construction settlement. For foundation soil less than competent, a ground improvement technique may be used prior to construction.

The British Standard BS 8006 design manual (BSI, 1995) for MSE walls also allows use of a truncated base. It states that a truncated base (or trapezoidal) wall should only be considered where foundations are formed by excavation into rock or other competent foundation conditions exist. The manual prescribes a minimum reinforcement length of $0.4H$ for the lower portion of the wall. Hong Kong's Geoguide 6 (GEO, 2002) essentially follows the FHWA NHI manual, except it stipulates that soil arching needs to be accounted for in the design.

Japan Railway allows reinforcement with truncated lengths in the lower part of a GRS wall be used when the costs of excavation to allow for full-length reinforcement are high (Tatsuoka et al., 1992, 2005; Morishima et al., 2005). The walls constructed with a truncated base have been reported to have performed satisfactorily during heavy rainfall and severe earthquake events.

Segrestin (1994) of Terre Armée International reported on the applications of truncated-base walls with steel-strip reinforcement in situations where constrained fill is present. It has been reported that wider metal strips, or more often an increased number of strips, have been employed in truncated base reinforced earth walls.

The Colorado Department of Transportation constructed a 7.6-m high GRS wall with a truncated base in the DeBeque Canyon along Interstate Highway 70 (Wu, 2001). For comparison purposes, a 10-m long control section was also constructed with full-length reinforcement throughout (termed "the control wall"). A road base material was used as backfill, and the wall was situated over a firm foundation. Lateral displacements of the control wall (reinforcement length = 5 m) and the truncated-base wall (reinforcement length at base = 1.1 m, or $0.14H$), taken 6 months after construction, were found to be very similar, both on the order of 3–6 mm, with the maximum displacement being 8 mm. Adams et al. (2011a) of the Federal Highway Administration also reported a number of GRS abutments with a truncated base. The GRS abutments have been reported to have performed satisfactorily.

Lee et al. (1994) conducted a forensic study on a series of failed walls founded on rock and concluded that the resistance against sliding failure is reduced by the truncated base due to a smaller base area. They noted that soil arching due to the rock behind the fill would reduce vertical stress above the back of the lower reinforcements, hence can lead to overestimation of resistance to pullout failure.

Thomas and Wu (2000) conducted a finite element study on the behavior of GRS walls with a truncated base. The major findings of the study are as follows:

- When designing a GRS wall with a truncated base, external stability should be thoroughly checked; truncated base walls are more likely to experience lateral sliding failure and lift-off failure. The length of reinforcement at the lowest level of a truncated base wall should be at least $0.35H$ or 0.9 m.

- Fill type and fill compaction play an especially important role in the performance of GRS walls with a truncated base. The use of cohesive backfill should be completely avoided for truncated base walls.

- The foundation soil needs to be sufficiently stiff (with little post-construction settlement) to permit construction of truncated-base walls.

(B)   *Discussions and Concluding Remarks*

A minimum reinforcement length of 0.6–$0.7H$ (H = height of the wall at the wall face) has been used in most design guides for GRS/GMSE walls with a uniform reinforcement length throughout. However, walls with uniform reinforcement lengths as small as 0.3–$0.4H$ have consistently been shown to be stable, and studies have suggested that GRS walls with reinforcement length between $0.4H$ and $0.7H$ behave approximately the same. For situations where a uniform reinforcement length is employed, a minimum reinforcement length of $0.6H$ is well justified. A minimum reinforcement length of $0.7H$ is overly conservative, especially for closely spaced reinforcement (spacing not more than 0.3 m). Care must be exercised to prevent tension crack on the wall crest if the wall is to carry significant vertical loads near the wall face (e.g., bridge abutments). Extending the top 1 to 2 layers of reinforcement well beyond the assumed failure plane would help in eliminating the tension cracks.

Use of shorter reinforcement for reinforced soil walls, with a uniform reinforcement length of 0.35–$0.5H$, will result in larger lateral movement than those with a uniform reinforcement length of 0.6*H*, and is recommended only if the following four conditions are met:

1) The wall will not be subject to heavy edge loads, such as a bridge abutment wall.

2) Free-draining granular backfill is employed and well compacted.

3) The foundation is competent (little post-construction settlement).

4) External stability, especially against lift-off (eccentricity) failure and lateral sliding failure, is thoroughly checked.

Use of a truncated base is a viable measure when the cost of excavation to allow for uniform reinforcement length is impractical. There is strong evidence based on a large number of case histories and research studies that a truncated base wall will perform satisfactorily as long as the base of a reinforced soil wall is founded on competent foundation, and external stability, especially sliding and eccentricity (i.e., lift-off) stability, is ensured. The reinforcement length at the lowest level should generally be at least 0.3*H*. However, it can be significantly smaller (as small as 0.1*H* has been used routinely by the Japan Railway) if a constrained fill is involved (i.e., there is firm existing overconsolidated soil within a short distance behind the wall).

5.4 The U.S. Forest Service (USFS) Design Method

This section describes a design method, known as the U.S. Forest Service (USFS) method (developed in 1977 and revised in 1983). The method was the very first design method developed for reinforced soil walls with geosynthetics as reinforcement, and laid the groundwork for all subsequently developed design methods. A design example is given to illustrate how to use the USFS design method. The method assumes the wall is with a wrapped geotextile face. However, the design concept is applicable to geosynthetic reinforced soil walls of different types of flexible facing. The method is known to be very conservative and is very reliable for routine designs.

5.4.1 Design Procedure: The U.S. Forest Service Method

Step 1: Establish wall profile and check design assumptions

A wall profile should be established from the grading plan of the wall site. The following design assumptions should be verified:

- wall face is vertical or near vertical; i.e., no more than 10° off the vertical, or steeper than a batter of 1 (horizontal) to 7 (vertical)
- wall crest is essentially horizontal
- backfill is granular and free-draining
- wall is constructed over a firm foundation; i.e., very small post-construction settlement
- live loads are essentially vertical
- no concern for seismic loads

If any of the design assumptions are not satisfied or if the wall geometry cannot be approximated as an equivalent system that satisfies the design assumptions, the design method should not be used.

Step 2:   Determine backfill properties ϕ and γ

The drained friction angle of the fill material, ϕ (more precisely ϕ'), can be estimated conservatively by an experienced soils engineer or determined by performing appropriate direct shear or triaxial tests. The unit weight, γ, can be estimated or determined by a test. Generally, the unit weight at 95% Standard Proctor relative compaction (i.e., 95% of AASHTO T-99 maximum unit weight) is specified. However, other densities may also be specified as long as the friction angle ϕ is consistent with the density.

Step 3:   Develop a lateral earth pressure diagram due to overburden plus surcharge

The friction angle ϕ determined in Step 2 can be used to calculate the coefficient of earth pressure at rest, $K_0 = 1 - \sin\phi$, and determine the lateral earth pressure diagram along the height of the wall due to overburden (soil self-weight) plus surcharge (see Figure 5.18). The lateral earth pressure at depth z (measured from the crest) is:

$$\sigma_{h(0)} = K_0 \left(\gamma z + q \right) \tag{5-16}$$

where K_0 is the coefficient of earth pressure at rest, and q is the vertical surcharge pressure uniformly applied on the wall crest.

Step 4:   Develop a lateral earth pressure diagram due to live loads

Boussinesq theory allows determination of stress at any point within a semi-infinite elastic soil mass due to a single point vertical load applied on the crest (the top surface of the soil mass). Boussinesq theory also allows lateral stress within a soil mass to be determined, which can be regarded as being equal to lateral earth pressure on a virtual wall face. Boussinesq theory is used in the USFS design method to determine lateral earth pressure due to loads applied on the crest of a reinforced soil wall.

For a concentrated vertical live load P applied on the crest, the lateral earth pressure at depth z on the wall face can be evaluated by the Boussineq equation:

$$\sigma_{h(l)} = \frac{P x^2 z}{R^5} \tag{5-17}$$

where $\sigma_{h(l)}$ is the lateral earth pressure on the wall face at depth z due to a concentrated vertical live load P applied at a point with a horizontal distance x from the wall face (distance measured in the orientation perpendicular to wall face), and R is the radial distance from the point where P is applied to the point where $\sigma_{h(l)}$ is being calculated. The lateral pressure profile due to live loads, $\sigma_{h(l)}$, is commonly evaluated at 0.6-m (2-ft) vertical intervals (i.e., $z = 0.6$ m, 1.2 m, or 2 ft, 4 ft, etc.) along the height of the wall. The lateral pressure profile due to live loads (see Figure 5.18) can be obtained from the influence diagram shown in Figure 2.11. The diagrams are *conservative* (see Section 2.2.6) for evaluation of $\sigma_{h(l)}$. When multiple live loads are present, $\sigma_{h(l)}$ can be obtained by superimposing earth pressure due to each live load. An example for calculating lateral earth pressure due to an eight-wheel 40-kip (178 kN) dual tandem axle truck can be found in Steward et al. (1977) and Koerner (2005), and is given (with modifications) in Section 5.4.2. Normally, more than one section along the wall should be checked to

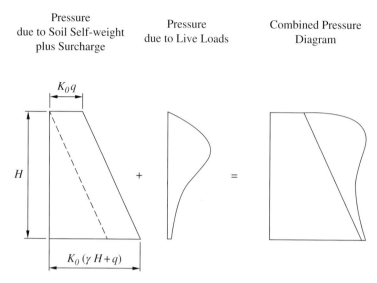

Figure 5.18 Combined lateral earth pressure on facing due to soil self-weight plus surcharge and live loads, the USFS design method

determine the most critical section. The values of live loads P can be determined as the larger value between 1.5 × (legal loads) and 1.2 × (heavy loads).

Step 5:   **Develop a combined lateral earth pressure diagram**
The lateral earth pressure diagrams determined from Steps 3 and 4 are superimposed to form a combined lateral earth pressure diagram, similar to that shown in Figure 5.18.

Step 6:   **Determine the vertical spacing of reinforcement layers**
The vertical spacing between reinforcement layers, s_v, is determined as:

$$s_v = \frac{T_{ult}}{F_s \sigma_h} \tag{5-18}$$

where T_{ult} is the ultimate strength of the reinforcement; F_s is the factor of safety, which should at least be 1.2–1.5, depending on the confidence level in the value of T_{ult}; and σ_h is lateral earth pressure at the mid-depth of a layer, as obtained from the combined pressure diagram (Step 5). The value of T_{ult} may be specified by either Method A, the wide cut strip tensile test, or Method B, a combination of grab tensile and 25-mm (1-in.) cut strip tests. The tests should be performed with the reinforcement in its weakest principal direction, and Method A is generally preferred over Method B.

To account for long-term creep potential, a reduction factor should be applied to the tensile strength obtained from the tests. The value of the reduction factor depends on the test method, the polymer type, and style of the reinforcement, as follows:

Polymer Type and Style	Method A	Method B
Polyester needled	0.7	1.0
Polypropylene needled	0.55	0.8
Polypropylene bonded	0.4	0.6
Polypropylene woven	0.25	0.4

When Method A is used for determination of T_{ult}, the following conditions should be observed:

- The aspect ratio (the ratio of width to gauge length) of the test specimen should be 2 or greater. The minimum gauge length (between grips of test specimen) should be 100 mm (4 in.).
- The test should be performed at a constant strain rate of 10% per minute.
- The test should be performed at 65 ± 2% relative humidity and 21 ± 1°C (70 ± 2°F) temperature.
- The test specimen should be soaked in water for at least 12 hours and maintained surface damp during the test.
- The grips used in the test should not weaken the specimen and should be able to hold the specimen without slippage. Tests which fail at the grips should be disallowed. If slippage cannot be sufficiently limited, elongation must be measured between points on the specimen rather than between the grips.
- Test results should include applied tensile force per unit width of specimen vs. strain curve, failure load per unit width, and strain at failure.

When using Method B to determine T_{ult}, the smaller value of the following two strengths should be used: (i) 90% of the 25-mm (1-in.) cut strip strength or (ii) 33% of the grab tensile strength.

Step 7: Determine the length of reinforcement required to develop pullout resistance

As shown in Figure 5.19, the total length of reinforcement, L, required to prevent pullout failure from occurring is equal to the sum of the anchored length behind the potential failure plane, L_e, and the length within the potential failure zone, L_f. For a reinforcement located at depth z below the crest,

$$L = L_e + L_f \tag{5-19}$$

in which

$$L_f = (H - z)\tan\left(45° - \frac{\phi}{2}\right) \tag{5-20}$$

$$L_e = \frac{K_0\, s_v\, F_s}{2\tan\left(\frac{2}{3}\phi\right)} \tag{5-21}$$

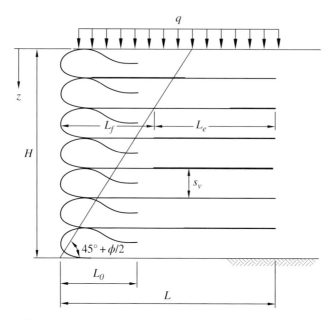

Figure 5.19 Notations for describing the USFS design method

where H is the wall height and z is the depth of the reinforcement layer being considered. The safety factor against pullout failure, F_s, should be at least 1.5–1.75. A minimum value of $L_e = 0.9$ m (3 ft) should be used.

When different soils are used above and below a reinforcement layer, the equation for calculating L_e is modified as:

$$L_e = \frac{K_0\, s_v\, F_s}{\tan\left(\dfrac{2}{3}\phi_a\right) + \tan\left(\dfrac{2}{3}\phi_b\right)} \qquad (5\text{-}22)$$

where ϕ_a and ϕ_b are the friction angles of the soils above and below the reinforcement layer, respectively.

Following the equations above, the reinforcement layers near the base can be shorter than near the top to satisfy the internal stability requirement of a reinforced soil structure. However, due to external stability considerations (Step 9), particularly with respect to sliding and bearing capacity, and to avoid confusion in the actual rein-forcement placement sequence in the field, all reinforcement layers are usually taken as being of a uniform length.

Step 8: **Determine the wrapped length of reinforcement into the wall face**
The wrapped length of reinforcement into the wall face, L_o (see Figure 5.19) can be determined as:

$$L_o = \frac{\sigma_h s_v F_s}{2(\gamma z + q)\tan\left(\frac{2}{3}\phi\right)}$$

(5-23)

The minimum value of safety factor, F_s, is 1.2–1.5. The minimum value of L_o is 0.9 m (3 ft). It is very important to abide by this minimum L_o value.

Step 9:   Check external stability
The external stability of the reinforced soil wall against overturning failure, base sliding, bearing failure, and overall slope failure should be checked. The checks should be carried out by treating the reinforced soil mass as a rigid wall.

- For overturning failure check, stabilizing and destabilizing overturning moments about the toe of the wall should be calculated to determine the safety factor against overturning failure. Typically, a minimum safety factor of 1.5 is needed.

- For base sliding failure check, stabilizing and destabilizing forces along the base of the wall should be calculated to determine the safety factor against sliding failure. Typically, a minimum safety factor of 1.5 is needed.

- For bearing failure check, the applied resultant load (typically an inclined resultant load) at the base of the wall (due to self-weight of the wall, externally applied loads on the crest, and lateral earth thrust against the reinforced soil mass) and the load-carrying capacity of the ground beneath the base of the wall should be calculated (by using the bearing capacity equations for shallow foundation, e.g., NAVFAC, 1986; Terzaghi et al., 1996) to determine the safety factor against bearing failure. Typically, a minimum safety factor of 3.0 is needed.

- For overall slope failure check, slope stability software is usually needed to evaluate the safety factors of the different assumed slip planes of the wall, involving soil behind and below the reinforced soil mass, and determine the minimum safety factor against overall slope failure. Typically, a minimum safety factor of 1.3 is needed.

If any of the required minimum safety factors are not met, one of the following two measures (or combination of the measures) could be taken to achieve required safety margins: (i) revise the design, such as decrease reinforcement spacing, increase reinforcement length, decrease wall height, etc., or (ii) improve the soil, including use of better reinforced fill material, improve fill compaction, improve the foundation soil, etc.

5.4.2   Design Example: The U.S. Forest Service Method

Given Conditions:

A typical cross-section of a wrapped-faced geotextile-reinforced soil retaining wall is shown in Figure 5.20.

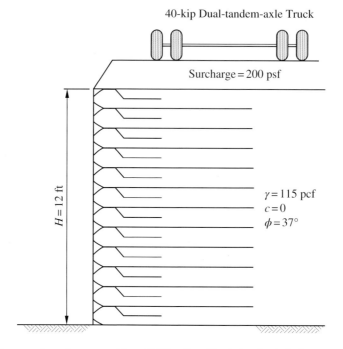

Figure 5.20 Cross-section of a wrapped-faced GRS wall and loads involved, design example of the USFS design method

Features of wall:

- uniform wall height: H = 12 ft
- near-vertical wall
- all geosynthetic reinforcement layers have the same length
- no concern for erosion or other unfavorable events at the wall base during design life, no embedment (wall constructed directly on the ground surface).

Features of fill material:

- same soil to be used in reinforced zone and *retained earth* (i.e., same soil in and behind the reinforced zone)
- soil is granular and free-draining, with moist unit weight γ = 115 lb/ft^3, c = 0, and ϕ = 37° at 95% of AASHTO T-99 maximum density; placement density and moisture of the backfill in the reinforced zone are similar to those in retained earth
- surface drainage and subsurface drainage are to be properly provided.

Features of foundation soil:

- dense sandy gravel with a uniform blow count (N) of 45
- deep free water level.

Features of *trial* reinforcement type:

- polyester needled geotextile with ultimate strength, T_{ult} = 2,520 lb/ft, determined by the wide cut strip tensile test (Method A, Step 6 of Section 5.4.1).

Features of loading:

- vertical surcharge is uniformly distributed over the crest, q = 200 lb/ft^2
- 40-kip live load on the crest due to a dual-tandem-axle truck whose wheel dimensions and positions with respect to the wall face are shown in Figure 5.21
- no concern for seismic loads.

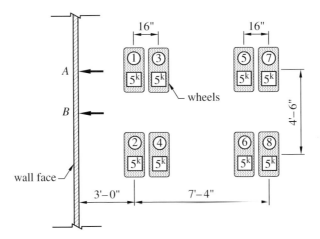

Figure 5.21 Footprint of live loads on a wall crest due to a 40-kip dual-tandem-axle truck (modified after Koerner, 2005)

Design Computations:

Step 1: **Establish wall profile and check design assumptions**

A typical cross-section of the wall is shown in Figure 5.20. The design assumptions listed in Step 1 of Section 5.4.1 are verified, including granular backfill and a firm foundation.

Step 2: **Determine backfill properties ϕ and γ**

The backfill is granular, with moist unit weight γ = 115 lb/ft^3, c' = 0 and ϕ' = 37° at 95% of AASHTO T-99 maximum density.

Step 3: **Develop a lateral earth pressure diagram due to soil self-weight plus surcharge**

The coefficient of earth pressure at-rest, $K_0 = 1 - \sin\phi = 1 - \sin(37°) = 0.4$

The lateral earth pressure at depth z due to self-weight and surcharge is:

$$\sigma_{h(o)} = K_0(\gamma z + q) = 0.4(115 \times z + 200) \quad (\text{unit}: \text{lb/ft}^2)$$

This lateral earth pressure is tabulated under $\sigma_{h(o)}$ in Table 5.4 and plotted in Figure 5.22.

Table 5.4 Calculations of required reinforcement length for the design example of the USFS design method

Depth (ft)	$\sigma_{h(o)}$ (psf)	$\sigma_{h(l)}$* (psf)	σ_h** (psf)	S_v (ft)	L_e (ft)	L_f (ft)	L*** (ft)	L_o (ft)
0	80	0	80	14.7	3.0	6.0	9.0	0.8
2	172	137	309	3.8	3.0	5.0	8.0	1.4
4	264	181	445	2.6	3.0	4.0	7.0	1.3
6	356	134	490	2.4	3.0	3.0	6.0	1.1
8	448	85	533	2.2	3.0	2.0	5.0	0.9
10	540	46	586	2.0	3.0	1.0	4.0	0.9
12	632	24	656	1.8	3.0	0	3.0	0.8

Note: *Obtained from Table 5.5 (Section A in Figure 5.21).
$**\sigma_h = \sigma_{h(o)} + \sigma_{h(l)}$
$***L = L_e + L_f$

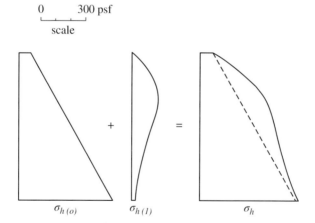

Figure 5.22 Profiles of lateral earth pressures, design example of the USFS design method

Step 4: Develop a lateral earth pressure diagram due to live loads

The lateral earth pressure due to the eight-wheel live load of 40 kips can be determined conservatively with the aid of the lateral earth pressure influence diagram shown in Figure 2.11. The earth pressure due to the live load in 2-ft increments of the wall height is evaluated at two selected vertical sections: Sections A and B (see Figure 5.21; Table 5.5). The earth pressure is found to be larger in Section A than in Section B, thus the pressure at Section A is used in the design. The earth pressure due to the live load is tabulated under $\sigma_{h(l)}$ in Table 5.4 and plotted in Figure 5.22.

Table 5.5 Lateral earth pressure due to live load for the design example of the USFS design method

(a) Section A (Figure 5.21)

Depth (ft)	Wheel Number								$\sigma_{h(l)}$ (psf)
	1	2	3	4	5	6	7	8	
0	0	0	0	0	0	0	0	0	0
2	45.1	9.9	45.1	18.0	6.9	5.3	3.5	2.8	136.6
4	55.5	12.2	59.0	23.6	10.4	7.9	6.9	5.6	181.1
6	34.7	7.6	34.7	13.9	13.8	10.5	10.4	8.4	134.0
8	20.8	4.6	20.8	8.3	10.4	7.9	6.9	5.6	85.3
10	10.4	2.3	10.4	4.2	6.9	5.2	3.5	2.8	45.7
12	3.5	0.8	5.2	2.1	3.4	2.6	3.5	2.8	23.9

Note: σ_h (from wheel 2) = $0.22\sigma_h$ (from wheel 1); $\theta = 56.3°$
σ_h (from wheel 4) = $0.40\sigma_h$ (from wheel 3); $\theta = 46.1°$
σ_h (from wheel 6) = $0.76\sigma_h$ (from wheel 5); $\theta = 26.5°$
σ_h (from wheel 8) = $0.81\sigma_h$ (from wheel 7); $\theta = 23.5°$

(b) Section B (Figure 5.21)

Depth (ft)	Wheel Number				$\sigma_{h(l)}$ (psf)
	1 & 2	3 & 4	5 & 6	7 & 8	
0	0	0	0	0	0
2	23.4	31.6	6.4	3.3	129.4
4	28.9	41.3	9.7	6.4	172.6
6	18.0	24.3	12.8	9.7	129.6
8	10.8	14.6	9.7	6.4	83.0
10	5.4	7.3	6.4	3.3	44.8
12	1.8	3.6	3.2	3.3	23.8

Note: σ_h (from wheels 1 & 2) = $0.52\sigma_h$ (from wheel 1 @ Section A); $\theta = 39.8°$
σ_h (from wheels 3 & 4) = $0.70\sigma_h$ (from wheel 3 @ Section A); $\theta = 30.2°$
σ_h (from wheels 5 & 6) = $0.93\sigma_h$ (from wheel 5 @ Section A); $\theta = 15.5°$
σ_h (from wheels 7 & 8) = $0.93\sigma_h$ (from wheel 7 @ Section A); $\theta = 13.6°$

Step 5: **Develop a combined lateral earth pressure diagram**

A combined earth pressure diagram is obtained by superimposing the earth pressures obtained in Steps 3 and 4. The combined earth pressure along the wall height is tabulated under σ_h in Table 5.4 and plotted in Figure 5.22.

Step 6: Determine the vertical spacing of reinforcement layers

For the trial reinforcement (a polyester needled geotextile), the ultimate strength is reduced by a factor of 0.7, per Step 6 of Section 5.4.1, thus $T_{ult} = 0.7\,(2{,}520) = 1{,}760$ (lb/ft). Using a safety factor F_s of 1.5, the vertical spacing, s_v, is:

$$s_v\,(\text{ft}) = \frac{1{,}760}{(1.5)\sigma_h}$$

The required minimum vertical spacing along the wall height is tabulated in Table 5.4. A uniform vertical spacing of 1.5 ft (or 18 in.) is selected.

Step 7: Determine the length of reinforcement

With $F_s = 1.5$, the required anchored length behind the potential failure plane, L_e, and the length within the potential failure wedge, L_f, are:

$$L_e = \frac{K_o\,s_v\,F_s}{2 \times \tan\left(\dfrac{2}{3}\phi\right)} = \frac{0.4 \times 1.5 \times 1.5}{2 \times \tan\left(\dfrac{2}{3} \times 37°\right)} = 1.0\,(\text{ft})$$

$$L_f = (H-z)\tan\left(45° - \frac{\phi}{2}\right) = (12-z)\tan\left(45° - \frac{37°}{2}\right)$$

Note that the length L_e must at least be 0.9 m (3 ft), thus $L_e = 0.9$ m (3 ft) is selected. The values of L_e and L_f as well as the total reinforcement length L ($L = L_e + L_f$) along the wall height are listed in Table 5.4. A uniform length of 2.7 m (9 ft) is selected for all reinforcement layers.

Step 8: Determine the wrapped length of reinforcement

The wrapped length, L_o, into the wall face is:

$$L_o = \frac{\sigma_h\,s_v\,F_s}{2(\gamma z + q)\tan\left(\dfrac{2}{3}\phi\right)} = \frac{\sigma_h \times 1.5 \times 1.2}{2\,(115 \times z + 200)\tan\left(\dfrac{2}{3} \times 37°\right)}$$

The values of L_o along the wall height are also listed in Table 5.4. All the L_o values are found to be less than the minimum required length of 0.9 m (3 ft), therefore L_o of 0.9 m (3 ft) is selected for all reinforcement layers.

Step 9: Check external stability

External stability against overturning failure, base sliding, bearing failure, and overall slope failure need to be checked before accepting the design.

5.5 The AASHTO Allowable Stress Design (ASD) Method

The Standard Specifications for Highway Bridges published by the American Association of State Highway and Transportation Officials (AASHTO) provides design guidelines for mechanically stabilized earth (MSE) walls (AASHTO 2002).

Different assumptions of earth pressure distribution are made for extensible (polymeric) and inextensible (metallic) reinforcements. For walls with extensible reinforcements such as geosynthetics, the earth pressure distribution is assumed to follow the active Rankine theory. Other than the earth pressure distribution, the allowable stress design (ASD) approach in the AASHTO design guidelines is similar to the U.S. Forest Service method described in Section 5.4. Of note is that the AASHTO Standard Specifications give only general design guidelines, rather than clearly defined design procedure. The design procedure is to some extent subject to the user's interpretation. The method presented in this section should be considered as a design method that is based on the AASHTO guidelines. The design method presented here is of the ASD approach.

5.5.1 Design Procedure: The AASHTO ASD Method

<u>Step 1:</u> **Determine the friction angle (ϕ) and unit weight (γ) of the fill in and behind the reinforced zone**

The friction angle (ϕ) should be estimated conservatively by an experienced soils engineer or determined by performing direct shear tests (AASHTO T236-72) or triaxial tests (AASHTO T234-74) on project-specific fill materials. The soil strength should be evaluated at residual stress levels. In the absence of reliable project-specific soil data, maximum friction angles of 30° and 34° should be used for external stability analysis and internal stability analysis, respectively.

<u>Step 2:</u> **Determine the depth of embedment**

Minimum embedment at the front face of the wall (measured from the adjoining finished ground surface to the bottom of footings) should be at least 2 ft (0.6 m), and greater than the maximum frost penetration depth at the wall site. In addition, the following requirements must be satisfied (H = wall height, measured from the bottom of the lowest block to the top of the fill behind the facing):

- For a horizontal ground surface in front of the wall: minimum embedment = H(ft)/20 (for walls) and H(ft)/10 (for abutments).

- For a sloping ground surface in front of the wall: minimum embedment = H(ft)/10 for a 3H:1V slope, H(ft)/7 for a 2H:1V slope, and H(ft)/5 for a 1.5H:1V slope, and there needs to be a minimum horizontal bench of 4 ft (1.2 m) in front of the wall.

For walls constructed along rivers and streams, the foundation depth must be at least 2 ft (0.6 m) below the potential scour depth.

<u>Step 3:</u> **Select a tentative reinforcement length**

A tentative reinforcement length of at least $0.7H$ (H = wall height measured from the bottom of the lowest block to the top of the fill behind the facing near the top of the wall) should be selected, and no less than 8 ft (2.4 m) for strip- or grid-type reinforcement. Unless substantial evidence is presented to justify variation in reinforcement length, the length of reinforcement should be uniform throughout the entire wall.

Step 4: Determine the forces acting on the reinforced soil mass

The reinforced soil mass is treated as a rigid body for external stability computations. All forces acting on the reinforced soil mass, including body force due to self-weight of the reinforced mass and forces acting on the reinforced soil mass, should be determined. The forces for walls with a level crest and an inclined crest are shown in Figures 5.23(a) and (b), respectively.

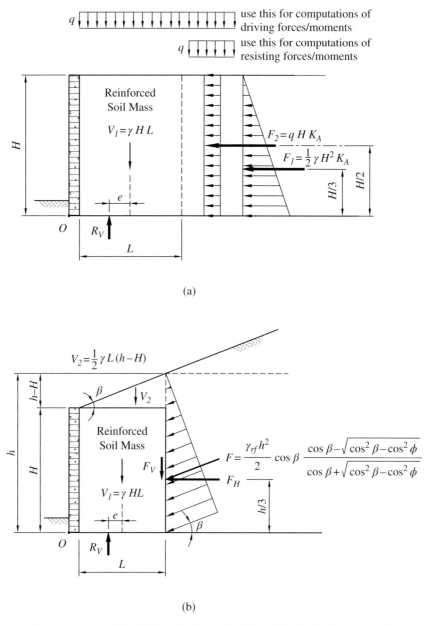

(a)

(b)

Figure 5.23 Forces acting on the reinforced soil mass for (a) a wall with a level crest and (b) a wall with an inclined crest (modified after AASHTO, 2002)

Step 5: **Check stability against overturning**

For walls with a level crest, the safety factor against overturning (about point O, Figure 5.23(a)) is calculated as:

$$FS_{overturning} = \frac{\text{resisting moment }(M_r)}{\text{driving moment }(M_d)} = \frac{V_1\left(\dfrac{L}{2}\right)}{F_1\left(\dfrac{H}{3}\right) + F_2\left(\dfrac{H}{2}\right)} \tag{5-24}$$

As shown in Figure 5.23(a), L is the length of the reinforced zone, H is the wall height, V_1 is the body force (self-weight) of the reinforced soil mass, F_1 is lateral force due to the fill behind the reinforced mass, and F_2 is lateral force due to a uniform traffic surcharge load, q, which is assumed to act only beyond the reinforced zone in computations involving resisting forces/moments for external stability check.

For walls with an inclined crest, the safety factor against overturning (about point O) can be calculated as:

$$FS_{overturning} = \frac{\text{resisting moment }(M_r)}{\text{driving moment }(M_d)} = \frac{V_1\left(\dfrac{L}{2}\right) + V_2\left(\dfrac{2L}{3}\right) + F_V\left(L\right)}{F_H\left(\dfrac{h}{3}\right)} \tag{5-25}$$

As shown in Figure 5.23(b), L is the length of the reinforced zone, H is the wall height, h is the height of the fill above the end of the reinforcement, V_1 and V_2 are body forces of the reinforced soil mass, and F_H and F_V are, respectively, the horizontal and vertical components of the lateral thrust F due to fill behind the reinforced mass.

The minimum value of $FS_{overturning}$ should be at least 2.0 for all walls. If the safety margin is found insufficient, the safety factor may be increased by increasing the reinforcement length.

Step 6: **Check stability against base sliding**

For walls with a level crest, the safety factor against sliding is calculated as:

$$FS_{sliding} = \frac{V_1 \tan\delta}{(F_1 + F_2)} \tag{5-26}$$

Note that the uniform traffic surcharge load, q, is assumed to act only beyond the reinforced zone in the computation of resistance to base sliding in the external stability check.

For walls with an inclined crest, the safety factor against sliding is calculated as:

$$FS_{sliding} = \frac{(V_1 + V_2)\tan\delta}{F_H} \tag{5-27}$$

where parameter δ is the friction angle between the backfill and foundation. The value of δ can be assumed to be the internal friction angle of the backfill or the foundation soil, whichever is smaller. The value of $FS_{sliding}$ must be at least 1.5. If the safety margin is found to be insufficient, the safety factor may be increased by increasing the roughness at the base of the wall (i.e., at the backfill–foundation contacting surface) or by increasing the reinforcement length.

Step 7:   Check overall stability of slopes in the vicinity of the wall

Overall slope stability encompassing the entire wall and the stability of the inclined sloping crest (if present) should be checked by using the limiting equilibrium methods of analysis, such as the modified Bishop, simplified Janbu, or Spencer methods of analysis. The recommended minimum factors of safety are listed in Table 5.6. If the safety margin is found to be insufficient, the slope should be stabilized by an appropriate method (e.g., TRB, 1996).

Table 5.6 Minimum factors of safety for overall stability of slopes in the vicinity of the wall

Determination of Soil and Rock Parameters and Ground Water Levels	Minimum Safety Factor	
	Retaining Wall	Abutment Wall
Determined based on In Situ and/or Laboratory Tests	1.3	1.5
Not Determined Based on In Situ and/or Laboratory Tests	1.5	1.8

Step 8:   Check bearing capacity of wall foundation

Bearing capacity of the wall foundation should be checked by considering the bottom of the reinforced soil mass as a rigid footing. The footing has a width of L, which is the length of the reinforcement at the foundation level, and is subjected to a concentrated eccentric load. The eccentricity, e, measured from the centerline of the reinforced soil mass (see Figure 5.23), can be calculated as:

$$e = \frac{L}{2} - \frac{M_R - M_d}{R_V} \tag{5-28}$$

where M_r and M_d are, respectively, resisting moment and driving moment for overturning of the reinforced soil mass (as determined in Step 5), and R_V is the resultant of vertical forces ($R_V = V_1 + V_2 + F_V$; for walls with a level crest, $V_2 = F_V = 0$). The eccentricity must stay within the middle third of the reinforced soil mass, i.e.,

$$e \leq \frac{L}{6} \tag{5-29}$$

The bearing pressure, $\sigma_v = R_V/(L - 2e)$, must have a safety factor of at least 2, i.e.,

$$FS_{bearing} = \frac{q_{ult}}{\sigma_v} \geq 2.0 \tag{5-30}$$

If the safety margin is found to be insufficient, the safety factor may be increased by improving the foundation soil (using techniques such as compaction, staged construction, preloading, wick drain, grouting, etc.) or by increasing the reinforcement length.

Step 9: **Select reinforcement and determine the limit state tensile load and serviceability tensile load**

Upon selecting a particular reinforcement for the wall, a series of controlled laboratory creep tests will need to be conducted. The tests should be performed for a minimum duration of 10,000 hours for a range of load levels on samples of the selected reinforcement. The samples should be tested in the anticipated loading direction, in either a confined or an unconfined condition, and typically at an assumed in-ground temperature of 70°F. The test results are to be extrapolated over the required design life using the procedures outlined in ASTM D2837.

From the laboratory creep tests, two tensile loads may be determined: limit state tensile load (T_{limit}) and serviceability state tensile load ($T_{service}$). The limit state tensile load is the highest load level at which the log time vs. creep-strain rate relationship continues to decrease with time within the design life without inducing either brittle or ductile failure. The serviceability state tensile load, on the other hand, is the load level at which total strain will not exceed 5% within the design life. The design life is typically 75 years. Designated critical walls, however, may be designed for a 100-year service life.

Step 10: **Determine the allowable tensile load of reinforcement**

The allowable tensile load in the reinforcement, T_a, is the lesser of two allowable tensile loads at limit state (T_{al}) and at serviceability state (T_{as}). The allowable limit state tensile load, T_{al}, is determined as:

$$T_{al} = \frac{T_{limit}}{F_D \cdot F_C \cdot F_S} \tag{5-31}$$

where T_{limit} is the limit state tensile load determined in Step 9; F_D is a durability safety factor to account for such effects as aging, chemical degradation, biological degradation, temperature variations, environmental stress cracking, and plasticization; F_C is a construction damage safety factor; and F_S is an overall safety factor to account for uncertainties in structure geometry, fill properties, reinforcement manufacturing variations, and externally applied loads. The minimum value of F_S is 1.5, whereas F_D and F_C should be determined by tests simulating the anticipated field conditions. If the facing of a wall is composed of dry stacked concrete blocks and facing connection strength is lower than T_{al}, the allowable tensile load at limit state (T_{al}) should be reduced further to account for possible connection failure.

On the other hand, the allowable serviceability state tensile load, T_{as}, is determined as:

$$T_{as} = \frac{T_{service}}{F_C \cdot F_D} \tag{5-32}$$

where $T_{service}$ is determined in Step 9, F_D is the durability safety factor, and F_C is the construction damage safety factor.

Step 11:   Develop a combined lateral earth pressure diagram due to overburden pressure, surcharge, and live loads

Similar to Step 5 of the U.S. Forest Service method (Section 5.4.1), the lateral earth pressure diagrams due to gravity, uniform normal surcharge, and live loads are superimposed to form a combined lateral earth pressure diagram.

Step 12:   Determine vertical spacing of reinforcement layers

Vertical spacing between reinforcement layers, s_v, is determined as:

$$s_v = \frac{T_a}{\sigma_h}$$

(5-33)

where T_a is the allowable tensile load determined in Step 10 and σ_h is the lateral earth pressure at the mid-depth of the reinforcement layer, obtained from the combined pressure diagram determined in Step 11.

For walls of low to medium height (say, 10–15 ft or 3.0–4.5 m), a uniform vertical spacing is generally selected for all reinforcement layers. In such a case, the value of s_v only needs to be evaluated at the depth where σ_h is largest. If the value of s_v turns out to be too small (say, less than the typical block height of 8 in. or 0.2 m), a different reinforcement with a larger allowable tensile load may be selected.

Step 13:   Check stability against pullout failure

Similar to the USFS method shown in Figure 5.19, the anchored length behind the potential failure plane (L_e) needed to prevent pullout failure from occurring is equal to the total reinforcement length (L) minus the length within the potential failure zone. For a reinforcement layer at depth z below the crest,

$$L_e = L - (H - z)\tan\left(45° - \frac{\phi}{2}\right)$$

(5-34)

where H is wall height and z is the depth of the reinforcement layer being considered.

The safety factor against pullout failure for the reinforcement layer being considered is evaluated as:

$$FS_{pullout} = \frac{\text{resisting shear force along } L_e}{\text{lateral earth thrust of the layer}} = \frac{2f\sigma_v L_e}{K_A \sigma_v s_v} = \frac{2f L_e}{K_A s_v}$$

(5-35)

where σ_v is the vertical stress at the depth of the reinforcement and f is the friction coefficient between backfill and reinforcement; $f = \tan\delta$, in which δ = friction angle at the backfill–reinforcement interface. In the absence of reliable test data, δ can be taken as 2/3 of the internal friction angle of the backfill, ϕ. The safety factor against pullout for every reinforcement layer as determined from Eqn. (5-35) must be at least 1.5, and the anchored length L_e must be at least 3 ft. If the safety factor is less than 1.5, the reinforcement length may be increased.

5.5.2 Design Example: The AASHTO ASD Method

Given Conditions:

A geotextile reinforced soil wall is situated near Chicago, Illinois. The cross-section of the wall is shown in Figure 5.24.

Features of wall:
- uniform wall height of 12 ft above finished ground surface
- vertical wall
- all geosynthetic reinforcement layers are to have the same length
- ground surface is horizontal in front of the wall.

Features of backfill and retained earth:
- same soil to be used for the reinforced zone and the *retained earth* (the soil behind reinforced zone)
- soil is granular, with moist unit weight $\gamma = 115$ lb/ft^3, $c' = 0$, and $\phi' = 37°$ at 95% of AASHTO T-99 maximum density; placement density and moisture of the backfill in the reinforced zone are similar to those in the retained earth
- surface drainage on top of the wall and subsurface drainage behind and beneath fill material will be properly provided during construction and during service life.

Features of foundation soil:
- dense sandy gravel with a uniform blow count (N) of 30
- deep free water level
- maximum frost penetration depth = 4 ft.

Features of a *trial* reinforcement:
- polypropylene woven geotextile
- for a design life of 75 years, the geotextile reinforcement has a limit state tensile load of 2,500 lb/ft and a serviceability state tensile load of 1,500 lb/ft, as determined from a series of laboratory creep tests.

Features of loading:
- surcharge load (including traffic load), $q = 250$ lb/ft^2
- no concern for seismic loads.

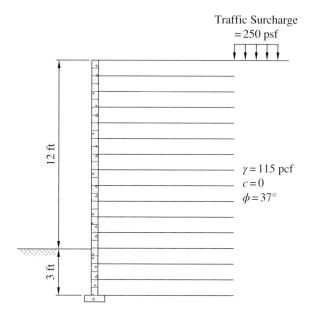

Figure 5.24 Cross-section of a reinforced soil wall, design example of the AASHTO ASD method, for calculations of resisting forces/moments

Design Computations

Step 1: **Determine friction angle, ϕ', and unit weight, γ, for the fill in and behind the reinforced zone**

The fill has a moist unit weight $\gamma = 115$ lb/ft³ and $c' = 0$, $\phi' = 37°$.

Step 2: **Determine depth of embedment**

Consider the following three conditions: (i) minimum embedment depth, measured from the finished ground surface to the bottom of the wall footings, should be at least 2 ft; (ii) minimum embedment for horizontal ground surface in front of the wall = $H/20$, i.e., 0.6 ft; and (iii) the embedment depth of the wall, measured from the finished ground surface to the bottom of the wall footings, should not be less than the maximum frost penetration depth at the wall site (4 ft). For 1-ft thick footings, the wall height (H), measured from the leveling pad of the footings, will be $12 + 4 - 1 = 15$ (ft).

Step 3: **Select a tentative reinforcement length**

A tentative length of $0.7H$, or 10.5 ft, is selected for the entire wall.

Step 4: **Determine the forces acting on the reinforced soil mass**

As depicted in Figure 5.25, the reinforced soil mass is subjected to three resultant forces: (i) weight of the reinforced soil mass, $V_1 = 18,100$ lb/ft, (ii) lateral earth thrust due to retained earth, $F_1 = 3,234$ lb/ft, and (iii) lateral force due to traffic surcharge, $F_2 = 938$ lb/ft.

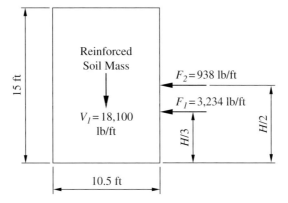

Figure 5.25 Forces acting on the reinforced soil mass, design example of the AASHTO ASD method

Step 5: **Check stability against overturning**
The factor of safety against overturning is calculated as:

$$FS_{overturning} = \frac{\text{resisting moment} \left(M_r \right)}{\text{driving moment} \left(M_d \right)} = \frac{18,100 \left(\dfrac{10.5}{2} \right)}{938 \left(\dfrac{15}{3} \right) + 3,234 \left(\dfrac{15}{2} \right)} = \frac{95,000}{28,900} = 3.29$$

The safety factor is greater than the minimum required value of 2.0 (OK).

Step 6: **Check stability against base sliding**

$$FS_{sliding} = \frac{18,100 \times \tan 37°}{938 + 3,234} = 3.27$$

The safety factor is greater than the minimum required value of 1.5 (OK).

Step 7: **Check overall stability of slopes in the vicinity of the wall**
Using the simplified Bishop method for stability analysis of slopes encompassing the entire wall, the minimum safety factor against overall slope failure can be determined. The safety factor needs to be at least 1.5.

Step 8: **Check the bearing capacity of wall foundation**
The eccentricity measured from the centerline of the reinforced soil mass is:

$$e = \frac{10.5}{2} = \frac{95,000 - 28,900}{18,100} = 1.60 \left(\text{ft} \right)$$

Alternatively,

$$e = \frac{M_d}{V_1} = \frac{28,900}{18,100} = 1.60 \text{ (ft)}$$

The value e is less than $L/6 = 1.75$ ft (OK).

The applied bearing pressure $\sigma_v = 18,100/[10.5/2 - 2(1.60)] = 2,480$ (lb/ft^2). For blow count (N) = 30, $\phi = 36°$, hence $N_\gamma = 50$ and $N_q = 32$. Therefore, the ultimate bearing capacity is:

$$q_{ult} = \frac{\gamma B'}{2} N_\gamma + \gamma D_f N_q$$

$$= \left(\frac{1}{2}\right) 115 \left[10.5 - 2(1.6)\right](50) + 115(3)(32) = 32,000 \text{ (psf)}$$

Thus,

$$FS_{bearing} = \frac{q_{ult}}{\sigma_v} = 12.9$$

The bearing capacity safety factor is greater than the minimum required value of 2.0 (OK).

Step 9: **Select reinforcement and determine the limit state tensile load and serviceability tensile load**
A polypropylene woven geotextile is selected as the reinforcement. For a design life of 75 years, the geotextile reinforcement has a limit state tensile load $T_{limit} = 2,500$ lb/ft and a serviceability state tensile load $T_{service} = 1,500$ lb/ft, as determined from a series of controlled laboratory creep tests.

Step 10: **Determine the allowable tensile load of reinforcement**
The allowable limit state tensile load, T_{al}, is determined as:

$$T_{al} = \frac{T_{limit}}{F_D \cdot F_C \cdot F_S} = \frac{2,500}{(2.5)(1.2)(1.5)} = 560 \text{ (lb/ft)}$$

The allowable serviceability state tensile load, T_{as}, is determined as:

$$T_{as} = \frac{T_{service}}{F_C \cdot F_D} = \frac{1,500}{(2.5)(1.2)} = 500 \text{ (lb/ft)}$$

Thus, the allowable tensile load, $T_a = 500$ lb/ft.

Step 11: **Develop a combined lateral earth pressure diagram due to overburden pressure, surcharge, and live loads**
The lateral earth pressure at depth z is:

$$\sigma_h = K_A\left(\gamma\,z+q\right)=0.25\left(115\times z+250\right) \qquad \left(\text{unit}:\text{lb/ft}^2\right)$$

where the coefficient of active earth pressure, $K_A = \dfrac{1-\sin\phi}{1+\sin\phi} = \dfrac{1-\sin\left(37^\circ\right)}{1+\sin\left(37^\circ\right)} = 0.25.$

In the absence of live loads, the maximum lateral earth pressure will occur at the base of the wall (i.e., at $z = 15$ ft), which is equal to 494 lb/ft^2.

Step 12: **Determine the vertical spacing of reinforcement layers**
Select uniform spacing for all reinforcement layers,

$$S_v = \frac{T_a}{\sigma_h} = \frac{500}{494} = 1.0 \ \left(\text{ft}\right)$$

Step 13: **Check stability against pullout failure**
The smallest anchored length behind the potential failure surface is at the uppermost reinforcement layer, for which $L_e = 10.5-(15-1)[\tan(45^\circ-37^\circ/2)] = 3.5$ (ft). This L_e value is greater than the minimum required value of 3.0 ft (OK).

The minimum safety factor against pullout failure is:

$$FS_{pullout} = \frac{2\times\tan\left(\dfrac{2\phi}{3}\right)\times 3.5}{\left(0.25\right)\left(1\right)} = 12.8$$

The safety factor is greater than the minimum required value of 1.5 (OK).

Design Summary:

> Reinforcement: a polypropylene woven geotextile with $T_{limit} = 2{,}500$ lb/ft and $T_{service} = 1{,}500$ lb/ft
>
> Reinforcement length (at all depths) = 10.5 ft
> Reinforcement vertical spacing = 1.0 ft
> Total height of wall measured from leveling pad (from the top of footing) = 15 ft
> Embedment depth (to the bottom of footing) = 4 ft

5.6 The NCHRP Design Method for GRS Bridge Abutments

The NCHRP design method is specifically for design of GRS bridge abutments with flexible facing (Wu et al., 2006). It was developed by using the FHWA NHI reference manual (Elias et al., 2001) as the basic framework, and modified for GRS abutments. The modifications were based on the results of finite element analysis as well as experience gained from actual construction of GRS abutments. The finite element model used in the study was calibrated and validated by measured results of two full-scale experiments conducted for the study. Measured results of a number of other full-scale loading experiments of GRS structures were also used for validation of the analytical model.

The allowable load-carrying pressure of GRS bridge sills in the NCHRP design method (Design Procedure Step 4, Section 5.6.1) were developed based on results of finite element analyses with two performance criteria. One criterion involves a limiting sill settlement, where the allowable bearing pressure is corresponding to a sill settlement of 1.0% of the load-carrying wall height (i.e., 1.0% × H). The other criterion involves distribution of critical shear strain in the reinforced soil mass, where the allowable bearing pressure is corresponding to a limiting condition in which a triangular critical shear strain distribution *reaches* the heel of the sill. The allowable load-carrying pressures were also validated by full-scale experiments.

The NCHRP design method is applicable only to GRS abutments that satisfy the following conditions:

- Total abutment wall height is not greater than 10 m.

- Facing may comprise dry-stacked concrete blocks, timber, natural rocks, wrapped geosynthetic sheets, or gabions—with or without facing connection enhancement elements (pins, lips, or keys) between vertically adjacent facing units.

- No prop or any temporary bracing is used during construction of the abutment wall.

- The backfill meets the following criteria: 100% passing 10 cm (4 in.) sieve, 0–60% passing 0.425 mm (No. 40) sieve, and 0–15% passing 0.075 mm (No. 200) sieve, free from organic material, plasticity index (*PI*) not greater than 6.

- The backfill has an internal friction angle of at least 34°, as determined by the standard direct shear test on the portion finer than 2 mm (No. 10) sieve, using soil samples compacted to 95% of AASHTO T-99, Method C or D, with oversize correction, at the optimum moisture content.

- The backfill in construction is compacted to at least 100% of AASHTO T-99 (i.e., 100% of standard Proctor maximum dry density) or 95% of AASHTO T-180 (i.e., 95% of modified Proctor maximum dry density), and the placement moisture is within ±2% of the optimum.

- The foundation soil is *competent*. The term *competent* is relative to the abutment height and loads on the sill. For a medium height GRS abutment (say, with a total height of 5–8 m) under a maximum allowable sill pressure defined in Step 4 of the design procedure (Section 5.6.1), the foundation is considered *competent* if the in situ undrained shear strength is at least 140 kPa (3,000 lb/ft^2) for a clayey foundation, or the standard penetration blow count is 25 or greater for a non-prestressed granular foundation. A specific check of foundation bearing pressure for a given bridge abutment is performed in Step 7 of the design procedure in Section 5.6.1.

5.6.1   Design Procedure: The NCHRP Method for GRS Abutments

Before using the design method, the conditions described above (immediately before this section) should be checked.

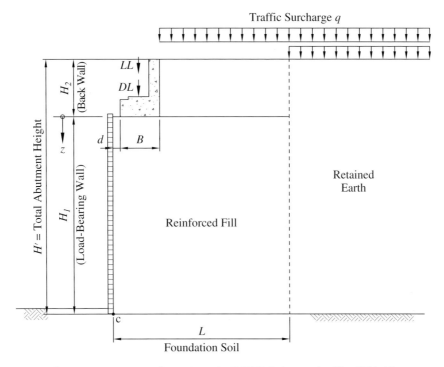

Figure 5.26 Abutment geometry and notations, the NCHRP design method for GRS bridge abutments

Step 1:   Establish design parameters

- Establish abutment geometry and loads (ref. Figure 5.26):
  - the wall is composed of a lower wall (load-bearing wall) and an upper wall (back wall), with concrete block facing
  - total abutment height, H', is the sum of lower wall height and upper wall height
  - load-bearing wall (lower wall) height, H_1, is measured from the base of the embedment to the top of the load-bearing wall
  - back wall (upper wall) height, H_2
  - equivalent traffic surcharge, q
  - bridge vertical dead load, DL
  - bridge vertical live load, LL
  - bridge horizontal load
  - bridge span and type (simple or continuous span)
  - length of approach slab
  - If there is a strong likelihood that erosion or other untoward events at the wall base will not be an issue throughout the design life, embedment of a GRS abutment wall needs only be a nominal depth (say, one

block height, typically 8 in.). If the foundation contains frost susceptible soils (i.e., silts, silty tills, etc.), they should be excavated to at least the maximum frost penetration line and replaced with a non-frost susceptible soil. If the GRS abutment is situated in a stream environment, scour/abrasion/channel protection measures should be undertaken.

- Establish trial design parameters:
  ○ sill width, B (a minimum sill width of 0.6 m (12 in.) is recommended)
  ○ clear distance between the backface of facing and the front edge of sill, d (a clear distance $d = 0.3$ m (6 in.) is recommended; the fill compaction within the clear distance is generally less than the rest of the fill)
  ○ sill type (*isolated sill* or *integrated sill*, the former refers to a sill that is a separate unit from the upper wall of the abutment, whereas the latter refers to a sill that is integrated with the upper wall as a combined unit)
  ○ facing type (dry-stacked concrete blocks, timber, natural rocks, wrapped geosynthetics, or gabions) and facing block size (for concrete block facing)
  ○ vertical facing is recommended, although batter of facing with a minimum front batter of 1/35 to 1/40 is commonly used (if a batter is employed, a typical minimum setback of 5–6 mm between successive courses of facing blocks is common for block height of 200 mm (or 8 in.))
  ○ reinforcement spacing (the default value of reinforcement spacing is 0.2 m (8 in.). For wrapped-faced geotextile walls, temporary walls or walls where facing may be added subsequently, a reinforcement spacing of 0.15 m (6 in.) is recommended. Reinforcement spacing greater than 0.4 m (16 in.) is not recommended under any circumstances).

Step 2: Determine geotechnical engineering properties

- Check to ensure that the selected fill satisfies the following criteria: 100% passing 100 mm (4 in.) sieve, 0–60% passing No. 40 (0.425 mm) sieve, and 0–15% passing No. 200 (0.075 mm) sieve; and plasticity index (PI) ≤ 6.
- Determine reinforced fill parameters, including:
  ○ moist unit weight of the reinforced fill
  ○ the *design friction angle* of the reinforced fill, ϕ_{design}, should be taken as one (1) degree lower than the friction angle obtained from direct shear tests, $\phi_{design} = \phi_{test} - 1°$, where ϕ_{test} is determined by a single set of the standard direct shear tests performed on the portion passing 2 mm (No. 10) sieve, with the specimen compacted to 95% of AASHTO T-99, Methods C or D, at the optimum moisture content. If multiple sets of direct shear tests are performed, the smallest friction angle could be used in design. For instance, if two sets of tests are performed; one showing a friction angle of 35° and the other 37°, the *design friction angle* will be 35°. On the other hand, if a single set of tests shows that a soil has a friction angle of 35°, then the *design friction angle* will be taken as 34°.

- Determine retained earth parameters:
  - friction angle of the retained earth
  - moist unit weight of the retained earth
  - coefficient of active earth pressure of the retained earth.
- Determine foundation soil parameters:
  - friction angle of the foundation soil
  - moist unit weight of the foundation soil
  - allowable bearing pressure of the foundation soil, q_{af}.
- Determine allowable bearing pressure of the reinforced fill:

  The allowable bearing pressure of the reinforced fill, q_{allow}, can be determined by a 3-step procedure as follows:

  1) Use Table 5.7 to determine the allowable bearing pressure for an abutment in the following conditions: (i) the sill is an *integrated sill*, (ii) sill width = 1.5 m, (iii) a sufficiently strong reinforcement is employed (with minimum required values of stiffness and strength to be determined in Step 9, below), and (iv) the abutment is constructed over a competent foundation (satisfying the bearing pressure requirement in Step 7, below).

  2) Select a sill width for the abutment and use Figure 5.27 to determine a *sill width correction factor* for the selected sill width. Calculate a corrected allowable bearing pressure, which is equal to the allowable pressure determined in Step 1 above, and multiply by the correction factor. A minimum sill width of 0.6 m is recommended.

  3) If an *isolated sill* is used, a reduction factor of 0.75 should be applied to the corrected allowable bearing pressure in Step 2, above. No correction is needed for an integrated sill.

Table 5.7 Recommended allowable bearing pressures of a GRS abutment with an integrated sill of sill width = 1.5 m on a competent foundation (Wu et al., 2006)

	Design Friction Angle of Fill Material[*][**]						
	$\phi = 34°$	$\phi = 35°$	$\phi = 36°$	$\phi = 37°$	$\phi = 38°$	$\phi = 39°$	$\phi = 40°$
Reinforcement Spacing = 0.2 m (8 in.)	180 kPa (26 psi)	190 kPa (27.5 psi)	200 kPa (29 psi)	220 kPa (32 psi)	235 kPa (34 psi)	255 kPa (37 psi)	280 kPa (40.5 psi)
Reinforcement Spacing = 0.4 m (16 in.)	125 kPa (18 psi)	140 kPa (20 psi)	155 kPa (22.5 psi)	175 kPa (25 psi)	195 kPa (28 psi)	215 kPa (31 psi)	240 kPa (34.5 psi)

[*]The internal friction angle should be determined by the standard direct shear test on the portion finer than 2 mm (No. 10) sieve, using specimens compacted to 95% of AASHTO T-99, Methods C or D, at optimum moisture content.

[**]If multiple sets of direct shear tests are performed, the lowest friction angle judged as reasonable should be used as the *design friction angle*. If a single set of shear tests is performed, the *design friction angle* will be 1° lower than the value obtained from the tests.

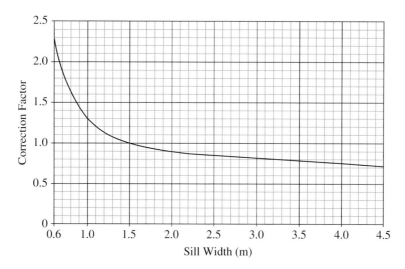

Figure 5.27 Relationship between sill width and correction factor, the NCHRP design method for GRS bridge abutments (Wu et al., 2006)

The 3-step procedure for determination of allowable bearing pressure is illustrated in Example 1 and Example 2 below. Note that the allowable bearing pressure determined by the procedure described above is applicable only to GRS abutments founded on a competent foundation and with a sufficiently strong reinforcement.

Example 1   The Founders/Meadows abutment (Abu-Hejleh et al., 2000)

Given Conditions:

> Fill: ϕ_{design} = 39° (Note: ϕ_{test} = 40.3°, determined from a single set of standard direct shear tests)
>
> Reinforcement spacing = 0.4 m
>
> Integrated sill, sill width = 3.8 m

Allowable bearing pressure:

1) From Table 5.7, for ϕ = 39° and reinforcement spacing = 0.4 m, allowable pressure = 215 kPa

2) From Figure 5.27, for sill width = 3.8 m, the correction factor is 0.77, thus corrected allowable bearing pressure = 215 kPa × 0.77 = 166 kPa

3) No reduction for an integrated sill, thus q_{allow} = 166 kPa

Example 2   The NCHRP test abutments (Wu et al., 2008)

Given Conditions:

Fill: ϕ_{design} = 34° (Note: ϕ_{test} = 34.8° determined from a single set of standard direct shear tests)

Reinforcement spacing = 0.2 m

Isolated sill, sill width = 0.9 m

Allowable bearing pressure:

1) From Table 5.7, for ϕ_{design} = 34° and reinforcement spacing = 0.2 m, allowable pressure = 180 kPa

2) From Figure 5.27, for sill width = 0.9 m, the correction factor is 1.4, thus corrected allowable bearing pressure = 180 kPa × 1.4 = 252 kPa

3) Reduction factor for an isolated sill = 0.75, thus q_{allow} = 252 × 0.75 = 189 kPa

Step 3:   **Establish design requirements**

- Establish external stability design requirements:
  - factor of safety against reinforced fill base sliding ≥ 1.5
  - eccentricity ≤ L/6 (L = length of reinforcement at base of reinforced zone)
  - average sill pressure ≤ allowable bearing pressure of the reinforced fill, q_{allow}
  - average contact pressure at the foundation level ≤ allowable bearing pressure of the foundation soil, q_{af}
- Establish internal stability design requirements:
  - factor of safety against reinforcement pullout, $FS_{pullout}$ of *every* reinforcement layer should be no less than 1.5
  - connection strength is not a design concern if (i) the reinforcement spacing is kept no greater than 0.2 m, (ii) the fill is compacted to the specified value (see Section 6.2.3), and (iii) the average applied pressure on the sill does not exceed the recommended allowable pressure determined in Step 2 above
  - for reinforcement spacing of 0.4 m, facing connection failure should be checked to ensure facing stability (see Section 5.3.5); moreover, for walls with light-weight concrete block facing, the contact surfaces of the top 2 to 3 courses of the facing block should be bonded together by pouring concrete mix through the course or using adhesives to improve facing stability (see Section 6.2.4 and Figure 6.1)
  - studies have suggested that concrete reinforced soil walls will be much stronger if the facing blocks are interconnected through the hollow cells (by use of rebar and cement) *after* all facing units are in place.

Step 4:   **Choose facing type and reinforcement spacing**

- Facing type, including wrapped geotextile facing, dry-stacked concrete block facing, timber facing, natural rock facing and gabion facing, may be selected.

- Reinforcement spacing between 0.15 m (6 in.) and 0.4 m (16 in.) can be selected. However, smaller reinforcement spacing is highly recommended. For concrete block facing, reinforcement spacing should be a multiple of a single block height.

Step 5:   **Determine a trial reinforcement length**

- A trial reinforcement length, L, can be taken as 0.7 × total abutment wall height ($L = 0.7 \times H'$, see Figure 5.26).

- The reinforcement length may be *truncated* in the bottom portion of the wall (see Section 5.3.6) provided that the foundation is *competent* (as defined earlier in this section). The recommended configuration for the truncated base is: reinforcement length = $0.35H'$ (H' = total abutment height) at the foundation level and increases upward at about a 45° angle.

- The allowable bearing pressure of the sill (reinforced fill), as determined in Step 2, should be reduced by 10% for a wall with a truncated base.

- When reinforcement is truncated in the bottom portion, the external stability of the wall (i.e., sliding failure, overall slope failure, eccentricity failure, and foundation bearing failure) in Steps 6 and 7 must be thoroughly examined.

Step 6:   **Size abutment footing/sill**

- Establish a trial sill configuration (e.g., establish the value of B, d, H_2, t, b, fw and fh, Figure 5.28).

- Determine the forces acting on the sill (e.g., see Figure 5.28) and calculate the factor of safety against sliding, $FS_{sliding}$, which needs to be ≥ 1.5.

- Check the sill eccentricity requirement: The load eccentricity at the base of the sill, e', needs to be $\leq B/6$ (B = width of sill).

- Check the allowable bearing pressure of the reinforced fill; the applied contact pressure on the base of the sill (reinforced fill) should be $\leq q_{allow}$, as determined in Step 2.

Step 7:   **Check external stability of reinforced fill (for the trial reinforcement length selected in Step 5)**

- Determine the forces needed for evaluating the external stability of the abutment, e.g., V_4, V_5, V_q, F_3, F_4 (see Figure 5.29) and l_1, the influence depth due to the horizontal forces in the upper wall (see Figure 5.30).

- Calculate the factor of safety against sliding of the reinforced zone, $FS_{sliding}$, which needs to be ≥ 1.5.

- Check the eccentricity requirement for the reinforced volume, e, which needs to be $\leq L/6$ (L = length of reinforcement).

- Check the allowable bearing pressure of the foundation soil:

  o Determine the *influence length* D_1 at the foundation level ($D_1 = d + (B - 2e') + H_1/2$, see Figure 5.30) and compare it to the effective reinforcement length, $L' = L - 2e$.

  o Determine the contact pressure on the foundation level, $p_{contact}$, calculated by dividing the total vertical load in the reinforced volume by D_1 or L', whichever is smaller.

  o The contact pressure $p_{contact}$ needs to be $\leq q_{af}$.

If the bearing pressure of the foundation soil supporting the bridge abutment is only marginally acceptable or slightly unacceptable, a *reinforced soil foundation* (RSF) may be employed to increase the bearing capacity and reduce potential settlement. A typical RSF is formed by excavating a pit at the planned location of bridge abutment and backfilled with compacted road base material that is reinforced by the same reinforcement for the reinforced abutment wall on 0.3 m vertical spacing. The pit should typically be 0.5L deep (L = reinforcement length), and at least cover the footprint of the reinforced fill, and should extend no less than 0.25L in front of the wall face. A generic design protocol for RSF has recently been proposed by Wu (2018).

Step 8: Determine internal stability at each reinforcement level

The internal stability is evaluated by computing the factor of safety against the reinforcement pullout failure at each reinforcement level. Rupture failure is addressed later in Step 9. The factor of safety against pullout failure, $FS_{pullout}$, at a given reinforcement level is equal to pullout resistance at the reinforcement level divided by T_{max}, which is the reinforcement tensile thrust at the reinforcement level where the pullout safety factor is being evaluated.

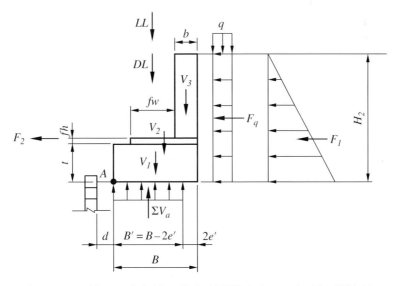

Figure 5.28 Geometry and forces of a bridge sill, the NCHRP design method for GRS bridge abutments

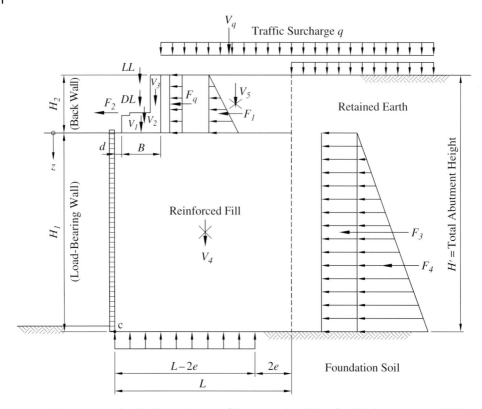

Figure 5.29 Forces involved in the evaluation of the external stability of a GRS abutment, the NCHRP design method for GRS bridge abutments

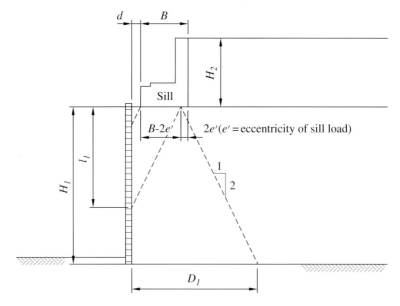

Figure 5.30 Assumed distribution of vertical stress from sill load and definitions of influence depth l_1 and influence length D_1, the NCHRP design method for GRS bridge abutments

The value of T_{max} can be calculated as the product of the average Rankine active lateral earth pressure of the reinforcement level and reinforcement vertical spacing. The pullout resistance, on the other hand, derives from the frictional resistance at the soil–reinforcement interface along the portion of reinforcement beyond the potential failure plane. The potential failure plane is taken as the active Rankine failure plane with a uniform vertical surcharge. $FS_{pullout}$ at all reinforcement levels need to be ≥ 1.5.

Step 9: **Determine the required minimum reinforcement stiffness and strength**

A required minimum ultimate tensile stiffness and a required minimum tensile strength are needed for specifying the geosynthetic reinforcement to be used in the design of a GRS abutment to ensure sufficient tensile resistance at the service load and a sufficient safety margin against rupture failure.

The tensile stiffness is defined as the tensile resistance at a working strain, namely, the *secant* modulus of reinforcement at a selected strain under in-service condition. It is recommended that a tensile strain of 1.0% be taken as the working strain for GRS abutment walls. The required minimum reinforcement stiffness in the direction perpendicular to abutment wall face, $T_{@\varepsilon=1.0\%}$, is determined as:

$$T_{@\varepsilon=1.0\%} \geq \sigma_{h(max)} \cdot s_v \qquad (5\text{-}36)$$

where $\sigma_{h(max)}$ is the largest lateral stress in the reinforced fill and s_v is the vertical reinforcement spacing. For non-uniform reinforcement spacing, s_v = (½ distance to reinforcement layer above) + (½ distance to reinforcement layer below).

The required minimum value of the ultimate reinforcement strength in the direction perpendicular to abutment wall face, T_{ult}, is determined by imposing a combined safety factor on $T_{@\varepsilon=1.0\%}$, i.e.,

$$T_{ult} \geq F_s \cdot T_{@\varepsilon=1.0\%} \qquad (5\text{-}37)$$

The combined safety factor F_s is used to (i) ensure satisfactory long-term performance, (ii) ensure sufficient ductility of the abutment, and (iii) account for various uncertainties. Recommended values for the combined safety factor are: F_s = 5.5 for reinforcement spacing ≤ 0.2 m and F_s = 3.5 for reinforcement spacing of 0.4 m (Note: F_s is smaller for s_v = 0.4 m than for s_v ≤ 0.2 m because the influence of spacing is compensated for in Eqn. (5-36)). These combined safety factors are applicable only to the condition where the backfill material and placement conditions satisfy those given in Section 6.2.3.

Step 10: **Design the back/upper wall**

If the back wall (upper wall) of a bridge abutment is to be a reinforced soil wall, it should be designed in a manner similar to the load-bearing wall (lower wall). In most cases, the same fill, same reinforcement, and same fill placement conditions as those of the load-bearing wall should be employed in the back wall, although the default reinforcement spacing in the approach fill may be increased somewhat, say from 0.2 m to 0.3 m. The length of all the layers of reinforcement should be extended approximately

1.5 m beyond the end of the approach slab to produce a *smoother* subsidence profile over the design life of the abutment and help eliminate potential *bridge bumps*. If there is serious space constraint, it is recommended that at least the top two to three layers of reinforcement should undertake this measure.

Step 11: Check angular distortion between abutments or piers

Angular distortion between any two bridge supporting structures (abutments or piers) should be checked to ensure ride quality and structural integrity. Angular distortion = (difference in total settlements between bridge supporting structures)/(span between the bridge supporting structures). The angular distortion should be limited to 0.005 (or 1:200) for simple span structures, and 0.004 (or 1:250) for continuous span structures.

The total settlement of a bridge supporting structure is the sum of foundation settlement (i.e., settlement occurs beneath the abutment) and settlement within the bridge supporting structure. Foundation settlement due to the self-weight of a GRS abutment/pier and subjected to bridge loads can be estimated by conventional settlement computation methods in soils engineering textbooks or reports (e.g., Terzaghi et al., 1996; Perloff, 1975; Poulos, 2000). Settlement of bridge supporting structures, under the recommended allowable bearing pressure determined in Step 2 (above), can be conservatively estimated as 1.5% of H_1 (H_1 = height of the load-bearing wall).

In situations where the heights of two load-bearing walls at the two ends of a bridge are significantly different, preloading or prestressing the load-bearing walls may be an effective measure to alleviate problems associated with differential settlement. The level of preloading/prestressing and probable reduction in differential settlement due to preloading of a reinforced soil mass can be evaluated by a procedure proposed by Ketchart and Wu (2001, 2002).

If differential settlement is due to differences in foundation settlement, a number of remedial measures (e.g., vibroflotation for granular deposits, wick drain with surcharge for cohesive deposits) for reducing foundation settlement can be employed to mitigate the problem.

5.6.2 Design Example: The NCHRP Method for GRS Abutments

Given Conditions:

A design example is given here to illustrate how to carry out design computations for the NCHRP design method. A spreadsheet-based software named BAWFF (bridge abutments with flexible facing) has been prepared to help with design of bridge abutments by the NCHRP method.

Design Computations:

Step 1: Establish design parameters

Wall heights and external loads (ref. Figure 5.31):

- Total abutment height, H' 9.7 m
- Load-bearing wall height, H_1 7.5 m

- Back wall height, H_2 2.2 m
- Traffic surcharge, q 9.4 kN/m^2
- Bridge vertical dead load, DL 45 kN/m
- Bridge vertical live load, LL 50 kN/m
- Bridge horizontal load, F_2 2.25 kN/m
- Length of concrete approach slab   4.25 m

Trial design parameters (ref. Figure 5.31):

- Sill width, B 1.5 m
- Clear distance, d 0.3 m
- Sill type   integrated sill
- Facing   dry-stacked concrete blocks
- Facing block size   200 mm × 200 mm × 400 mm
- Batter of facing   1/35 (6 mm setback for each block)

Note: Since a batter of 1/35 corresponds to an angle of 1.6°, which is far less than 10°, the abutment wall can be designed as a vertical wall and the coefficient of earth pressure is to follow the Rankine active earth pressure theory for walls with a vertical backface.

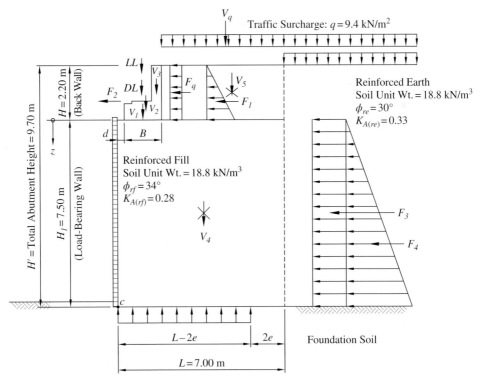

Figure 5.31 Configuration of abutment and forces involved in the evaluation of external stability, design example of the NCHRP design method for GRS bridge abutments

Step 2: **Determine geotechnical engineering properties**

Reinforced fill:

- The selected fill satisfies the following criteria: 100% passing 100 mm (4 in.) sieve, 0–60% passing No. 40 (0.42 mm) sieve, and 0–15% passing No. 200 (0.074mm) sieve; $PI \leq 6$, and maximum soil particle size = 20 mm (3/4 in.).

- The friction angle of the fill is 35°, as determined by one set of standard direct shear test on the portion finer than 2 mm (No. 10) sieve, using a specimen compacted to 95% of AASHTO T-99, Methods C or D, at optimum moisture content.

- $\phi_{test} = 35°, \gamma_{rf} = 18.8 \text{kN/m}^3, K_{A(rf)} = \tan^2(45° - \phi_{rf}/2) - 0.28$

  Note: The *design friction angle* is taken as one degree lower than ϕ_{test}, i.e., $\phi_{design} = 34°$ (see Design Procedure Step 2, Section 5.6.1).

Determine allowable bearing pressure of the reinforced fill, q_{allow}, for following conditions:

- $\phi_{rf} = \phi_{design}; \gamma_n = 34°$

- Reinforcement spacing = 0.2 m (uniform reinforcement length with no truncation)

- Integrated sill, sill width = 1.5 m

  1) From Table 5.7, for $\phi = 34°$ and reinforcement spacing = 0.2 m, allowable bearing pressure is 180 kPa.

  2) From Figure 5.27, correction factor for sill width of 1.5 m is 1.0, thus corrected allowable bearing pressure = 18kPa × 1.0 = 180 kPa.

  3) No reduction for an integrated sill, thus
     $q_{allow} = 180$ kPa

Retained earth:

- $\phi_{re} = 30°, \gamma_{re} = 18.8 \text{ kN / m}^3, K_{A(re)} = \tan^2(45° - \phi_{re}/2) = 0.33$

Foundation soil:

- $\phi_{fs} = 30°, \gamma_{fs} = 20.0 \text{ kN/m}^3, q_a = 300$ kPa

Step 3: **Establish design requirements**

External stability design requirements:

- Design life = 75 years
- Factor of safety against sliding, $FS_{sliding} \geq 1.5$
- Eccentricity $\leq L/6$
- Sill pressure $\leq$ allowable bearing pressure of the reinforced fill, $q_{allow} = 180$ kPa
- Average contact pressure at the foundation level $\leq$ allowable bearing pressure of the foundation soil, $q_a = 300$ kPa

Internal stability design requirements:

- Factor of safety against pullout, $FS_{pullout} \geq 1.5$
- Facing connection strength is OK with reinforcement spacing = 0.2 m (see Step 3, Section 5.6.1)

Step 4: **Choose facing type and reinforcement spacing**

Concrete blocks, 200 mm (depth) × 200 mm (height) × 400 (width)

Reinforcement spacing in the load-bearing wall (the lower wall) = 0.2 m

Step 5: **Determine a trial reinforcement length**

Trial reinforcement length = 0.7 × total abutment height

$$L = 0.7\,H' = 0.7 \times 9.7 \text{ m} = 6.8 \text{ m (use 7.0 m)}$$

Step 6: **Size abutment footing/sill**

Preliminary sill configuration and forces acting on the sill are shown in Figure 5.32. The dimensions of the sill are:

$$B = 1.5 \text{ m}$$
$$d = 0.3 \text{ m}$$
$$H_2 = 2.2 \text{ m}$$
$$t = 0.65 \text{ m}$$
$$b = 0.4 \text{ m}$$
$$fw = 0.8 \text{ m}$$
$$fh = 0.1 \text{ m}$$

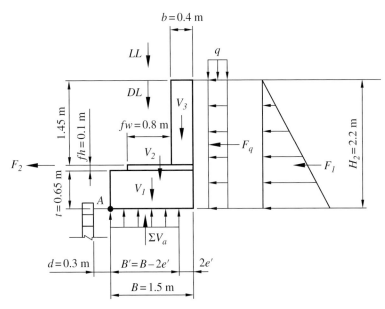

Figure 5.32 Configuration of a bridge sill and loads on a bridge sill, design example of the NCHRP design method for GRS abutments

With the unit weight of concrete, $\gamma_{concrete} = 23.6 \text{ kN/m}^3$, the forces acting on the sill are determined as:

$$V_1 = (B \cdot t) \, \gamma_{concrete}$$
$$= (1.5 \text{ m} \times 0.65 \text{ m}) \times 23.6 \text{ kN/m}^3$$
$$= 23.01 \text{ kN/m}$$

$$V_2 = \left[(fw + b) f h \right] \gamma_{concrete}$$
$$= \left[(0.8 \text{ m} \times 0.4 \text{ m}) \times 0.1 \text{ m} \right] \times 23.6 \text{ kN/m}^3$$
$$= 2.83 \text{ kN/m}$$

$$V_3 = \left[b \left(H_2 - fh - t \right) \right] \gamma_{concrete}$$
$$= \left[0.4 \text{ m} \times \left(2.2 \text{ m} - 0.1 \text{ m} - 0.65 \text{ m} \right) \right] \times 23.6 \text{ kN/m}^3$$
$$= 13.69 \text{ kN/m}$$

$$DL = 45 \text{ kN/m} \left(\text{from Step 1} \right)$$
$$LL = 50 \text{ kN/m} \left(\text{from Step 1} \right)$$
$$F_q = K_{A(rf)} \cdot q \cdot H_2$$
$$= 0.28 \times 9.4 \text{ kN/m}^2 \times 2.2 \text{ m}$$
$$= 5.79 \text{ kN/m}$$

$$F_1 = \frac{1}{2} K_{A(rf)} \cdot \gamma_{rf} \cdot H_2^2 a$$
$$= \frac{1}{2} \times 0.28 \times 18.8 \text{ kN/m}^3 \times \left(2.2 \text{ m} \right)^2$$
$$= 12.74 \text{ kN/m}$$

$$F_2 = 2.25 \text{ kN/m} \left(\text{from Step 1} \right)$$

Check factor of safety against sliding:

ΣV_a = sum of vertical forces acting on the sill
$$= V_1 + V_2 + V_3 + DL + LL$$
$$= 23.01 \text{ kN/m} + 2.83 \text{ kN/m} + 13.69 \text{ kN/m} + 45 \text{ kN/m} + 50 \text{ kN/m}$$
$$= 134.53 \text{ kN/m}$$

ΣF_a = sum of horizontal forces acting on the sill
$$= F_q + F_1 + F_2$$
$$= 5.79 \text{ kN/m} + 12.74 \text{ kN/m} + 2.25 \text{ kN/m}$$
$$= 20.78 \text{ kN/m}$$

$$FS_{sliding} = \frac{\left(\Sigma V_a - LL \right) \tan \phi_{rf}}{\Sigma F_a}$$
$$= \frac{\left(134.53 \text{ kN/m} - 50 \text{ kN/m} \right) \times \tan 34°}{20.78 \text{ kN/m}}$$
$$= 2.74 > 1.5 \left(\text{OK} \right)$$

Check eccentricity requirement:

ΣM_{OA} = sum of overturning moments about point A

$$= F_q\left(\frac{H_2}{2}\right) + F_1\left(\frac{H_2}{3}\right) + F_2(t + fh)$$

$$= 5.79\,\frac{kN}{m} \times \left(\frac{2.2\text{ m}}{2}\right) + 12.74\,\frac{kN}{m} \times \left(\frac{2.2\text{ m}}{3}\right) + 2.25\,\frac{kN}{m} \times \left(0.65\text{ m} + 0.1\text{ m}\right)$$

$$= 17.40\text{ kN/m}\cdot\text{m}$$

ΣM_{RA} = sum of resisting moments about point A

$$= V_1\left(\frac{B}{2}\right) + V_2\left[\frac{fw + b}{2} + (B - b - fw)\right] + V_3\left[\frac{b}{2} + (B - b)\right]$$

$$+ (DL + LL)\left[\frac{fw}{2} + (B - b - fw)\right]$$

$$= 23.01\,\frac{kN}{m}\left(\frac{1.5\text{ m}}{2}\right) + 2.83\,\frac{kN}{m}\left[\frac{0.8\text{ m} + 0.4\text{ m}}{2} + \left(1.5\text{ m} - 0.4\text{ m} - 0.8\text{ m}\right)\right]$$

$$+ 13.69\,\frac{kN}{m}\left[\frac{0.4\text{ m}}{2} + \left(1.5\text{ m} - 0.4\text{ m}\right)\right]$$

$$+ \left(45\,\frac{kN}{m} + 50\,\frac{kN}{m}\right)\left[\frac{0.8\text{ m}}{2} + \left(1.5\text{ m} - 0.4\text{ m} - 0.8\text{ m}\right)\right]$$

$$= 104.10\text{ kN/m}\cdot\text{m}$$

$$e' = \text{eccentricity at the base of sill} = \frac{B}{2} - \frac{\Sigma M_{RA} - \Sigma M_{OA}}{\Sigma V_a}$$

$$= \frac{1.5\text{ m}}{2} - \frac{104.10\text{ kN/m}\cdot\text{m} - 17.40\text{ kN/m}\cdot\text{m}}{134.53\text{ kN/m}}$$

$$= 0.11\text{ m}$$

$$\frac{B}{6} = \frac{1.5\text{ m}}{6} = 0.25\text{ m}$$

$$e' < \frac{B}{6}\text{ (OK)}$$

Check allowable bearing pressure of the reinforced fill:

$$p_{sill} = \text{applied pressure from the sill} = \frac{\Sigma V_a}{B - 2e'}$$

$$= \frac{134.53\text{ kN/m}}{1.5\text{ m} - \left(2 \times 0.11\text{ m}\right)}$$

$$= 105.1\text{ kN/m}^2 < q_{allow} = 180\text{ kPa (OK)}$$

Step 7: **Check external stability of reinforced fill (for the trial reinforcement length selected in Step 5)**

The forces involved in evaluation of external stability of the abutment are shown in Figure 5.31. These forces are calculated as follows.

$$V_4 = (L \cdot H_1) \, \gamma_{rf}$$
$$= \left(7 \text{ m} \times 7.5 \text{ m}\right) \times 18.8 \text{ kN/m}^3$$
$$= 987 \text{ kN/m}$$

$$V_5 = \left[(L - d - B) H_2 \right] \gamma_{rf}$$
$$= \left[\left(7 \text{ m} - 0.3 \text{ m} - 1.5 \text{ m}\right) \times 2.2 \text{ m} \right] \times 18.8 \text{ kN/m}^3$$
$$= 215.07 \text{ kN/m}$$

$$V_q = (L - d - B) \cdot q$$
$$= \left(7 \text{ m} - 0.3 \text{ m} - 1.5 \text{ m}\right) \times 9.4 \text{ kN/m}^2$$
$$= 48.88 \text{ kN/m}$$

$$F_3 = \left[K_{A(re)} \left(q + \gamma_{re} \cdot H_2 \right) \right] H_1$$
$$= \left[0.33 \times \left(9.4 \text{ kN/m}^2 + 18.8 \text{ kN/m}^3 \times 2.2 \text{ m}\right) \right] \times 7.5 \text{ m}$$
$$= 125.63 \text{ kN/m}$$

$$F_4 = \frac{1}{2} K_{A(re)} \cdot \gamma_{re} \cdot H_1^2$$
$$= \frac{1}{2} \times 0.33 \times 18.8 \text{ kN/m}^3 \times \left(7.5 \text{m}\right)^2$$
$$= 174.49 \text{ kN/m}$$

$$\Sigma V_a = 134.53 \text{ kN/m} \text{ (from Step 6)}$$
$$\Sigma F_a = 20.78 \text{ kN/m} \text{ (from Step 6)}$$

l_1 is the influence depth due to the horizontal forces in the back wall (see Figure 5.33, cf. Figure 5.30):

$$l_1 = (d + B - 2e') \tan\left(45° + \frac{\phi_{rf}}{2} \right)$$
$$= \left(0.3 \text{ m} + 1.5 \text{ m} - 2 \times 0.11 \text{ m}\right) \times \tan\left(45° + \frac{34°}{2} \right)$$
$$= 2.97 \text{ m}$$

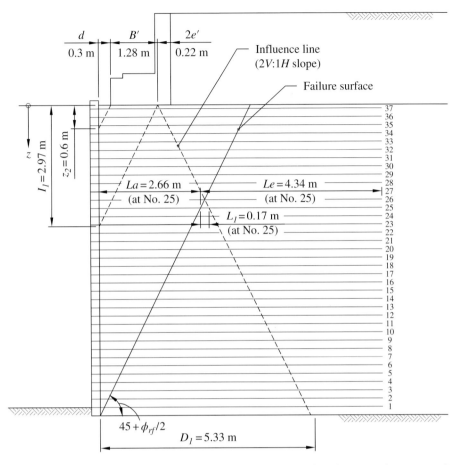

Figure 5.33 Parameters involved in evaluating the internal stability of an abutment, design example of the NCHRP design method for GRS abutments

Check the factor of safety against sliding for the reinforced volume:

ΣV = sum of vertical forces acting on the foundation soil

$= V_4 + V_5 + V_q + \Sigma V_a$

$= 987 \text{ kN/m} + 215.07 \text{ kN/m} + 48.88 \text{ kN/m} + 134.53 \text{ kN/m}$

$= 1{,}385.48 \text{ kN/m}$

ΣF_a = sum of horizontal forces acting on the foundation soil

$= F_3 + F_4 + \Sigma F_a$

$= 125.63 \text{ kN/m} + 174.49 \text{ kN/m} + 20.87 \text{ kN/m}$

$= 320.90 \text{ kN/m}$

$$FS_{sliding} = \frac{(\Sigma V - LL - Vq)\tan\phi_{fs}}{\Sigma F}$$

$$= \frac{\left(1{,}385.48 \text{ kN/m} - 50 \text{ kN/m} - 48.88 \text{ kN/m}\right) \times \tan 30°}{320.90 \text{ kN/m}}$$

$$= 2.31 > 1.5 \;(\text{OK})$$

Check the eccentricity requirement for the reinforced volume:

ΣM_{OC} = sum of overturning moments about point C

$$= F_3\left(\frac{H_1}{2}\right) + F_4\left(\frac{H_1}{3}\right) + \Sigma F_a\left(H_1 - \frac{l_1}{3}\right)$$

$$= 125.63\,\frac{\text{kN}}{\text{m}} \times \left(\frac{7.5 \text{ m}}{2}\right) + 174.49\,\frac{\text{kN}}{\text{m}} \times \left(\frac{7.5 \text{ m}}{3}\right) + 20.78\,\frac{\text{kN}}{\text{m}} \times \left(7.5 \text{ m} - \frac{2.97 \text{ m}}{3}\right)$$

$$= 1{,}042.62 \text{ kN/m} \cdot \text{m}$$

ΣM_{RC} = sum of resisting moments about point C

$$= V_4\left(\frac{L}{2}\right) + (V_5 + V_q)\left[\frac{L-d-B}{2} + (d+B)\right] + \left(\Sigma M_{RA} + \Sigma V_a d\right)$$

$$= 987\,\frac{\text{kN}}{\text{m}}\left(\frac{7 \text{ m}}{2}\right) + \left(215.07\,\frac{\text{kN}}{\text{m}} + 48.88\,\frac{\text{kN}}{\text{m}}\right)\left[\frac{7 \text{ m} - 0.3 \text{ m} - 1.5 \text{ m}}{2}\right.$$

$$\left. + \left(0.3 \text{ m} + 1.5 \text{ m}\right)\right] + \left[104.10 \text{ kN/m} \cdot \text{m} + 134.53 \text{ kN/m} \times \left(0.3 \text{ m}\right)\right]$$

$$= 4{,}760.34 \text{ kN/m} \cdot \text{m}$$

ΣM_{SC} = moments about point C due to traffic surcharge

$$= V_q\left[\frac{L-d-B}{2} + (d+B)\right]$$

$$= 48.88\,\frac{\text{kN}}{\text{m}}\left[\frac{7 \text{ m} - 0.3 \text{ m} - 1.5 \text{ m}}{2} + \left(0.3 \text{ m} + 1.5 \text{ m}\right)\right]$$

$$= 215.07 \text{ kN/m} \cdot \text{m}$$

e = eccentricity at the base of the reinforced volume

$$= \frac{L}{2} - \frac{(\Sigma M_{RC} - \Sigma M_{SC}) - \Sigma M_{OC}}{\Sigma V - V_q}$$

$$= \frac{7 \text{ m}}{2} - \frac{\left(4{,}760.34 \text{ kN/m} \cdot \text{m} - 215.07 \text{ kN/m} \cdot \text{m}\right) - 1{,}042.62 \text{ kN/m} \cdot \text{m}}{1{,}385.48 \text{ kN/m} - 48.88 \text{ kN}/\text{m}}$$

$$= 0.88 \text{ m}$$

$$\frac{L}{6} = \frac{7 \text{ m}}{6} = 1.17 \text{ m}$$

$$e < \frac{L}{6} \;(\text{OK})$$

Check the allowable bearing pressure of the foundation soil:

Calculate the *effective length* D_1 at the foundation level and compare with the effective reinforcement length, L'.

$$D_1 = d + B' + \frac{H_1}{2} = d + (B - 2e') + \frac{H_1}{2}$$

$$= 0.3 + (1.5\,\text{m} - 2 \times 0.11\,\text{m}) + \frac{7.5\,\text{m}}{2}$$

$$= 5.33\,\text{m}$$

$$L' = L - 2e$$
$$= 7.0\,\text{m} - 2 \times 0.88\,\text{m}$$
$$= 5.24\,\text{m}$$

As D_1 at the foundation level is greater than L' $(L' = L - 2e)$, the contact pressure on the foundation level, $p_{contact}$, is calculated as follows.

$$p_{contact} = \frac{\Sigma V}{L - 2e}$$

$$= \frac{1{,}385.48\,\text{kN/m}}{7.0\,\text{m} - 2 \times 0.88\,\text{m}}$$

$$= 264.40\,\text{kN/m}^2$$

$$q_a = 300\,\text{kN/m}^2 \ (\text{from Step 2})$$

$$p_{contact} = 264.40\,\text{kN/m}^2 < q_a = 300\,\text{kN/m}^2 \ (\text{OK})$$

Step 8: **Determine internal stability at each reinforcement level**

With geosynthetic reinforcement, the coefficient of lateral earth pressure is constant for the entire wall height. The internal stability is evaluated by checking stability against reinforcement pullout failure (Note: rupture failure is to be addressed later in Step 9).

Check reinforcement pullout failure:

$$P_r = \text{pullout resistance} = F_p \cdot \alpha \cdot (\sigma_v \cdot L_e) \cdot m \cdot R_C$$

$$F_p = \text{pullout resistance factor} = \frac{2}{3}\tan\phi_{rf} = \frac{2}{3}\tan(34°) = 0.45$$

α is a scale effect correction factor, ranging from 0.6 to 1.0 for geosynthetic reinforcement; for geotextile, α is defaulted to 0.6.

$\sigma_v \cdot L_e$ is the normal force at the soil–reinforcement interface at depth z (excluding traffic surcharge).

$$\sigma_v \cdot L_e = \sigma_{vs} \cdot L_e + \Delta\sigma_v \cdot L_i$$

L_e = length of reinforcement in the resistant zone behind the failure surface at depth z

$$= L - L_a$$

L_a = length of reinforcement in the active zone at depth z

$$= (H_1 - z) \tan (45° - \phi_{rf} / 2)$$

L_i = length of reinforcement within the influence area inside the resistant zone; this length can be measured directly from the design drawing

m = reinforcement effective unit perimeter; m = 2 for strips, grids, and sheets

R_c = coverage ratio; R_c = 1.0 for 100% coverage of reinforcement

σ_h = horizontal pressure at depth $z = K_{A(rf)} (\sigma_{vs} + \Delta\sigma_v + q) + \Delta\sigma_h$

σ_{vs} = vertical soil pressure at depth $z = \gamma_{rf} \cdot H_2 + \gamma_{rf} \cdot z$

$\Delta\sigma_v$ = distributed vertical pressure from sill = $\Sigma V_a/D$

D = effective width of applied load at depth z

For $z \leq z_2$, $D = (B - 2e') + z$

For $z > z_2$, $D = d + (B - 2e') + \dfrac{z}{2}$

$z_2 = 2 \cdot d$

$\Delta\sigma_h$ = supplement horizontal pressure at depth z

For $z \leq l_1$, $\Delta\sigma_h = 2 \cdot \Sigma F_a \cdot \dfrac{l_1 - z}{l_1^2}$

For $z > l_1$, $\Delta\sigma_h = 0$

T_{max} = maximum tensile force in the reinforcement at depth $z = \Delta\sigma_h \cdot s_v$

s_v = vertical reinforcement spacing

$FS_{pullout}$ = factor of safety against reinforcement pullout

$$FS_{pullout} = \dfrac{P_r}{T_{max}}$$

Let depth z be the depth measured from the top of the load-bearing wall. We will use reinforcement layer no. 25 at z = 2.5 m (see Figure 5.33) as an example for determination of $FS_{pullout}$.

$$\sigma_{vs} = \gamma_{rf} \cdot H_2 + \gamma_{rf} \cdot z$$

$$= \left(18.8\,\text{kN/m}^3 \times 2.2\,\text{m}\right) + \left(18.8\,\text{kN/m}^3 \times 2.5\,\text{m}\right)$$

$$= 88.36\,\text{kN/m}^2$$

$$z_2 = 2 \times d = 2 \times 0.3\,\text{m} = 0.6\,\text{m}$$

Since $z = 2.5$ m $> z_2 = 0.6$ m, therefore $D = d + (B - 2e') + \dfrac{z}{2}$

$$D = 0.3\text{ m} + \left(1.5\text{ m} - 2 \times 0.11\text{ m}\right) + \frac{2.5\text{ m}}{2} = 2.83\text{ m}$$

$$\Delta\sigma_v = \frac{\Sigma V_a}{D} = \frac{134.53\text{ kN/m}}{2.83\text{ m}} = 47.54\text{ kN/m}^2$$

Since $z = 2.5$ m $< l_1 = 2.97$ m, therefore $\Delta\sigma_h = 2\,\Sigma F_a \cdot \dfrac{(l_1 - z)}{l_1^2}$

$$\Delta\sigma_h = 2 \times 20.78\text{ kN/m} \times \frac{\left(2.97\text{ m} - 2.5\text{ m}\right)}{\left(2.97\text{ m}\right)^2} = 2.21\text{ kN/m}^2$$

$$\begin{aligned}
\sigma_h &= K_{A(rf)}\left(\sigma_{vs} + \Delta\sigma_v + q\right) + \Delta\sigma_h \\
&= 0.28 \times \left(88.36\text{ kN/m}^2 + 47.54\text{ kN/m}^2 + 9.41\text{ kN/m}^2\right) + 2.21\text{ kN/m}^2 \\
&= 42.89\text{ kN/m}^2
\end{aligned}$$

$$T_{max} = \sigma_h \cdot s_v = 42.89\text{ kN/m}^2 \times 0.2\text{ m} = 8.58\text{ kN/m}$$

$$\begin{aligned}
L_a &= \left(H_1 - z\right)\tan\left(45° - \frac{\phi_{rf}}{2}\right) \\
&= \left(7.5\text{ m} - 2.5\text{ m}\right) \times \tan\left(45° - \frac{34°}{2}\right) = 2.66\text{ m}
\end{aligned}$$

$$L_e = L - L_a = 7\text{ m} - 2.66\text{ m} = 4.34\text{ m}$$

Calculate the normal force at $z = 2.5$ m:

$$\begin{aligned}
\sigma_v \cdot L_e &= \sigma_{vs} \cdot L_e + \sigma_{vs} \cdot L_i \\
&= \left(88.36\text{ kN/m}^2 \times 4.34\text{ m}\right) + \left(47.54\text{ kN/m}^2 \times 0.17\text{ m}\right) \\
&= 391.56\text{ kN/m}
\end{aligned}$$

$$\begin{aligned}
P_r &= F_p \cdot \alpha \cdot \left(\sigma_v \cdot L_e\right) \cdot m \cdot Rc \\
&= 0.45 \times 0.6 \times \left(391.56\text{ kN/m}^2\right) \times 2 \times 1 \\
&= 211.44\text{ kN/m}
\end{aligned}$$

$$FS_{pullout} = \frac{P_r}{T_{max}} = \frac{211.44\text{ kN/m}}{8.58\text{ kN/m}} = 24.64 > 1.5 \quad (OK)$$

The values of $FS_{pullout}$ for all reinforcement layers in the load-bearing wall are summarized in Table 5.8.

Table 5.8 Design example for the NCHRP design method: $FS_{pullout}$ at each reinforcement level

No.	Depth z (m)	s_v (m)	σ_{vs} (kN/m²)	D (m)	$\Delta\sigma_v$ (kN/m²)	$\Delta\sigma_h$ (kN/m²)	σ_h (kN/m²)	T_{max} (kN/m)	L_a (m)	L_e (m)	L_i (m)	$(\sigma_v \cdot L_e)$ (kN/m)	Pr (kN/m)	$FS_{pullout}$
1	7.3	0.2	178.6	5.2	25.7	0.0	59.8	12.0	0.1	6.9	5.1	1,363	735	61.5
2	7.1	0.2	174.8	5.1	26.2	0.0	58.9	11.8	0.2	6.8	4.9	1,316	710	60.2
3	6.9	0.2	171.1	5.0	26.8	0.0	58.0	11.6	0.3	6.7	4.7	1,267	685	59.0
4	6.7	0.2	167.3	4.9	27.3	0.0	57.1	11.4	0.4	6.6	4.5	1,223	660	57.8
5	6.5	0.2	163.6	4.8	27.9	0.0	56.2	11.3	0.5	6.5	4.3	1,178	635	56.5
6	6.3	0.2	159.8	4.7	28.4	0.0	55.3	11.1	0.6	6.4	4.1	1,133	611	55.2
7	6.1	0.2	156.0	4.6	29.1	0.0	54.5	10.9	0.7	6.3	3.9	1,089	588	54.0
8	5.9	0.2	152.3	4.5	29.7	0.0	53.6	10.7	0.9	6.2	3.7	1,046	564	52.7
9	5.7	0.2	148.5	4.4	30.4	0.0	52.7	10.5	1.0	6.0	3.5	1,003	541	51.3
10	5.5	0.2	144.8	4.3	31.1	0.0	51.9	10.4	1.1	5.9	3.3	961	518	50.0
11	5.3	0.2	141.0	4.2	31.8	0.0	51.0	10.2	1.2	5.8	3.1	919	496	48.6
12	5.1	0.2	137.4	4.1	32.6	0.0	50.2	10.0	1.3	5.7	2.9	879	474	47.2
13	4.9	0.2	133.5	4.0	33.4	0.0	49.4	9.9	1.4	5.6	2.7	838	452	45.8
14	4.7	0.2	129.7	3.9	34.2	0.0	48.5	9.7	1.5	5.5	2.4	798	431	44.4
15	4.5	0.2	126.0	3.8	35.1	0.0	47.7	9.6	1.6	5.4	2.2	759	410	42.9
16	4.3	0.2	122.2	3.7	36.1	0.0	47.0	9.4	1.7	5.3	2.0	721	389	41.4
17	4.1	0.2	118.4	3.6	37.1	0.0	46.2	9.2	1.8	5.2	1.8	682	368	39.9
18	3.9	0.2	114.7	3.5	38.1	0.0	45.4	9.1	1.9	5.1	1.6	645	345	38.3
19	3.7	0.2	110.9	3.4	39.2	0.0	44.7	8.9	2.0	5.0	1.4	608	328	36.7
20	3.5	0.2	107.2	3.3	40.4	0.0	44.0	8.8	2.1	4.9	1.2	571	308	35.0
21	3.3	0.2	103.4	3.2	41.7	0.0	43.3	8.7	2.2	4.8	1.0	534	288	33.3

22	3.1	0.2	99.6	3.1	43.0	0.0	42.6	8.5	2.3	4.7	0.8	498	269	31.6
23	2.9	0.2	95.9	3.0	44.4	0.3	42.3	8.5	2.5	4.6	0.6	463	250	29.5
24	2.7	0.2	92.1	2.9	45.9	1.3	42.6	8.5	2.6	4.5	0.4	427	230	27.1
25	2.5	0.2	88.4	2.8	47.5	2.2	42.9	8.6	2.7	4.3	0.2	392	211	24.6
26	2.3	0.2	84.6	2.7	49.3	3.2	43.3	8.7	2.8	4.2	0.0	358	193	22.3
27	2.1	0.2	80.8	2.6	51.2	4.1	43.7	8.7	2.9	4.1	0.0	334	180	20.6
28	1.9	0.2	77.1	2.5	53.2	5.0	44.2	8.8	3.0	4.0	0.0	310	167	19.0
29	1.7	0.2	73.3	2.4	55.4	6.0	44.7	8.9	3.1	3.9	0.0	287	155	17.4
30	1.5	0.2	69.6	2.3	57.7	6.9	45.2	9.0	3.2	3.8	0.0	265	143	15.8
31	1.3	0.2	65.8	2.2	60.3	7.9	45.8	9.2	3.3	3.7	0.0	244	131	14.4
32	1.1	0.2	62.0	2.1	63.2	8.8	46.5	9.3	3.4	3.6	0.0	223	120	13.0
33	0.9	0.2	58.3	2.0	66.3	9.8	47.3	9.5	3.5	3.5	0.0	203	110	11.6
34	0.7	0.2	54.5	1.9	69.7	10.7	48.1	9.6	3.6	3.4	0.0	185	100	10.4
35	0.5	0.2	50.8	1.8	75.6	11.6	49.6	9.9	3.7	3.3	0.0	166	90	9.0
36	0.3	0.2	47.0	1.6	85.2	12.6	52.2	10.4	3.8	3.2	0.0	149	80	7.7
37	0.1	0.2	43.2	1.4	97.5	13.5	55.6	11.1	3.9	3.1	0.0	133	72	6.4

Step 9: **Determine the required minimum reinforcement stiffness and strength**

The required minimum *working reinforcement stiffness* is determined as:

$$T_{@ \, \varepsilon=1.0\%} \geq \sigma_{h(max)} \cdot s_v$$

The required minimum *ultimate reinforcement strength* is determined as:

$$T_{ult} \geq F_s \cdot T_{@ \, \varepsilon=1.0\%}$$

The recommended combined safety factor is $F_s = 5.5$ for reinforcement ≤ 0.2 m, and $F_s = 3.5$ for reinforcement spacing of 0.4 m.

From Table 3.2, $\sigma_{h(max)}$ is 59.84 kN/m², at reinforcement no. 1 at depth $z = 7.3$ m and with $s_v = 0.2$ m.

$$T_{@ \, \varepsilon=1.0\%} = \sigma_{h(max)} \cdot s_v$$
$$T_{@ \, \varepsilon=1.0\%} = 59.84 \text{ kN/m}^2 \times 0.2 = 12.0 \text{ kN/m}$$

The uniform reinforcement spacing is 0.2 m, hence $F_s = 5.5$.

$$T_{ult} \geq F_s \cdot T_{@ \, \varepsilon=1.0\%}$$
$$T_{ult} = 5.5 \times 12 \text{ kN/m} = 66.0 \text{ kN/m}$$

A reinforcement with minimum working stiffness, $T_{@ \, \varepsilon=1.0\%} = 12.0$ kN/m and minimum ultimate strength (per ASTM D4595 for geotextiles and D6637 for geogrids), $T_{ult} = 66.0$ kN/m is required.

Step 10: **Design of back/upper wall**

Reinforced fill:	same as that of the load-bearing wall
Reinforcement:	same as that of the load-bearing wall
Reinforcement length:	4.25 m (length of approach slab) + 1.5 m = 5.75 m (see Step 10, Section 5.6.1)
Reinforcement layout:	vertical spacing = 0.3 m
	Wrapped-face with wrapped return at least 0.5 m (18 in.) in the horizontal direction and anchored in at least 100 mm of fill material.
	A compressible layer of approximately 50-mm in thickness installed between the wrapped face and the rigid back wall is recommended.

Step 11: **Check angular distortion between abutments or piers**

If the magnitude of settlement between any span of a bridge (i.e., between any adjacent abutments or piers) is expected to be different, angular distortion will need to be checked. For demonstration purposes, the following calculations for angular distortion are made with an assumed value of foundation settlement.

$\delta_{abutment}$ = abutment settlement

$= 1.5\% \times H_1$

$= 0.015\,(7.5\,\text{m}) = 0.1125\,\text{m}$

$\delta_{foundation}$ = foundation settlement (as determined by conventional settlement computation methods)

Assume $\delta_{foundation} = 0.01\,\text{m}$

δ_{total} = total settlement

$= \delta_{abutment} + \delta_{foundation}$

$= 0.1125\,\text{m} + 0.01\,\text{m} = 0.1225\,\text{m}$

Assuming the other end of the bridge segment is zero,
angular distortion = $(\delta_{total} - 0)$/span length = 0.1225 m/24 m = 0.0051

Largest tolerable angular distortion for simple span = 0.005

Angular distortion = 0.0051 ≈ 0.005 (OK)

Design Summary:

The configuration for the final design is shown in Figure 5.34.

Abutment configuration:

Load-bearing wall height, H_1	7.5 m
Back wall (upper wall) height	2.2 m
Facing	dry-stacked concrete blocks (200 × 200 × 400 mm)
Front batter	1/35
Sill type	integrated sill
Sill width	1.5 m
Sill clear distance	0.3 m
Embedment	200 mm (one facing block height)

Reinforcement:

Minimum working stiffness, $T_{@\varepsilon=1.0\%}$ = 12.0 kN/m
Minimum ultimate strength, T_{ult} = 66.0 kN/m
Length in load-bearing wall = 7.0 m
Length of top three layers in load-bearing wall = 7.5 m
Vertical spacing in load-bearing wall = 0.2 m
Length in back wall = 5.75 m
Vertical reinforcement spacing in back wall = 0.3 m

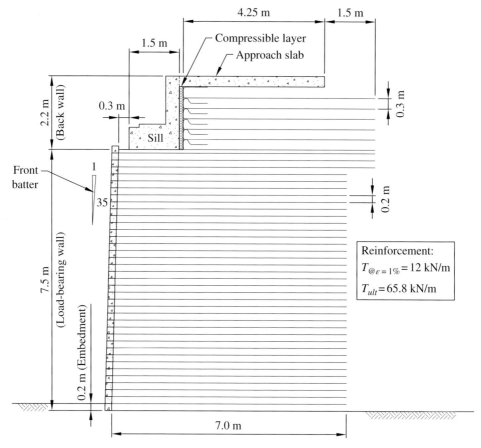

Figure 5.34 Configuration of the final design, design example of the NCHRP design method for GRS bridge abutments

5.7 The GRS Non-Load-Bearing (GRS-NLB) Walls Design Method

Wu (2012) developed a generic design method for GRS walls in non-load-bearing (NLB) applications where external loads are limited to common traffic loads. The design method, referred to as the GRS-NLB method, adopts an allowable stress design (ASD) approach. Another version of the GRS-NLB design method based on the *load and resistance factor design* (LRFD) approach has been presented by Beauregard (2016). A step-by-step procedure of the GRS-NLB ASD design method is presented in this section. All six major design features described in Section 5.3 are incorporated in the GRS-NLB design method.

The GRS-NLB design method is primarily for concrete block GRS walls in non-load-bearing applications under the following conditions:

- The facing comprises dry-stacked concrete blocks.

- The reinforcement spacing is not more than 0.3 m.
- The wall face is vertical or near-vertical, i.e., no more than 10° off the vertical (or steeper than a batter of 1 (horizontal) to 7 (vertical)); vertical wall face is recommended for walls of low to medium height (i.e., height not over 20 ft or 6 m).
- No prop or temporary bracing is used during construction.
- The fill material is predominantly granular and free-draining; the fill typically satisfies the following criteria: 100% passing 100 mm (4 in.) sieve, 0–60% passing No. 40 (0.425 mm) sieve, and no more than 20% of fines (passing No. 200 sieve), with liquid limit (*LL*) not more than 30, and plasticity index (*PI*) not more than 6.
- The backfill is compacted to at least 95% of the Standard Proctor maximum density or 90% of the Modified Proctor maximum density.
- The backfill, when compacted to 95% of the Standard Proctor at the optimum moisture content, should have an angle of internal friction of 34° or greater, as determined by the standard direct shear test on portion finer than 2 mm (No. 10) sieve.
- The slope of the wall crest is not more than $\tan^{-1}\left(\dfrac{\tan\phi}{1.3}\right)$ for granular fill.
- The foundation soil is competent (i.e., with insignificant post-construction settlement in the foundation soil); approximately level within $0.3H$ in front of the wall face (H = height of wall face, measured from the bottom of the lowest facing block).
- The wall is only subject to common traffic loads, no high-intensity edge load (as in bridge abutments) is present. Vertical surcharge/traffic load on the crest, q, when present, is no more than an equivalence of 3 ft (0.9 m) of overburden.
- Seismic loads are not a major concern.

For applications of the GRS-NLB design method, the following should be noted:

- Concrete block GRS walls with closely spaced reinforcement (i.e., spacing not more than 0.2 m), although designed to withstand only static loads, have been shown to be capable of withstanding earthquake motions up to $0.2g$ peak ground acceleration without visible distress, especially with measures taken on two ends of a bridge (Helwany et al., 2012).
- Even though there are practical height limitations in considerations of ease of construction, visual perception, and aesthetics, there is no theoretical limit to the height of closely spaced GRS walls in non-load-bearing applications (see Section 5.3.5).
- The GRS-NLB design method is primarily for GRS walls with concrete block facing. For applications of the design method to other types of wall facing, adequate mobilization of reinforcement loads must be permitted during construction, also the stability of facing and connection between facing and reinforced soil mass must be ensured.

5.7.1   Design Procedure: The GRS-NLB Method

Step 1:   Establish wall profile and verify design conditions

A wall profile for design should be established in consideration of the grading plan at the wall site. When a project involves variation in wall height along the longitudinal direction, the designer should choose proper height intervals to allow for a gradual change in height. An example is given in Figure 5.35.

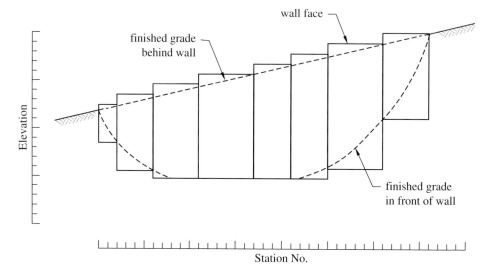

Figure 5.35 Profile of a GRS wall with variation in wall height

The conditions for non-load-bearing applications listed at the beginning of Section 5.7 should be checked. If the conditions are not met, the design method is not recommended.

Step 2:   Determine design parameters for reinforced fill, retained fill, and foundation

Soil parameters needed for design include:

- moist unit weight (γ), Mohr–Coulomb angle of internal friction (at peak strength) (ϕ), and angle of dilation (ψ) of the compacted fill in the reinforced soil zone

- moist unit weight (γ_{rf}) and Mohr–Coulomb angle of internal friction (at peak strength) (ϕ_{rf}) of the fill in the retained fill zone (i.e., immediately behind the reinforced soil zone)

- moist unit weight (γ_f) and Mohr–Coulomb strength parameters (at peak strength) (c_f and ϕ_f) of the foundation soil; for purely cohesive foundation: $\phi_f = 0$ and $c_f =$ undrained shear strength, which can be determined by appropriate field or laboratory tests; for granular foundation: $c_f = 0$, and ϕ_f can be estimated from standard penetration blow count or other in situ shear strength tests.

For a granular fill material, the value of ϕ in a set of direct shear tests should be determined by drawing a *best-fit straight line* through the data points and ignoring the cohesion.

Step 3: **Determine forces acting on reinforced soil mass**

The reinforced soil mass together with block facing (where the block facing is regarded as a façade) should be considered as a rigid body when evaluating external stability. Figures 5.36(a) and (b) show forces acting on and in the reinforced soil mass for external stability evaluation of a GRS wall with a level crest (also referred to as horizontal backslope) and with a sloping crest (also referred to as sloping backslope), respectively. The figure includes body force (e.g., self-weight), traffic surcharge, and earth thrusts from the retained fill on the reinforced soil mass. The coefficient of Rankine active earth pressure, K_A, in Figure 5.36(a) is calculated as: $K_A = \tan^2\left(45° - \dfrac{\phi_{rf}}{2}\right)$.

GRS walls with a break in the crest slope (see Figure 5.37), referred to as a broken crest or broken backslope, can be analyzed as an equivalent sloping crest (shown by a dashed line in Figure 5.37) if the break in the slope is located within $2H$ from the back of the wall face (H = total wall height measured from the bottom of the lowest block), as suggested by AASHTO (2002, 2014). Alternatively, the actual slope geometry of the broken crest can be analyzed by the Coulomb theory, with the soil–wall friction angle δ as discussed in Section 2.4.

In subsequent steps, the design computations will involve resisting forces/moments and driving forces/moments for evaluation of safety factors. To be on the conservative side, for calculations of *resisting* forces/moments, the traffic surcharge in the crest area is assumed to begin at the end of the reinforced zone (see Figure 5.36(a)). For calculations of *driving* forces/moments, on the other hand, the surcharge is assumed to extend over the entire crest area.

Step 4a (for walls with a sloping crest, skip to Step 4b): **Determine a tentative reinforcement length of a GRS wall with a level crest**

A tentative reinforcement length is obtained by considering two potential failure modes: (i) lateral sliding failure and (ii) eccentricity failure. If all reinforcement layers are to have the same length, the tentative reinforcement length should be the greater of the two lengths obtained from the above two modes of failure. Should construction of the wall involve excavation into rock outcrop or heavily overconsolidated clay, the costs of excavation to accommodate uniform reinforcement length may be impractically high. In such a situation, the use of shorter reinforcement in the lower portion of the reinforced zone (referred to as the *truncated base*) is allowed, especially if the fill material is free-draining and the foundation soil is firm. Note that the provision of a truncated base (see Step 5 of the NCHRP design method, Section 5.6.1) will affect external stability, movement of the wall face, etc. Note also that *overturning* failure check is not needed because it is always less critical than *eccentricity* failure (*lift-off* of reinforced soil mass) as long as the safety factor against overturning is smaller than 3.0. For added conservatism, all reinforcement lengths are to be measured from the back of the wall face.

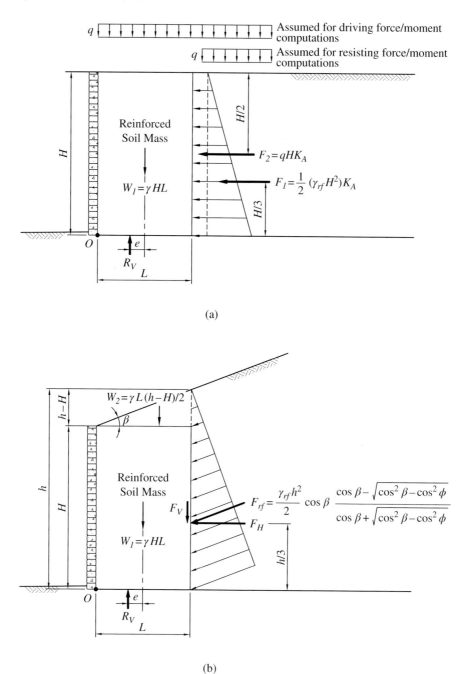

(a)

(b)

Figure 5.36 Forces acting on the reinforced zone of a GRS wall with (a) a level crest with traffic surcharge and (b) a sloping crest, the GRS-NLB design method

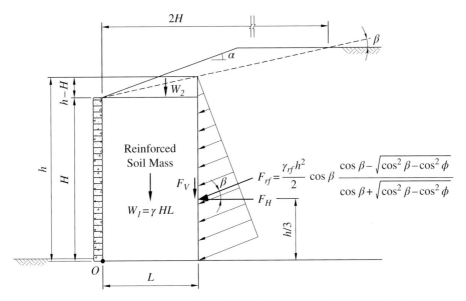

$$F_{rf} = \frac{\gamma_{rf}h^2}{2}\cos\beta\ \frac{\cos\beta - \sqrt{\cos^2\beta - \cos^2\phi}}{\cos\beta + \sqrt{\cos^2\beta - \cos^2\phi}}$$

Figure 5.37 Equivalent sloping crest (denoted by the dashed line) for a wall with a broken crest

(i) *Lateral sliding failure*

To obtain a tentative reinforcement length against lateral sliding failure, an initial trial value of $L_t = 0.5H$ may be assumed. For a GRS wall of uniform reinforcement length (see Figure 5.36(a)), the safety factor against lateral sliding is:

$$FS_{sliding} = \frac{\gamma H L_t \tan\delta}{0.5 K_A H(\gamma_{rf} H + 2q)} \tag{5-38}$$

Therefore, the required length of reinforcement (L_t) to resist lateral sliding failure is:

$$L_t = \frac{FS_{sliding}\ K_A H(\gamma_{rf} H + 2q)}{2\gamma H \tan\delta} \tag{5-39}$$

In Eqns. (5-38) and (5-39), H is the height of the wall at wall face measured from the bottom of the lowest facing block, and δ is the friction angle between the base of the reinforced zone and the foundation soil. If the lowest layer of reinforcement is on the ground surface, δ will be the friction angle between the geosynthetic reinforcement and the foundation soil, which can be obtained by performing interface shear tests of the two materials. If δ is greater than either ϕ or ϕ_f, the lesser value of ϕ and ϕ_f should be used for δ. In the absence of interface shear test results, $\delta = (2/3) \cdot$ (lesser value between ϕ and ϕ_f) may be assumed. The recommended minimum safety factor $FS_{sliding}$ against lateral sliding failure is 1.5.

(ii) Eccentricity failure

The reinforced soil mass in a GRS wall is typically subject to both vertical and horizontal forces acting above the base, hence the resultant force at the base of the reinforced soil mass without introducing a bending moment will usually be eccentric. The line of action of the eccentric resultant force needs to stay within the middle third of the base of the reinforced soil mass to prevent lift-off of the reinforced soil mass from the foundation soil (i.e., to prevent tension at the contact surface). Using the tentative reinforcement length L_t determined from the lateral sliding failure check, the eccentricity e, measured from the centerline of the reinforced soil mass, is:

$$e = \frac{F_1 \left(\dfrac{H}{3} \right) + F_2 \left(\dfrac{H}{2} \right)}{\gamma H L_t} \tag{5-40}$$

The parameters in Eqn. (5-40) are as defined in Figure 5.36(a). Note that surcharge q is not included in the denominator of Eqn. (5-40) as it is part of the resisting forces. The e value must not be greater than $L_t/6$. If $e > L_t/6$, a larger value of L_t should be selected, and the corresponding value of e should be re-calculated. This procedure needs to be repeated until a value of L_t that satisfies $e \le L_t/6$ is obtained.

Step 4b (for walls with a level crest, skip to Step 5): **Determine a tentative reinforcement length of a GRS wall with a sloping crest**

A tentative reinforcement length for GRS walls with a sloping crest needs to be obtained by an iterative procedure. A trial reinforcement length, L_t, is first assumed for determination of the safety factor against the lateral sliding failure at the base of the reinforced soil mass. The safety factor against lateral sliding, $FS_{sliding}$, needs to be at least 1.5. If $FS_{sliding}$ is less than 1.5, a longer trial length should be analyzed until $FS_{sliding} \ge 1.5$ is obtained. The tentative reinforcement length L_t that satisfies the lateral sliding failure criterion is then used to check against eccentricity failure (i.e., lift-off of reinforced soil mass). The reinforcement length needs to be increased again if it fails to meet the eccentricity failure criterion. When construction of a wall involves excavation into rock outcrop or heavily overconsolidated clay, use of shorter reinforcement in the lower portion of reinforced zone (referred to as the *truncated base*) to minimize excavation is allowed, especially if the fill material is free-draining and the foundation soil is firm. Note that the provision of a truncated base (see Step 5 of the NCHRP design method, Section 5.6.1) may affect external stability, movement of wall face, etc. Note also that there is no need to check overturning failure in design because it is always less critical than eccentricity failure as long as the safety factor against overturning is smaller than 3.0. For added conservatism, all reinforcement lengths are measured from the back of the wall face.

(i) Lateral sliding failure

To obtain a tentative reinforcement length against lateral sliding failure, an initial trial value of $L_t = 0.5H$ may be assumed. Using this trial value for L_t, a projected vertical height above the end of reinforced zone, h, is determined as $h = H + L_t \tan \beta$, where β is the angle of the sloping crest with the horizontal plane (see Figure 5.36(b)). The resultant active force F_{rf}, exerted by retained fill on the reinforced soil mass, inclining counterclockwise at an angle β from the horizontal plane, is:

$$F_{rf} = \frac{\gamma_{rf} h^2}{2} (\cos \beta) \frac{\cos \beta - \sqrt{\cos^2 \beta - \cos^2 \phi}}{\cos \beta + \sqrt{\cos^2 \beta - \cos^2 \phi}} \tag{5-41}$$

Once F_{rf} is determined, its horizontal and vertical components F_H and F_V (see Figure 5.36(b)) can be calculated as $F_H = F_{rf} (\cos \beta)$, and $F_V = F_{rf} (\sin \beta)$.

For a GRS wall of uniform reinforcement length (see Figure 5.36(b)), the safety factor against lateral sliding failure, $FS_{sliding}$, is:

$$FS_{sliding} = \frac{\left[0.5\gamma L_t (h - H) + \gamma H L_t + F_V \right] \tan \delta}{F_H} \tag{5-42}$$

In Eqn. (5-42), δ is the friction angle between the base of the reinforced zone and the foundation soil. If the lowest layer of reinforcement is on the ground surface, δ will be the friction angle between the geosynthetic and foundation soil, and the value of δ can be obtained by performing foundation soil–geosynthetic interface shear tests. If δ is greater than either ϕ or ϕ_f, the lesser of ϕ and ϕ_f should be used as δ in the calculation. In the absence of interface shear test results, $\delta = (2/3) \cdot$ (lesser value between ϕ and ϕ_f) may be assumed.

If $FS_{sliding} < 1.5$, the above procedure should be repeated with an increased L_t until a value of L_t that gives $FS_{sliding} \geq 1.5$ is obtained. This can be easily accomplished with the aid of a spreadsheet.

(ii) Eccentricity failure

The reinforced soil mass in a GRS wall is typically subject to both vertical and horizontal forces (acting above the base), hence the resultant force at the base of the reinforced soil mass without introducing a bending moment will be eccentric. The line of action of the eccentric resultant force needs to stay within the middle third of the base of the reinforced soil mass to prevent lift-off of the reinforced soil mass from the foundation soil. The length L_t determined from lateral sliding failure may be used to check eccentricity failure by calculating the eccentricity e (measured from the center line of the reinforced soil mass) to see if it satisfies the eccentricity requirement of $e \leq L_t /6$:

$$e = \frac{F_H \left(\dfrac{h}{3} \right) - F_V \left(\dfrac{L_t}{2} \right) - W_2 \left(\dfrac{L_t}{6} \right)}{R_V} \tag{5-43}$$

where R_V is the resultant of vertical forces ($R_V = W_1 + W_2 + F_V$). If $e \leq L_t/6$, the tentative length is said to satisfy the eccentricity failure criterion; otherwise, L_t needs to be increased, and a new value of e needs to be calculated until it meets the eccentricity requirement of $e \leq L_t/6$.

Step 5: Check stability against foundation bearing failure

Following the work of Meyerhof (1963), the safety factor against the foundation bearing failure, $FS_{bearing}$, of the reinforced mass can be determined by the following equation:

$$FS_{bearing} = \frac{0.5\gamma_f \left(L_t - 2e\right)^2 i_\gamma N_\gamma + c_f \left(L_t - 2e\right) i_c N_c}{\sigma_v} \tag{5-44}$$

where σ_v is the average reaction pressure at the base, which can be determined as follows.

For a level crest with uniform traffic surcharge q:

$$\sigma_v = \frac{W_1 + q L_t}{L_t - 2e} \tag{5-45}$$

For a sloping crest:

$$\sigma_v = \frac{W_1 + W_2 + F_V}{L_t - 2e} \tag{5-46}$$

The value of eccentricity, e, in Eqns. (5-45) or (5-46) has been determined in Step 4. Forces W_1, W_2, and F_V are as defined in Figure 5.36(b). The multipliers i_γ and i_c in Eqn. (5-44) are defined as:

$$i_r = \left(1 - \frac{\lambda}{\phi}\right)^2, \quad i_c = \left(1 - \frac{\lambda}{90°}\right)^2, \quad i_r = 0 \text{ if } \lambda \geq \phi \tag{5-47}$$

The angle of inclination λ is the angle between the vertical plane and the orientation of the resultant of all forces acting on the reinforced soil mass (i.e., the resultant of q, W_1, F_1 and F_2 for walls with a level crest, or the resultant of W_1, W_2, and F_{rf} for walls with a sloping crest, see Figure 5.36). The bearing capacity factors, N_γ and N_c are a function of friction angle (ϕ_f) of the foundation soil, as given in Figure 5.38. For a cohesive foundation, $N_\gamma = 0$ and $N_c = 5.14$.

The safety factor against foundation bearing failure, $FS_{bearing}$, should be at least 2.5. Note that the side shear of the reinforced zone is ignored for conservatism in the above computations, hence $FS_{bearing} = 2.5$ (slightly lower than the commonly used value of 3.0 for footings) is recommended. If the value of $FS_{bearing}$ is less than 2.5, L_t may be increased until $FS_{bearing} \geq 2.5$ is obtained. Alternatively, the embedment depth of the reinforced soil mass may be increased to increase resistance against foundation bearing failure. An increase in embedment is particularly effective if the foundation soil is granular. In this case, $\gamma_f D_f (L_t - 2e) N_q$ (where D_f = depth of embedment) should be added to the numerator in Eqn. (5-44) for the calculation of $FS_{bearing}$, i.e.,

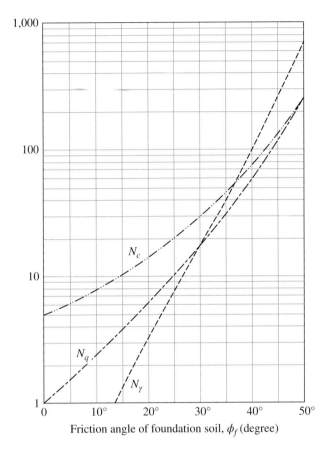

Figure 5.38 Bearing capacity factors N_c, N_γ, and N_q as a function of soil friction angle (modified after Perloff and Baron, 1976)

$$FS_{bearing} = \frac{0.5\,\gamma_f\left(L_t - 2e\right)^2 i_\gamma N_\gamma + c_f\left(L_t - 2e\right) i_c N_c + \gamma_f D_f\left(L_t - 2e\right)N_q}{\sigma_v} \qquad (5\text{-}48)$$

The values of N_q as a function of ϕ_f are given in Figure 5.38. If the reinforcement length becomes too large (say, $L_t > 0.7H$), especially if the foundation soil is cohesive, a reinforced soil foundation as described in Step 12 may be considered.

For GRS walls with a truncated base, the design computations are very similar to those described above. The detailed procedure is illustrated in Design Example 3 in Section 5.7.2.

Step 6: Select vertical reinforcement spacing and block height
Selecting a small value for vertical reinforcement spacing, in multiples of facing block height between 0.1 m (4 in.) and 0.3 m (12 in.), is recommended to take advantage of tight reinforcement spacing to gain significantly better performance and improved stability (Wu et al., 2010). Uniform spacing throughout is usually selected to avoid confusion during actual installation of geosynthetic reinforcement. For taller GRS walls

(say, $H \geq 6$ m), however, use of two or more values of vertical spacing may prove to be justifiable, especially for very long walls. When more than one value of reinforcement spacing is selected, smaller spacing is generally used in lower layers. Alternatively, if reinforcement spacing is to be kept constant for the entire wall, weaker reinforcement may be used for the upper part of the wall.

When reinforcement is not placed at every course of facing blocks, the *CTI tail* (shortened reinforcement in place of full-length reinforcement) on every other block may be employed. The CTI tail typically extends only 1.0 m (3 ft) behind the back of facing blocks (see Section 6.1.1). Great care needs to be exercised to avoid confusion on sequencing of reinforcement placement during construction when the CTI tail is used.

Step 7: **Determine required tensile stiffness and strength for reinforcement**
As noted in Section 5.3.1, both tensile stiffness and tensile strength are needed for the specification and design of geosynthetic reinforcement. Tensile stiffness is defined as the tensile resistance at a reference reinforcement strain. For GRS-NLB, a reference strain of 3.0% is recommended. Specification of a minimum value for tensile stiffness ensures that the selected reinforcement will be stiff enough to provide satisfactory performance under service loads. The required minimum reinforcement stiffness in the direction perpendicular to the wall face, $T_{@\varepsilon=3.0\%}$, can be determined by the following equation (see Section 5.3.1 for background information):

$$
T_{@\,\varepsilon=3.0\%} = \left[\frac{\sigma_h - \sigma_c}{0.7^{\left(\frac{s_v}{6d_{max}}\right)}} \right] s_v
\tag{5-49}
$$

where $T_{@\,\varepsilon=3.0\%}$ = required minimum tensile stiffness at $\varepsilon = 3\%$ on the load–deformation curve (as obtained from the wide-width tensile test, ASTM D4595 for geotextiles and D6637 for geogrids), σ_h = horizontal stress in GRS mass at the depth of reinforcement being considered, σ_c = lateral confining pressure of GRS mass at the depth of reinforcement being considered, s_v = reinforcement spacing, and d_{max} = maximum particle size of the backfill.

The largest value of σ_h typically occurs at the base of the wall and can be estimated by the Rankine active earth pressure theory. For a GRS wall with a level crest and subjected to a uniform surcharge, $\sigma_h = (\gamma H + q) K_A = (\gamma H + q) \tan^2(45° - \phi/2)$, where γ is the unit weight of the backfill, H is the height of the wall, q is the uniform surcharge, and K_A is the coefficient of Rankine active earth pressure. For a GRS wall with a sloping crest (slope angle = β, Figure 5.36(b)), the bottom layer of reinforcement is usually subject to the highest horizontal stress, and the largest vale of σ_h is $\gamma(H + L_t \tan \beta) \tan^2(45° - \phi/2)$.

For a GRS wall with dry-stacked concrete block facing, the value of σ_c can be estimated as $\sigma_c = \gamma_b \cdot D \cdot \tan \delta_b$, where γ_b is the *bulk* unit weight of the facing block, D is the depth of the facing block (in the direction perpendicular to the wall face), and δ_b is the friction angle between facing blocks (or between the geosynthetic reinforcement and the facing block, if geosynthetic reinforcement is installed between adjacent blocks).

The value of σ_c is very small for light-weight dry-stacked facing, and can be conservatively assumed to be zero.

In addition to specifying the stiffness requirement, the strength requirement for reinforcement is needed as well. The required minimum value of the ultimate strength of reinforcement in the direction perpendicular to the wall face, T_{ult}, can be determined by imposing a *ductility and long-term factor, F_{dl}*, on $T_{@\,\varepsilon=3.0\%}$ to ensure sufficient ductility (i.e., to prevent sudden collapse) and satisfactory long-term performance, i.e.,

$$T_{ult} = F_{dl} \cdot T_{@\,\varepsilon=3.0\%} \geq F_{dl} \left[\frac{\sigma_h - \sigma_c}{0.7^{\left(\frac{s_v}{6d_{max}}\right)}} \right] s_v \qquad (5\text{-}50)$$

From limited experience, the tensile strain corresponding to T_{ult} needs to be at least 5–7% (referred to as the ductility strain). The ductility and long-term factor F_{dl} is a function of soil properties and geosynthetic type. The following values of F_{dl} are recommended (Note: Plasticity index (*PI*) of any fill material greater than 6 is not recommended):

- If the plasticity index (*PI*) of fill material passing No. 40 sieve is 3 or less:

    F_{dl} = 3.5 for all geosynthetics

- If the plasticity index (*PI*) of fill material passing No. 40 sieve is between 4 and 6:

    F_{dl} = 5.5 for polypropylene geosynthetics

    F_{dl} = 5.0 for polyethylene geosynthetics

    F_{dl} = 4.0 for polyester geosynthetics

Note that if more than one reinforcement (i.e., reinforcements of different stiffnesses/ strengths) is selected, the σ_h-value in Eqn. (5-49) should be evaluated at the bottom of each group of reinforcement layers. Different values of required reinforcement stiffness and reinforcement strength for each group of reinforcement can then be determined.

Step 8: Select geosynthetic reinforcement

Select a geosynthetic product that satisfies the following conditions (cf. Figure 5.9):

a) Stiffness requirement: the tensile stiffness in the direction perpendicular to wall face, $T_{@\,\varepsilon=3.0\%}$ (tensile stiffness at 3% strain), must be equal to or greater than the value calculated by Eqn. (5-49).

b) Strength requirement: the ultimate strength, T_{ult}, in the direction perpendicular to wall face must be equal to or greater than the value calculated by Eqn. (5-50).

The load–deformation relationship of geosynthetic product, commonly expressed in terms of load per unit width versus tensile strain, can be determined by conducting a wide-width tensile test (ASTM D4595 for geotextiles, ASTM D6637 for geogrids). The T_{ult} value should be the minimum average roll value (MARV) of the product to account for statistical variance. For most geosynthetics that are not sensitive to

confining pressure (i.e., woven geotextiles and geogrids), the load–deformation relationship determined by in-isolation tests can be used for the selection of geosynthetic
reinforcement. For confining pressure-sensitive geosynthetics (i.e., nonwoven needle-
punched geotextiles), the intrinsic confined test proposed by Wu (1991), Ling et al.
(1992), and Ballegeer and Wu (1993) is recommended. Note that for pressure-sensitive
geosynthetics, the effect of confining pressure has been found to be much more pronounced on the tensile stiffness than on the ultimate strength. For example, the
increase in stiffness (secant modulus) for a nonwoven needle-punched geotextile
under a confining pressure of 80 kPa is found to be about 100%, whereas the corresponding increase in ultimate strength is only about 30%. Note that pressure-sensitive
geosynthetics are generally not selected in the design of GRS walls due to their relatively low stiffness values.

Step 9: **Estimate lateral movement of wall face**

The lateral displacement profile of the wall face of a block facing GRS wall can be
estimated by the following equation (Pham, 2009; Wu and Pham, 2010). The lateral
movement Δ_i at a given depth z_i of a GRS wall is:

$$
\Delta_i = 0.5 \left[\frac{K_A(\gamma z_i + q)s_v - (\gamma_b D s_v)(\tan \delta_b)}{K_{reinf}} \right] (H - z_i) \left[\tan\left(45° - \frac{\psi}{2}\right) + \tan(90° - \phi_{ds}) \right]
$$

$$(5-51)$$

where K_{reinf} = stiffness of the geosynthetic reinforcement ($K_{reinf} = E \cdot t$, see Figure 4.19),
ϕ_{ds} = friction angle of soil in the reinforced zone, as obtained from direct shear tests,
and ψ = angle of dilation of the soil in reinforced zone. Other parameters are as defined
in Section 5.3.3. In the absence of soil dilation data, the angles of dilation ψ can be estimated as 0°, 10°, and 20°, for loose, medium, and dense sands, respectively. The value
of ϕ_{ds} can be estimated from the plane-strain angle of friction, ϕ_{ps}, and the angle of
dilation, ψ, as (Jewell and Milligan, 1989):

$$
\tan \phi_{ds} = \frac{\sin \phi_{ps} \cos \psi}{1 - \sin \phi_{ps} \sin \psi}
$$

$$(5-52)$$

A vertical face without a batter is recommended for walls of low to medium height
(less than 20 ft or 6 m in height) because a batter usually offers little practical benefit in
these cases. For walls with a battered face, an empirical facing batter factor Φ_{fb}, suggested by Allen and Bathurst (2001), may be applied to account for the facing batter. For
a facing batter of 8°, Allen and Bathurst (2001) suggest an empirical equation for
determination of the lateral earth pressure on the wall face:

$$
\Phi_{fb} = \left(\frac{K_{ab}}{K_{av}} \right)^d = 0.88
$$

$$(5-53)$$

where K_{ab} and K_{av} are horizontal components of the active earth pressure coefficient for
a battered wall and a vertical wall, respectively, and d is a constant. Allen and Bathurst

(2001) found that $d = 0.5$ would yield the best fit for available T_{max} data, and recommended using $d = 0.5$ for determination of Φ_{fb}. For design of a battered wall, K_{ab} determined from Eqn. (5-53) can be used in place of K_A in Eqn. (5-51) for evaluation of the wall displacement profile. This procedure is illustrated in Design Examples 2 and 3 in Section 5.7.2.

For GRS walls with a sloping crest, the portion of soil above the top of the wall face can be regarded as an equivalent uniform overburden pressure of $0.5\gamma L(\tan\beta)$ for the term q in Eqn. (5-51). Note that Eqn. (5-51) ignores the down-drag force along the contact surface between the back of the wall face and the backfill behind. This will lead to a conservative (i.e., larger than actual) lateral displacement profile.

From Eqn. (5-51), the maximum lateral displacement of a wall, Δ_{max}, can readily be determined. If the maximum wall displacement exceeds a prescribed performance limit, measures should be taken to alleviate the problem. Measures to consider include decreasing reinforcement spacing and increasing reinforcement stiffness/strength. Strictly speaking, Eqn. (5-51) is applicable only to GRS walls with concrete block facing. For GRS walls with other types of facing, Δ_{max} would be different. Take geotextile wrapped-faced walls for instance, Δ_{max} will be about 15% larger than that calculated by Eqn. (5-51).

Step 10: Check stability against pullout failure
For GRS walls with a level crest and uniform reinforcement length, the safety factor against pullout failure for reinforcement at depth z_i (referred to as *reinforcement layer-i*) is:

$$Fs_i = \frac{2\gamma z_i \tan\delta\left[L_t - (H - z_i)\tan\left(45° - \frac{\phi}{2}\right)\right]}{\sigma_{h(max)} S_v} = \frac{Pr_i\left(pullout\ resisting\ force\ for\ layer\ i\right)}{Pd_i\left(pullout\ driving\ force\ for\ layer\ i\right)}$$

$$(5\text{-}54)$$

where L_t is the most updated value of reinforcement length (up to Step 9), $\sigma_{h(max)}$ is the maximum horizontal stress, and δ is the angle of friction at the soil–geosynthetic interface.

For GRS walls with a sloping crest, the sloping soil overburden can be regarded as an equivalent uniform surcharge, and the sloping crest wall considered as an equivalent level crest wall with an added wall height of $0.5 \cdot L_t \cdot (\tan\beta)$ or subjected to a vertical surcharge pressure of $0.5 \cdot \gamma \cdot L_t \cdot (\tan\beta)$.

The coefficient of interface friction, $f = \tan\delta$, can be determined from the results of a pullout test by the interface pullout formula (Eqn. (4-2), Section 4.2.4), i.e.,

$$\tan\delta = \frac{E \cdot t \cdot \ln\left(\frac{F_f}{E \cdot t} + 1\right)}{2 \cdot \sigma_n \cdot L}$$

$$(5\text{-}55)$$

where F_f = applied pullout force at failure (per unit width of reinforcement) in the pull-out test, L = total length of the reinforcement specimen used in the pullout test, E = Young's modulus per unit width of reinforcement, t = nominal thickness of reinforcement, and σ_n = normal stress (overburden) on the reinforcement specimen. In the absence of pullout test results, the angle δ can be assumed as 2/3 of the friction angle of the backfill, i.e., $\delta = (2/3)\phi$.

The maximum horizontal stress for a given layer of reinforcement at depth z_i (measured from the crest), $\sigma_{hi(max)}$, in Eqn. (5-54) can be determined as:

$$\sigma_{hi(max)} = K_A \cdot \gamma \cdot z_i \tag{5-56}$$

As in Step 9, for GRS walls with a battered wall face, an empirical facing batter factor, Φ_{fb}, suggested by Allen and Bathurst (2001), may be applied to account for the facing batter:

$$\Phi_{fb} = \left(\frac{K_{ab}}{K_{av}}\right)^d = 0.88 \tag{5-57}$$

where K_{ab} and K_{av} are the horizontal components of the active earth pressure coefficient for a battered wall and a vertical wall, respectively, and $d = 0.5$ for the best fit of available T_{max} data. For a battered wall, the value of K_{ab} determined from Eqn. (5-57) can be used in place of K_A in Eqn. (5-56). Again, wall batter is generally of little practical value for walls of low to medium heights. The maximum tensile force, i.e., the denominator of Eqn. (5-54), $\sigma_{h(max)} \cdot s_v$, should be calculated for each selected value of vertical spacing. To be conservative, the contribution of surcharge q is ignored in Eqn. (5-54). For a wall on uniform vertical spacing and with a uniform reinforcement length for all layers, the maximum horizontal stress ($\sigma_{h(max)}$) at the lowest reinforcement layer should be used for calculation of T_{max}. For a wall of more than one reinforcement spacing, T_{max} should be calculated at the lowest reinforcement layer of each spacing used in design.

In all prevailing design methods, stability against pullout failure is checked by requiring each and every layer of reinforcement to meet a minimum safety factor against pullout. As discussed in Section 5.3.2, it is not a realistic scenario that a single reinforcement layer is being pulled out while all the adjacent reinforcement layers remain immobilized in a closely spaced reinforced soil mass. In other words, pullout failure cannot occur in a single layer alone when reinforcement layers are tightly spaced. If pullout failure is to occur, it will only occur as a group. To evaluate pullout of reinforcement, a group safety factor as defined in Section 5.3.2 may be used to allow for a more rational evaluation of pullout stability.

It is recommended that the *group* safety factor be evaluated by the *rule of three*, i.e., pullout failure will occur if the group safety factor for any three consecutive reinforcement layers is below a prescribed value (see Section 5.3.2). The group safety factor $Fs_{(group)}$, with its mid-layer located at depth z_i, is defined as:

$$Fs_{i(group)} = \frac{Pr_{i-1} + Pr_i + Pr_{i+1}}{Pd_{i-1} + Pd_i + Pd_{i+1}} \tag{5-58}$$

For a level crest with surcharge q:

$$Pr_i = 2\,\gamma\,z_i \tan\delta \left[L_t - (H - z_i) \tan\left(45° - \frac{\phi}{2}\right) \right] \tag{5-59}$$

$$Pd_i = \left(\frac{1}{W}\right) K_A \left(\gamma\,z_i + q\right) s_v \tag{5-60}$$

For a sloping crest:

$$Pr_i = 2\gamma\left(z_i + 0.5 L_t \tan\beta\right) \tan\delta \left[L_t - (H - z_i) \tan\left(45° - \frac{\phi}{2}\right) \right] \tag{5-61}$$

$$Pd_i = \left(\frac{1}{W}\right) K_A \gamma \left(z_i + L_t \tan\beta\right) s_v \tag{5-62}$$

where $W = 0.7^{\left(\frac{s_v}{6d_{max}}\right)}$, and the subscripts $i-1$, i, and $i+1$ refer to reinforcement layers at depths z_{i-1}, z_i and z_{i+1}, respectively. Eqn. (5-58) states that the group safety factor against reinforcement pullout is the sum of resisting pullout forces for any three consecutive layers divided by the sum of the driving pullout forces for the same three layers. Note again that there is no reinforcement above the top layer or beneath the bottom layer of a wall, hence the values of the non-existent layers should be taken as zero (i.e., $Pr_{i-1} = Pd_{i-1} = 0$ for evaluation of $F_{s\,(group)}$ for the top layer, and $Pr_{i+1} = Pd_{i+1} = 0$ for evaluation of $F_{s\,(group)}$ for the bottom layer). Note also that surcharge q does not appear in Eqn. (5-59) as it is for calculation of a resisting force, but appears in Eqn. (5-60) as it is for calculation of a driving force (see Figure 5.23(a)). If the any of the $F_{s\,(group)}$ is less than 1.5, the reinforcement length L_t will need to be increased until all values of $F_{s\,(group)}$ are at least 1.5.

Step 11: Check stability against rotational slide-out failure

Using the updated reinforcement length determined in Step 10, stability against rotational slide-out failure should be evaluated. This can be carried out by using limiting-equilibrium based slope stability computer software such as SLOPE-W, STABL WV, ReSSA, etc.

The rotational slide-out analysis should include not only analysis of overall slide-out failure encompassing the entire reinforced soil wall, but also analysis of compound failure (see Figure 5.4), especially in situations where the wall is situated over sloping or soft ground. For a trial slip surface passing through the soil–geosynthetic composite mass, the strength parameters of the reinforced soil mass (c_R and ϕ_R) can be taken as (Wu et al., 2010):

$$\phi_R = \phi, \quad c_R = \left[0.7^{\left(\frac{s_v}{6d_{max}}\right)}\right] \frac{T_{ult}}{2s_v} \sqrt{K_P} + c \tag{5-63}$$

where K_P = coefficient of Rankine passive earth pressure, T_{ult} = tensile strength of the reinforcement, as determined in Step 8, s_v = vertical spacing of reinforcement, d_{max} = maximum particle size of the reinforced fill, and c = cohesion of the soil (c can be ignored if it is small, say less than 10 kPa).

The factor of safety against any trial of rotational slide-out failure should be at least 1.3. Otherwise, the reinforcement length may be increased to provide an adequate safety margin. Alternatively, a viable ground improvement technique may be employed to improve the strength of the foundation soil when warranted. The design reinforcement length can then be finalized at the completion of this design step. Recall that all reinforcement lengths referred to in the design computations are to measure from the back of facing.

Step 12: Determine facing requirements

Facing requirements involve three aspects: embedment, facing batter, and connection strength.

(a) Embedment

If there is no concern for erosion or other untoward events surrounding the wall base during design life (typically 75 years), the required embedment of a GRS-NLB wall can be a nominal depth of 0–200 mm, i.e., up to one typical concrete block height. Note that a nominal embedment of ½ to 1 block height can help to prevent outward movement of facing blocks during construction before the wall gets taller.

Design of shallow foundations requires that the base of the footing be placed below the maximum frost penetration depth to alleviate uneven heaving due to formation of ice lenses in the frost zone. Since heaving of the foundation soil can be *absorbed* by the reinforced soil mass of a GRS-NLB wall, and will usually not hinder wall performance, it can be argued that embedding the wall base below the maximum frost penetration depth is unnecessary. However, if the foundation soil contains frost-susceptible soils (e.g., silty soils, see Holtz et al., 2011 for details), it is recommended that the foundation soils be excavated to the maximum frost penetration depth and replaced with a non-frost-susceptible soil. The backfill in the excavated zone can be reinforced by geosynthetic reinforcement to form a reinforced soil foundation (RSF) (Adams et al., 2011a). A typical RSF for GRS abutments involves excavating a pit that is 0.5L deep (L = reinforcement length, determined in Step 11), backfilling it with compacted road base material, and reinforcing it by the same reinforcement used in the reinforced soil wall on 0.2–0.3 m vertical spacing. The lateral extent of the RSF should at least cover the footprint of the reinforced fill and should extend no less than 0.25L in front of the wall face. A generic design protocol for RSF has recently been proposed by Wu (2018).

For GRS walls constructed along rivers and streams, embedment depths should be at least 0.6 m (2 ft) below the potential scour depth, as determined based on subsurface investigation and hydraulic studies. For GRS walls founded on a slope, a minimum horizontal bench width of 0.3H (H = wall height) should be provided.

At the base of the wall face, a leveling pad made of compacted granular soil may be formed to help align the first course of facing blocks. Good alignment of the first course of facing blocks is essential to overall alignment of a concrete block GRS wall. An alternative is to use a concrete leveling pad (see Section 6.2.4 about leveling pads).

(b) Facing batter

A vertical face is recommended for low to medium height concrete block walls (not exceeding 20 ft or 6 m in height). The wall face may batter slightly, typically from 1:10 to 1:12 (horizontal to vertical) to help mask minor lateral movement of the wall face. For design computations, slightly battered walls can be considered as vertical walls with only small errors.

(c) Connection strength

The facing of GRS walls should be designed to resist lateral pressure from the reinforced soil mass and to resist down-drag force resulting from differential settlement between the reinforced soil mass and facing blocks. The connection strength required for a concrete block GRS wall is influenced by reinforcement spacing, constraints on facing movement (including weight of blocks and facing connection enhancement elements), surcharge on crest, reinforcement tensile properties, stress–strain–strength properties of the reinforced fill, relative settlement between facing blocks and reinforced mass, and construction sequence.

For walls that satisfy the four conditions for the bin pressure diagram (see Section 5.3.5), frictional resistance between concrete blocks has been found to be more than sufficient to satisfy connection stability requirements (i.e., resisting forces are always greater than driving forces at all depths). The four conditions are: (1) the facing offers insignificant restraint to lateral movement of the reinforced soil mass, (2) the fill material behind the facing is free-draining so that there is little hydraulic pressure involved, (3) the soil–reinforcement interface bonding is maintained under the design load, and (4) the reinforcement is sufficiently stiff. For walls which fail to satisfy any of the four conditions, use of the *connection force equations* (i.e., Eqns. (5-12) to (5-15)) is recommended to check if facing connection stability is satisfied. It is highly recommended that the reinforcement spacing is kept small, preferably on 0.2 m spacing for GRS-NLB. For closely spaced GRS walls with reinforcement at every facing block and reinforcement spacing not greater than 0.3 m, connection stability is satisfied regardless of wall height (see Section 5.3.5).

It is recommended that reinforcement sheets be placed in such a way that they cover the top surface area of the facing blocks as much as possible without protruding out of the wall face; in addition, when lightweight blocks are used, the top 2–3 courses of facing blocks be bonded together by concrete mix or mortar. This measure will offer added resistance to the lateral earth thrust under unexpected external loads.

5.7.2 Design Examples: The GRS-NLB Method

Three design examples for the GRS-NLB design method are given in this section, including (1) a GRS wall of a level crest, (2) a GRS wall of a broken crest, and (3) a GRS wall with a truncated base.

Design Example 1: To design a concrete block GRS wall of a level crest in a non-load-bearing application, as shown in Figure 5.39.

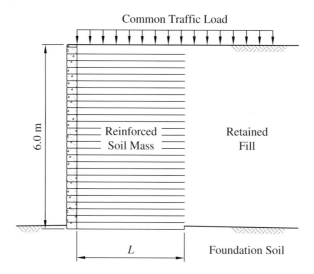

Figure 5.39 Cross-section of wall, Design Example 1 of the GRS-NLB design method

Given Conditions:

a) General:

- Uniform wall height: H = 6.0 m, vertical wall face.
- Level ground and a level crest.
- Subject to common traffic loads only.
- All geosynthetic reinforcement layers are to have the same length.

b) Backfill and retained fill:

- The backfill for the reinforced soil mass is a silty gravelly sand, with 12% of fines (by weight); maximum particle size = 25 mm (1 in.); liquid limit, LL = 10; plasticity index, PI = 2; moist unit weight, γ = 18.0 kN/m³; cohesion, c = 5 kPa; internal friction angle at peak strength, ϕ = 35°; and angle of dilation, ψ = 5°.

- Soil in the retained earth zone (i.e., soil behind the reinforced zone) is the same as that in the reinforced zone and with similar density and moisture.

- Surface runoff and subsurface drainage are to be provided properly.

c) Foundation soil:

- The foundation soil is uniform up to 8 m below the ground surface; moist unit weight, γ_f = 18.8 kN/m^3; cohesion, c_f = 9.5 kPa; angle of internal friction, ϕ_f = 38°.

- There is no concern for erosion or other untoward events surrounding the planned location of the wall base during design life.

- Deep free water level.

d) Loading:

- Common traffic loads only; vertical surcharge uniformly distributed over the crest, q = 12 kPa.

- No concern for seismic loads.

e) Performance limit:

- Maximum permissible lateral movement of wall face is 1% of the wall height, i.e., Δ_{max} = (6.0 m × 1,000 mm/m) (1%) = 60 mm.

f) Facing:

- Dry-stacked concrete blocks, with a vertical wall face.

- Facing blocks: split face CMU block (0.2 m × 0.2 m × 0.4 m) with hollow cells; bulk unit weight, γ_b = 12.6 kN/m^3 (80 lb/ft^3).

- Friction coefficient between block and geosynthetics, tan δ_b = 0.4.

- Interface friction and adhesion between backface of block and backfill: ignored (for conservatism).

Design Computations:

Step 1:   Establish wall profile and verify design conditions

The cross-section of the wall is as depicted in Figure 5.39. The conditions for non-load-bearing applications as listed at the beginning of Section 5.7 are verified, including fines content of reinforced backfill = 12% ≤ 20%, liquid limit = 10 ≤ 30, and plasticity index (*PI*) = 2 ≤ 6.

Step 2:   Determine design parameters for reinforced fill, retained fill, and foundation

The values of the soil parameters are:

$$\gamma = \gamma_{rf} = 18.0 \text{ kN/m}^3, \quad \gamma_f = 18.8 \text{ kN/m}^3$$

$$c = c_{rf} = 0 \text{ (ignored)}, \quad \phi = \phi_{rf} = 35°, \quad \psi = 5°, \quad c_f = 9.5 \text{ kPa}, \quad \phi_f = 38°$$

Step 3: Determine forces acting on reinforced soil mass

The forces acting on the reinforced soil mass for external stability checks are shown in Figure 5.40. The active Rankine earth pressure coefficient of the retained fill, $K_A = \tan^2 [45° - (35°/2)] = 0.27$.

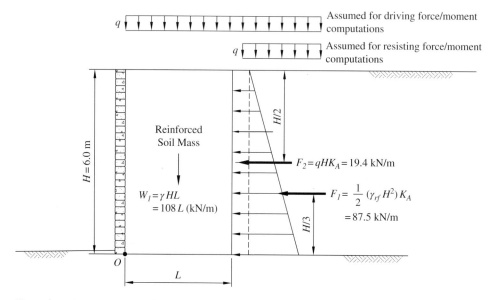

Figure 5.40 Forces acting on the reinforced soil mass for external stability evaluation, Design Example 1 of the GRS-NLB method

Step 4: Determine a tentative reinforcement length (for a GRS wall with a level crest)

Tentative reinforcement lengths (L_t) against lateral sliding failure and eccentricity failure are determined as follows:

(i) Lateral sliding failure

Determine L_t against lateral sliding failure:

$$L_t = \frac{F_s K_A H (\gamma_{rf} H + 2q)}{2\gamma H \tan\delta}$$

$$= \frac{(1.5)(0.27)(6.0)\left[(18.0)(6.0)+2(12)\right]}{2(18.0)(6.0)\tan\left[\left(\frac{2}{3}\right)35°\right]} = 3.4\,(\text{m})$$

(ii) Eccentricity failure

Determine L_t needed to prevent eccentricity failure:

Using L_t = 3.4 m (as determined in (i), above), the eccentricity, e, can be determined as (Eqn. (5-40)):

$$e = \frac{F_1(H/3) + F_2(H/2)}{\gamma H L_t}$$

$$= \frac{(87.5)(6.0/3) + (19.4)(6.0/2)}{(18.0)(6.0)(3.4)}$$

$$= 0.64 \text{ (m)} > \frac{L_t}{6} = 3.4 \text{ m}/6 = 0.57 \text{ m} \quad (\text{N.G.})$$

$$\text{Set } e = \frac{L_t}{6} = \frac{(87.5)(6.0/3) + (19.4)(6.0/2)}{(18.0)(6.0)(L_t)}$$

hence $L_t = 3.6$ m and $e = 0.6$ m

Therefore, the tentative reinforcement length needs to be increased from $L_t = 3.4$ m (determined from lateral sliding failure check) to $L_t = 3.6$ m (determined from eccentricity failure check).

Step 5: **Check stability against foundation bearing failure**
From Figure 5.38, for $\phi_f = 38°$, bearing capacity factors $N_c = 65$ and $N_\gamma = 80$

Vertical load $= (\gamma H + q) L_t = [(18.0)(6.0) + 12)](3.6) = 432$ (kN/m)

Horizontal load $= F_1 + F_2 = 87.5 + 19.4 = 106.9$ (kN/m)

Angle of inclination for the resultant load, $\lambda = \tan^{-1}(106.9/432) = 13.9°$ $(< \phi = 35°)$

Hence, $i_\gamma = (1 - \lambda/\phi)^2 = (1 - 13.9°/35°)^2 = 0.36$

$$i_c = \left(1 - \lambda/90°\right)^2 = \left(1 - 13.9°/90°\right)^2 = 0.72$$

From Eqns. (5-44) and (5-45)

$$FS_{bearing} = \frac{0.5\,\gamma_f\,(L_t - 2e)^2\,i_\gamma\,N_\gamma + c_f\,(L_t - 2e)\,i_c\,N_c}{(W_1 + q L_t)/(L_t - 2e)}$$

$$= \frac{0.5(18.8)(3.6 - 2\times0.6)^2(0.36)(80) + (9.5)(3.6 - 2\times0.6)(0.72)(65)}{\left[(18)(6.0)(3.6) + 12(3.6)\right]/(3.6 - 2\times0.6)}$$

$$= 2.8 \, (\geq 2.5, \, OK)$$

Since embedment of the wall is to be only nominal, its effect is not included in the calculation of $FS_{bearing}$ (Note: $FS_{bearing}$ determined by ignoring the effect of embedment has already been found adequate). The tentative reinforcement length of $L_t = 3.6$ m, therefore, remains unchanged.

Step 6: **Select vertical reinforcement spacing and block height**
Uniform spacing of 0.2 m (i.e., $s_v = 0.2$ m) is selected. The height of each facing block is to be 0.2 m.

Step 7: **Determine required tensile stiffness and strength for reinforcement**

The confining stress of the reinforced soil mass (σ_c) is:

$$\sigma_c = \gamma_b \cdot D \cdot \tan\delta_b = (12.6 \text{ kN/m}^3)(0.2 \text{ m})(0.4) = 1.0 \text{ (kPa)}$$

(for conservatism, use $\sigma_c = 0$ in design)

The maximum lateral stress (σ_h) is:

$$\sigma_h = (\gamma H + q)K_A = (18.0 \times 6.0 + 12)(0.27) = 32.4 \text{ (kPa)}$$

Required minimum reinforcement stiffness, $T_{@\varepsilon = 3.0\%}$, in the direction perpendicular to wall face is:

$$T_{@\varepsilon=3.0\%} = \left[\frac{\sigma_h - \sigma_c}{0.7\left(\frac{s_v}{6d_{max}}\right)}\right] s_v = \left[\frac{32.4 - 0}{0.7\left(\frac{0.2}{6 \times 0.025}\right)}\right](0.2) = 10.4 \text{ (kN/m)}$$

$PI = 2 \leq 3$, thus the ductility and long-term factor $F_{dl} = 3.5$, hence the required minimum value of ultimate reinforcement strength in the direction perpendicular to the wall face, T_{ult}, is:

$$T_{ult} = F_{dl} \cdot T_{@\varepsilon=3.0\%} = (3.5)(10.4) = 36.4 \text{ (kN/m)}$$

The tensile strain of the selected reinforcement corresponding to T_{ult} ($\geq 36.4 \text{ kN/m}$) needs to be at least 5–7%.

Step 8: **Select geosynthetic reinforcement**

Select a geosynthetic product with tensile stiffness $T_{@\varepsilon = 3.0\%} \geq 10.4 \text{ kN/m}$ and tensile strength $T_{ult} \geq 36.4 \text{ kN/m}$ at tensile strain ≥ 5–7%, both in the direction perpendicular to the wall face. A woven polypropylene geotextile with MARV = 38 kN/m (at $\varepsilon = 10\%$) and $T_{@\varepsilon = 3.0\%} = 12 \text{ kN/m}$ is selected.

Step 9: **Estimate lateral movement of wall face**

The lateral movement Δ_i at depth z_i is:

$$\Delta_i = 0.5\left[\frac{K_A(\gamma z_i + q)s_v - (\gamma_b D s_v)(\tan\delta_b)}{K_{reinf}}\right](H - z_i) \times$$

$$\left[\tan\left(45° - \frac{\psi}{2}\right) + \tan(90° - \phi_{ds})\right]$$

The reinforcement stiffness, K_{reinf}, is equal to $T_{@\varepsilon = 3.0\%}/(3\%) = 12 \text{ kN/m}/(3\%) = 400$ kN/m. The values of the other parameters are: $K_A = 0.27$, $\gamma = 18.0 \text{ kN/m}^3$, $q = 12 \text{ kN/m}^2$, $s_v = 0.2 \text{ m}$, $\gamma_b = 12.6 \text{ kN/m}^3$, $D = 0.2 \text{ m}$, $\tan\delta_b = 0.4$, $H = 6.0 \text{ m}$, and $\psi = 5°$. The direct shear friction angle ϕ_{ds} can be estimated as:

$$\phi_{ds} = \tan^{-1}\left[\frac{\sin\phi_{ps}\cos\psi}{1-\sin\phi_{ps}\sin\psi}\right] = \tan^{-1}\left[\frac{(\sin 35°)(\cos 5°)}{1-(\sin 35°)(\sin 5°)}\right] = 31°$$

The lateral movement profile for a vertical wall face is shown in Figure 5.41. From the profile, the maximum lateral movement is found to be 33 mm, smaller than the specified performance limit of 60 mm.

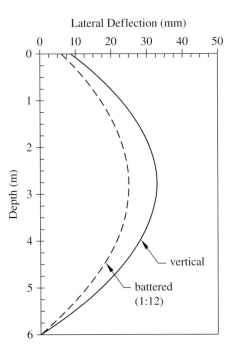

Figure 5.41 Calculated lateral movement profiles for a vertical wall face and 1:12 battered wall face: Design Example 1 of the GRS-NLB method

Should a facing batter of 1 (horizontal):12 (vertical) is employed (instead of a vertical wall specified in this example), the K_A value will need to be modified. The empirical correction suggested by Allen and Bathurst (2001) is:

$$\Phi_{fb} = \left(\frac{K_{ab}}{K_{av}}\right)^{0.5} = 0.88$$

Therefore the term K_A in the equation for Δ_i needs to be replaced by $K_{ab} = (0.88)^2(0.27) = 0.21$, and the maximum lateral movement will reduce from 33 mm to 25 mm (see Figure 5.41), both meeting the performance limit of 60 mm.

Step 10:   Check stability against pullout failure
The pullout resisting force (Pr_i) and driving force (Pd_i) for reinforcement *layer-i* are (Eqns. (5-59) and (5-60)):

$$Pr_i = 2\gamma\, z_i \tan\delta \left[L_t - (H - z_i)\tan\left(45° - \frac{\phi}{2}\right) \right]$$

$$Pd_i = \left(\frac{1}{W}\right) K_A \left(\gamma\, z_i + q\right) s_v$$

where $W = 0.7^{\left(\frac{s_v}{6d_{max}}\right)} = 0.7^{\left(\frac{0.2}{6\times(0.025)}\right)} = 0.62$. Note that surcharge q (=12 kPa) does not appear in the equation for Pr_i, but appears in the equation for Pd_i. The pullout resisting forces and driving forces for all layers are listed in Table 5.9.

Using the *rule of three*, i.e., the group pullout safety factor for *any three consecutive* reinforcement layers with its mid-layer located at depth z_i can be determined as:

$$Fs_{i(group)} = \frac{Pr_{i-1} + Pr_i + Pr_{i+1}}{Pd_{i-1} + Pd_i + Pd_{i+1}}$$

The group pullout safety factors of the wall are also listed in Table 5.9. Note that the group safety factor of *layer-i* in the table is the safety factor with *layer-i* being the mid-layer of the group. It is seen that the group safety factors for all layers are greater than 1.5, therefore the tentative reinforcement length will stay unchanged.

Step 11:   Check stability against rotational slide-out failure
A limiting-equilibrium based slope stability program (such as SLOPE-W, STABL WV, or ReSSA) can be used to evaluate different modes of rotational stability (including overall slide-out failure and compound failure). The analysis can be carried out with the reinforced soil zone having the following soil–geosynthetic composite strength parameters (because the cohesion of the soil is quite small, $c = 0$ is assumed):

$$\phi_R = \phi = 35°$$

$$c_R = \left[0.7^{\left(\frac{s_v}{6d_{max}}\right)} \right] \frac{T_{ult}}{2\, s_v} \sqrt{K_p}$$

$$= \left[0.7^{\left(\frac{0.2}{6\times0.025}\right)} \right] \left[\frac{38}{2(0.2)} \right] \sqrt{\frac{1}{0.27}} = 114\ (\text{kPa})$$

If the smallest safety factor against rotational slide-out failure is less than 1.3, the reinforcement length may be increased to raise the safety factor. Alternatively, if the critical rotational failure plane is passing through the foundation soil, a viable ground improvement technique may be employed to improve the strength of the foundation soil.

Step 12:   Determine facing requirements

a)   *Embedment*: The GRS wall is to be constructed with a nominal embedment of 0.1 m (half a block height) below the ground surface. This is based on the given information that there is very low probability of erosion or other unfavorable events occurring near the wall base through the design life of the wall.

Table 5.9 Resisting and driving pullout forces and group pullout safety factor for each reinforcement layer, $L_t = 3.6$ m, Design Example 1

z_i (m)	Pr_i (kN/m)	Pd_i (kN/m)	$Fs_{(group)}$
0.2	1.8	1.4	2.7
0.4	4.3	1.7	2.7
0.6	7.4	2.0	3.8
0.8	11.1	2.3	4.9
1.0	15.5	2.6	6.0
1.2	20.5	2.9	7.1
1.4	26.2	3.2	8.2
1.6	32.5	3.5	9.2
1.8	39.5	3.9	10.3
2.0	47.1	4.2	11.4
2.2	55.4	4.5	12.4
2.4	64.3	4.8	13.5
2.6	73.9	5.1	14.5
2.8	84.1	5.4	15.6
3.0	95.0	5.7	16.6
3.2	106.5	6.0	17.6
3.4	118.6	6.4	18.7
3.6	131.4	6.7	19.7
3.8	144.9	7.0	20.8
4.0	158.9	7.3	21.8
4.2	173.7	7.6	22.8
4.4	189.1	7.9	23.9
4.6	205.1	8.2	24.9
4.8	221.8	8.5	26.0
5.0	239.1	8.9	27.0
5.2	257.1	9.2	28.0
5.4	275.7	9.5	29.1
5.6	295.0	9.8	30.1
5.8	314.9	10.1	31.2
6.0	335.4	10.4	31.7

Note: The group safety factor against pullout is the sum of the resisting pullout forces of any consecutive three layers divided by the sum of the driving pullout forces of the same three layers. Note that there is no layer above the top layer (i.e., layer $i-1$ does not exist) or beneath the bottom layer (i.e., layer $i+1$ does not exist). The values of the non-existent layers (i.e., Pr_{i-1} and Pd_{i-1} for evaluation of the top layer, and Pr_{i+1} and Pd_{i+1} for evaluation of the bottom layer) are taken as zero.

b) *Facing batter*: The facing of the GRS wall is to be vertical, no batter.

c) *Connection strength requirements*: All facing blocks are to be dry-stacked. Reinforcement is at every course of facing blocks and reinforcement spacing is 0.2 m, hence connection stability is ensured. The top two courses of facing blocks are to be bonded by filling the hollow cells with concrete mix or mortar.

Design Summary for Reinforcement:

Reinforcement length (uniform for all depths), L = 3.6 m

Reinforcement spacing (same for all depths), s_v = 0.2 m

Reinforcement type: woven polypropylene with T_{ult} = 38 kN/m (@ ε = 10%) and $T_{@\varepsilon=3.0\%}$ = 12 kN/m

Design Example 2:   To design a concrete block GRS wall of a broken crest in a non-load-bearing application, as shown in Figure 5.42.

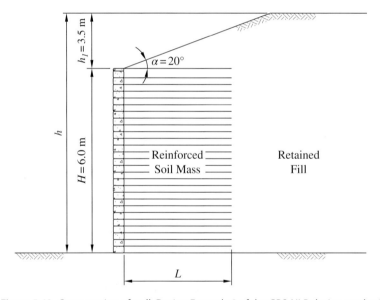

Figure 5.42 Cross-section of wall, Design Example 2 of the GRS-NLB design method

Given Conditions:

a) General:
- Uniform wall height: H = 6.0 m, with 1 (vertical) to 12 (horizontal) face batter.
- Level ground and a broken crest.
- All geosynthetic reinforcement layers are to have the same length.

b) Backfill and retained fill:

- The backfill for the reinforced soil mass is a silty gravelly sand, with 14% of fines (by weight); maximum particle size = 25 mm (1 in.); liquid limit, $LL = 10$; plasticity index, $PI = 2$; moist unit weight, $\gamma = 18.0$ kN/m^3; cohesion, $c = 5$ kPa; internal friction angle at peak strength, $\phi = 35°$; and angle of dilation, $\psi = 5°$.

- Same soil in the reinforced zone and retained earth zone (the soil behind reinforced zone) with similar placement density and moisture.

- Surface runoff and subsurface drainage are to be properly provided.

c) Foundation soil:

- The foundation soil is uniform up to 9 m below the ground surface, and moist unit weight, $\gamma_f = 18.8$ kN/m^3; cohesion, $c_f = 9.5$ kPa; angle of internal friction, $\phi_f = 38°$.

- There is no concern for erosion or other untoward events surrounding the planned location of the wall base during design life.

- Deep free water level.

d) Loading:

- A non-load-bearing wall.

- No concern for seismic loads.

e) Performance limit:

- Maximum permissible lateral movement of the wall face is 1% of wall height, i.e., Δ_{max} = (6.0 m × 1,000 mm/m) (1%) = 60 mm.

f) Facing:

- Dry-stacked concrete blocks.

- Facing blocks: split face CMU block (0.2 m × 0.2 m × 0.4 m) with hollow cells; bulk unit weight, $\gamma_b = 12.6$ kN/m^3 (80 lb/ft^3).

- Friction coefficient between block and geosynthetics, $\tan \delta_b = 0.4$.

- Interface friction and adhesion between backface of block and backfill: ignored (for conservatism).

Design Computations:

Step 1: **Establish wall profile and verify design assumptions**

The cross-section of the wall is shown in Figure 5.42. The conditions for non-load-bearing applications as listed at the beginning of Section 5.7 are verified, including fines content of reinforced backfill = 14% ≤ 20%, liquid limit = 10 ≤ 30, and plasticity index $(PI) = 2 ≤ 6$.

Step 2: **Determine design parameters for reinforced fill, retained fill, and foundation**

The values of the soil design parameters are:

$$\gamma = \gamma_{rf} = 18.0 \text{ kN/m}^3; \ \gamma_f = 18.8 \text{ kN/m}^3$$
$$c = c_{rf} = 0 \text{ (ignored)}; \ \phi = \phi_{rf} = 35°; \ \psi = 5°; \ c_f = 9.5 \text{ kPa}; \ \phi_f = 38°$$

Step 3: Determine forces acting on reinforced soil mass

The break in the slope is located within $2H$ from the back of the wall face, hence the broken crest can be converted into an equivalent sloping crest by following the procedure suggested by AASHTO (2002, 2014) (see Figure 5.37). The forces acting on the reinforced soil mass for external stability checks of the equivalent sloping crest are shown in Figure 5.43. The active Rankine earth pressure coefficient of the retained fill, $K_A = \tan^2[45° − (35°/2)] = 0.27$. The equivalent sloping crest is denoted by a dashed line in the figure. The angle of the equivalent sloping crest, $\beta = \tan^{-1}[h_1/(2H)] = 16.3°$.

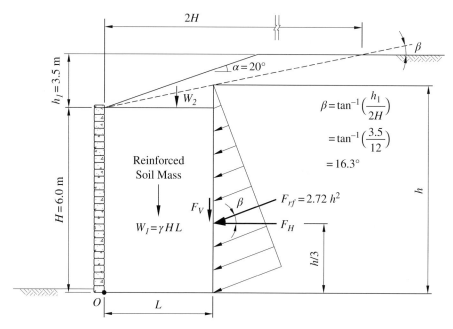

Figure 5.43 Forces acting on the reinforced soil mass for external stability evaluation, Design Example 2 of the GRS-NLB method (not to scale)

Step 4: Determine a tentative reinforcement length (for a GRS wall with a sloping crest)

A tentative reinforcement length (L_t) is determined by considering two failure modes: lateral sliding failure and eccentricity failure.

(i) Lateral sliding failure

Assume reinforcement length $L_t = 0.5H = 0.5\,(6.0\text{ m}) = 3.0$ m. For a GRS wall of uniform reinforcement length, the safety factor against lateral sliding at the lowest layer of reinforcement is (Eqn. (5-42)):

$$F_{sliding} = \frac{\left[0.5\,\gamma\,L_t\,(h-H) + \gamma\,H\,L_t + F_V\right]\tan\delta}{F_H}$$

where $h − H = L_t\,(\tan\beta) = (3.0\text{ m})\,(\tan 16.3°) = 0.9$ m, thus $h = 6.9$ m (see Figure 5.43). The resultant force exerted by the retained fill, F_{rf} (inclined at β counterclockwise from the horizontal) is (Eqn. (5-41)):

$$F_{rf} = \frac{(18.0)(h^2)}{2}(\cos 16.3°)\frac{\cos 16.3° - \sqrt{\cos^2 16.3° - \cos^2 35°}}{\cos 16.3° + \sqrt{\cos^2 16.3° - \cos^2 35°}}$$

$$= 2.72\,h^2 = 129\,(\text{kN/m})$$

$$F_H = F_{rf}\cos 16.3° = 2.6\,h^2 = 124\,\text{kN/m}$$

$$F_V = F_{rf}\sin 16.3° = 0.76\,h^2 = 36\,\text{kN/m}$$

Hence,

$$F_{sliding} = \frac{\left[0.5(18.0)(3.0)(0.9) + (18.0)(6.0)(3.0) + 36\right]\tan\left[\left(\frac{2}{3}\right)(35°)\right]}{124}$$

$$= 1.33 < 1.5\;(N.G.)$$

From trials that repeated the calculations with increasing value of L_t, or with the aid of a design spreadsheet, it was determined that $L_t = 3.5$ m would yield $F_s = 1.5$.

(ii) Eccentricity failure

To check eccentricity (lift-off) failure, the L_t as determined above is first used to calculate eccentricity e. For $L_t = 3.5$ m, $h = H + L_t (\tan \beta) = 6.0 + (3.5)$ $(\tan 16.3°) = 7.0$ (m). Thus, $W_1 = (18.0)(6.0)(3.5) = 378$ (kN/m), $W_2 = 0.5$ $(18.0)(3.5)(7.0 - 6.0) = 31.5$ (kN/m), $F_{rf} = 2.72 \cdot h^2 = 2.72\,(7.0)^2 = 133$ (kN/m), $F_H = F_{rf}(\cos 16.3°) = 128$ (kN/m), and $F_V = F_{rf}(\sin 16.3°) = 37$ (kN/m).

From Eqn. (5-43),

$$e = \frac{F_H\,(h/3) - F_V\,(L_t/2) - W_2\,(L_t/6)}{W_1 + W_2 + F_V}$$

$$= \frac{(128)(7.0/3) - (37)(3.5/2) - (31.5)(3.5/6)}{378 + 31.5 + 37}$$

$$= 0.48\,(\text{m})$$

Since $e = 0.48$ m $\le L_t/6 = 3.5/6 = 0.58$ m (OK), thus $L_t = 3.5$ m remains unchanged.

Step 5: Check stability against foundation bearing failure

From Figure 5.38, for $\phi_f = 38°$, bearing capacity factors are $N_c = 65$ and $N_\gamma = 80$

Vertical load $= W_1 + W_2 + F_v = 378 + 31.5 + 37 = 446.5$ (kN/m)

Horizontal load $= F_H = 128$ kN/m

Angle of inclination for the resultant load, $\lambda = \tan^{-1}(128/446.5) = 16.0°\;(< \phi = 35°)$

Hence, $i_\gamma = (1 - \lambda/\phi)^2 = (1 - 16.0°/35°)^2 = 0.29$
$i_c = (1 - \lambda/90°)^2 = (1 - 16.0°/90°)^2 = 0.68$ and
eccentricity $e = 0.48$ m (from Step 4)

From Eqns. (5-44) and (5-46),

$$FS_{bearing} = \frac{0.5\gamma_f \left(L_t - 2e\right)^2 i_\gamma N_\gamma + c_f \left(L_t - 2e\right) i_c N_c}{\left(W_1 + W_2 + F_V\right)/\left(L_t - 2e\right)}$$

$$= \frac{0.5(18.8)(3.5 - 2 \times 0.48)^2 (0.29)(80) + (9.5)(3.5 - 2 \times 0.48)(0.68)(65)}{446.5/(3.5 - 2 \times 0.48)}$$

$$= 14.1 \ (\geq 2.5, OK)$$

Step 6:   Select vertical reinforcement spacing and block height

Uniform spacing of 0.2 m (s_v = 0.2 m) is selected. The height of each facing block is to be 0.2 m also.

Step 7:   Determine required tensile stiffness and strength for reinforcement

The confining stress of the reinforced soil mass, σ_c, is:

$$\sigma_c = \gamma_b \cdot D \cdot \tan \delta_b = (12.6 \text{ kN/m}^3)(0.2 \text{ m})(0.4) = 1.0 \text{ (kPa)}$$

(for conservatism, use $\sigma_c = 0$ in design)

The maximum lateral stress, σ_h, at the wall base is

$$\sigma_{h(max)} = \left(\gamma H_{max} + q\right) \times K_A = \left[\gamma\left(H + L_t \tan \beta\right) + 0\right] K_A$$
$$= \left\{(18.0)\left[6.0 + (3.5)(\tan 16.3°)\right]\right\}(0.27) = 34.1 \text{ (kPa)}$$

The required minimum reinforcement stiffness, $T_{@\varepsilon = 3.0\%}$, in the direction perpendicular to the wall face, is:

$$T_{@\varepsilon=3.0\%} = \left[\frac{\sigma_h - \sigma_c}{\left(\frac{s_v}{6d_{max}}\right)^{0.7}}\right] s_v = \left[\frac{34.1 - 0}{\left(\frac{0.2}{6 \times 0.025}\right)^{0.7}}\right](0.2) = 11 \text{ (kN/m)}$$

$PI = 2 \leq 3$, thus $F_{dl} = 3.5$. The required minimum value of the ultimate reinforcement strength in the direction perpendicular to the wall face, T_{ult}, is:

$$T_{ult} = F_{dl} \cdot T_{@\varepsilon=3.0\%} = (3.5)(11) = 38.5 \text{ (kN/m)}$$

Also, the ductility strain of the selected reinforcement at T_{ult} ($\geq$ 38.5 kN/m) needs to be at least 5–7%.

Step 8:   Select geosynthetic reinforcement

Select a geosynthetic product with tensile stiffness $T_{@\varepsilon = 3.0\%} \geq 11$ kN/m and tensile strength $T_{ult} \geq 38.5$ kN/m (at $\varepsilon \geq 5$–7%), both in the direction perpendicular to the wall face. A woven polypropylene geotextile with MARV = 40 kN/m (at ε = 10%) and $T_{@\varepsilon = 3.0\%}$ = 12 kN/m is selected.

Step 9: **Estimate lateral movement of wall face**

The lateral movement of the wall face, Δ_i, at depth z_i is:

$$\Delta_i = 0.5 \left[\frac{K_A\left(\gamma z_i + q\right)s_v - \left(\gamma_b \, D \, s_v\right)\left(\tan\delta_b\right)}{K_{reinf}} \right] \left(H - z_i\right)\left[\tan\left(45° - \frac{\psi}{2}\right) + \tan\left(90° - \phi_{ds}\right) \right]$$

The reinforcement stiffness, K_{reinf} is equal to $T_{@\varepsilon=3.0\%}/(3\%) = 12 \text{ kN/m}/(3\%) = 400 \text{ kN/m}$. Other parameters are: $K_A = 0.27$, $\gamma = 18.0 \text{ kN/m}^3$, $q = 0.5\,\gamma\,L_t\,(\tan\beta) = 0.5(18.0)\,(3.5)(\tan 16.3°) = 9.2 \text{ (kN/m}^2)$, $s_v = 0.2 \text{ m}$, $\gamma_b = 12.6 \text{ kN/m}^3$, $D = 0.2 \text{ m}$, $\tan\delta_b = 0.4$, $H = 6.0 \text{ m}$, and $\psi = 5°$. The direct shear friction angle ϕ_{ds} can be estimated as:

$$\phi_{ds} = \tan^{-1}\left[\frac{\sin\phi_{ps}\cos\psi}{1 - \sin\phi_{ps}\sin\psi} \right] = \tan^{-1}\left[\frac{\left(\sin 35°\right)\left(\cos 5°\right)}{1 - \left(\sin 35°\right)\left(\sin 5°\right)} \right] = 31°$$

To account for facing batter,

$$\Phi_{fb} = \left(\frac{K_{ab}}{K_{av}} \right)^{0.5} = 0.88$$

The term K_A in the equation for calculating Δ_i is to be replaced by $K_{ab} = (0.88)^2 (0.27) = 0.21$ (as opposed to 0.27 for a vertical wall).

The lateral movement profile is shown in Figure 5.44. The maximum lateral movement is determined to be 24 mm, which is smaller than the specified performance limit of 60 mm. Note that a vertical wall without any batter would have sufficed.

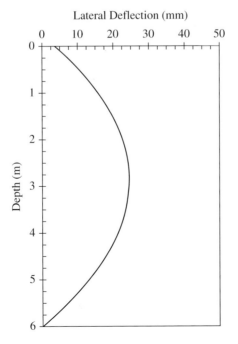

Lateral Deflection (mm)

Depth (m)

Figure 5.44 Calculated lateral movement profile, Design Example 2 of the GRS-NLB method

Step 10: **Check stability against pullout failure**

The pullout resisting force (Pr_i) and driving force (Pd_i) for reinforcement *layer-i* are (Eqns. (5-61) and (5-62)):

$$Pr_i = 2\gamma\left(z_i + 0.5L_t \tan\beta\right)\tan\delta\left[L_t - (H - z_i)\tan\left(45° - \frac{\phi}{2}\right)\right]$$

$$Pd_i = \left(\frac{1}{W}\right)K_A\gamma\left(z_i + L_t \tan\beta\right)s_v$$

where $W = 0.7^{\left(\frac{s_v}{6d_{max}}\right)} = 0.7^{\left(\frac{0.2}{(6\times 0.025)}\right)} = 0.62$. The pullout resisting force and driving force for each layer of the wall are listed in Table 5.10.

Using the *rule of three*, the group pullout safety factor for *any three consecutive* reinforcement layers with its mid-layer located at depth z_i can be determined as:

$$Fs_{i(group)} = \frac{Pr_{i-1} + Pr_i + Pr_{i+1}}{Pd_{i-1} + Pd_i + Pd_{i+1}}$$

The group pullout safety factors at all depths are also listed in Table 5.10. The group safety factor of *layer-i* is the safety factor with *layer-i* being the mid-layer of the group. It is seen that the group safety factors for all layers are greater than 1.5, therefore the tentative reinforcement length will stay unchanged.

Step 11: **Check stability against rotational slide-out failure**

A limiting-equilibrium based slope stability program (such as SLOPE-W, STABL WV, or ReSSA) can be used to evaluate different modes of rotational stability (including overall slide-out failure and compound failure). The analysis can be carried out with the following soil–geosynthetic composite strength parameters for the reinforced soil mass (because the cohesion of the soil is quite small, $c = 0$ is assumed):

$$\phi_R = \phi = 35°$$

$$c_R = \left[0.7^{\left(\frac{s_v}{6d_{max}}\right)}\right]\frac{T_{ult}}{2s_v}\sqrt{K_p}$$

$$= \left[0.7^{\left(\frac{0.2}{6\times 0.025}\right)}\right]\left[\frac{40}{2(0.2)}\right]\sqrt{\frac{1}{0.27}} = 120\ (kPa)$$

If the smallest safety factor against any modes of rotational slide-out failure is less than 1.3, the reinforcement length may be increased to raise the safety factor. Alternatively, if the critical rotational failure plane is passing through the foundation soil, a viable ground improvement technique may be employed to improve the strength of the foundation soil.

Table 5.10 Resisting and driving forces for pullout of each reinforcement layer and corresponding group pullout safety factor, $L_t = 3.5$ m, Design Example 2

z_i (m)	Pr_i (kN/m)	Pd_i (kN/m)	$F_{s(group)}$
0.2	5.3	1.9	2.9
0.4	8.3	2.2	3.8
0.6	11.9	2.5	4.8
0.8	16.2	2.9	5.7
1.0	21.1	3.2	6.7
1.2	26.6	3.5	7.7
1.4	32.8	3.8	8.7
1.6	39.7	4.1	9.7
1.8	47.2	4.4	10.7
2.0	55.3	4.7	11.7
2.2	64.1	5.0	12.8
2.4	73.5	5.4	13.8
2.6	83.6	5.7	14.8
2.8	94.3	6.0	15.8
3.0	105.7	6.3	16.8
3.2	117.7	6.6	17.9
3.4	130.4	6.9	18.9
3.6	143.7	7.2	19.9
3.8	157.7	7.5	20.9
4.0	172.3	7.9	22.0
4.2	187.5	8.2	23.0
4.4	203.4	8.5	24.0
4.6	220.0	8.8	25.0
4.8	237.2	9.1	26.1
5.0	255.0	9.4	27.1
5.2	273.5	9.7	28.1
5.4	292.6	10.0	29.2
5.6	312.4	10.4	30.2
5.8	332.8	10.7	31.2
6.0	353.9	11.0	31.7

Note: The group safety factor against pullout is the sum of the resisting pullout forces of any consecutive three layers divided by the sum of the driving pullout forces of the same three layers. Note that there is no layer above the top layer (i.e., layer $i-1$ does not exist) or beneath the bottom layer (i.e., layer $i+1$ does not exist). The values of the non-existent layers (i.e., Pr_{i-1} and Pd_{i-1} for evaluation of the top layer, and Pr_{i+1} and Pd_{i+1} for evaluation of the bottom layer) are taken as zero.

Step 12: **Determine facing requirements**

a) *Embedment*: Since there is no concern for erosion or other untoward events surrounding the planned location of the wall base during the design life, the GRS wall is to be constructed with a nominal embedment of 0.1 m below the ground surface.

b) *Facing batter*: The facing of the GRS wall is to have a batter of 1 (vertical) to 12 (horizontal); note that a vertical wall without a batter would suffice.

c) *Connection strength requirements*: All facing blocks are to be dry-stacked. Reinforcement is at every course of facing blocks and the reinforcement spacing is 0.2 m, hence connection stability is ensured (see Section 5.3.5). The top two courses of facing blocks are to be bonded by filling the hollow cells with concrete mix or mortar. To ensure facing stability under a sloping crest, the concrete mix may be reinforced with 0.2-m long steel pins through the cores. All other facing blocks are to be dry-stacked with their cells filled with compacted fill.

Design Summary for Reinforcement:

Reinforcement length (uniform for all depths), $L = 3.5$ m. [Note: The reinforcement length is slightly less than that in Design Example 1 because the soil above the reinforced zone in this design example is a dead weight (i.e., always present). In Design Example 1, however, the traffic surcharge above the reinforced zone is not considered present for evaluation of resisting force.]

Reinforcement spacing (same for all depths), $s_v = 0.2$ m

Reinforcement type: woven polypropylene with $T_{ult} = 40$ kN/m (at $\varepsilon = 10\%$) and $T_{@\varepsilon=3.0\%} = 12$ kN/m

Design Example 3: To design a truncated-base concrete block GRS wall of a level crest in a non-load-bearing application (i.e., subject to common traffic loads only), as shown in Figure 5.45.

Given Conditions:

a) General:

- Uniform wall height $H = 6.0$ m at the wall face, with 1 (vertical) to 10 (horizontal) face batter.
- Level ground and a level crest.
- Reinforcement lengths are truncated in the lower portion of the wall, with truncation inclining at approximately 45° from the horizontal ground surface.

b) Backfill and retained fill:

- The backfill for the reinforced soil mass is a silty gravelly sand, with 12% (by weight) of fines; maximum particle size = 25 mm (1 in.);

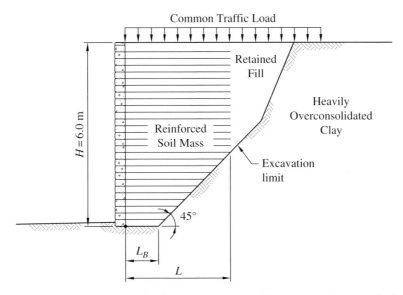

Figure 5.45 Cross-section of wall, Design Example 3 of the GRS-NLB design method

liquid limit, $LL = 10$; plasticity index, $PI = 2$; moist unit weight, $\gamma =$ 18.0 kN/m³; cohesion, $c = 5$ kPa; internal friction angle at peak strength, $\phi = 35°$; and angle of dilation, $\psi = 5°$.

- Same soil for the backfill in reinforced zone and retained fill (the soil behind the reinforced zone) with similar placement density and moisture.
- Surface runoff and subsurface drainage are to be provided properly.

c) Foundation soil:

- The foundation soil is a heavily overconsolidated clay, uniform up to 10 m below the ground surface, and moist unit weight, $\gamma_f = 18.8$ kN/m³; cohesion, $c_f = 9.5$ kPa; and angle of internal friction, $\phi_f = 30°$.
- There is no concern for erosion or other untoward events surrounding the planned location of the wall base during design life.
- Deep free water level.

d) Loading:

- A non-load-bearing wall, with common traffic loads only (an equivalent vertical surcharge, $q = 12$ kPa).
- No concern for seismic loads.

e) Performance limit:

- Maximum permissible lateral movement of wall face is 1% of the wall height, i.e., $\Delta_{max} = $ (6.0 m × 1,000 mm/m) (1%) = 60 mm.

f) Facing:

- Dry-stacked concrete blocks.
- Facing blocks: split face CMU block (0.2 m × 0.2 m × 0.4 m) with hollow cells; bulk unit weight, $\gamma_b = 12.6$ kN/m³ (80 lb/ft³).

- Friction coefficient between block and geosynthetics, tan δ = 0.4.
- Interface friction and adhesion between backface of block and backfill: ignored for conservatism.

Design Computations:

Step 1: Establish wall profile and check design assumptions

The cross-section of the wall is depicted in Figure 5.45. The conditions for non-load-bearing applications as listed at the beginning of Section 5.7 are verified, including fines content of reinforced backfill = 12% $\leq$ 20%, liquid limit = 10 $\leq$ 30, and plasticity index (PI) = 2 $\leq$ 6.

Step 2: Determine design parameters for reinforced fill, retained fill, and foundation

The values of the soil parameters needed for the design are:

$$\gamma = \gamma_{rf} = 18.0 \text{ kN/m}^3, \ \gamma_f = 18.8 \text{ kN/m}^3$$

$$c = c_{rf} = 0 \text{ (ignored)}, \ \phi = \phi_{rf} = 35°, \ \psi = 5°, \ c_f = 9.5 \text{ kPa}, \ \phi_f = 30°$$

Step 3: Determine forces acting on reinforced soil mass

With a trial reinforcement length of $0.5H$ ($L = 0.5H = 3.0$ m) in the upper portion of the wall and a trial reinforcement length at the base is set equal to $0.3H$ ($L_B = 0.3H = 1.8$ m), the forces acting on the reinforced soil mass of the GRS wall for external stability checks are as shown in Figure 5.46. Note that the active Rankine state can be mobilized in the zone of retained fill (see Section 2.4), hence the lateral earth

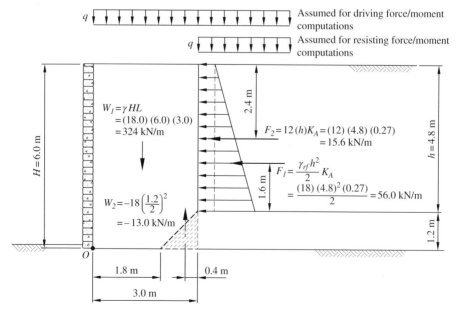

Figure 5.46 Forces acting on the reinforced soil mass for external stability checks, Design Example 3 of the GRS-NLB method

pressure in the retained fill above the truncated zone is assumed to be in the Rankine active state. The active Rankine earth pressure coefficient of the retained fill, $K_A = \tan^2 [45° − (35°/2)] = 0.27$.

Step 4: **Determine a tentative reinforcement length (for a GRS wall with a level crest)**

A tentative reinforcement length is obtained by considering two failure modes: lateral sliding failure and eccentricity failure.

(i) Lateral sliding failure

For the assumed reinforcement length $L = 0.5H = 3.0$ m, and reinforcement length at base $= 0.3H = 1.8$ m (an initial trial),

$$\text{Resisting force against lateral sliding} = (W_1 + W_2)\tan\delta$$
$$= (311)\tan[35° \times (2/3)] = 134\,(\text{kN/m})$$

Driving force for sliding $= F_1 + F_2 = 56 + 16 = 72$ (kN/m)

Factor of safety against lateral sliding $FS_{sliding} = 134/72 = 1.9 > 1.5$ (OK)

(ii) Eccentricity failure

To check eccentricity failure (i.e., lift-off at wall base), the tentative reinforcement length can be used for calculation of resultant forces and bending moment at the base of the reinforced soil mass. The calculation can be performed by treating the wall base as a projected horizontal plane, in the same way as for a wall without base-truncation, except the stabilizing force due to soil weight is now reduced due to truncation. The calculation of the resultant forces and bending moment on the projected wall base for $L = 3.0$ m and $L_B = 1.8$ m is shown in Figure 5.47.

The eccentricity of the resultant force R_V due to the bending moment (M) is

$$e\left(\text{for } L_t = 3.0 \text{ m}\right) = \frac{M_{centerline}}{P} = \frac{227}{311}$$
$$= 0.73\,(\text{m}) > \frac{L_t}{6} = \frac{3.0}{6} = 0.5\,\text{m}\,(\text{N.G.})$$

The initial trial tentative reinforcement length of $L_t = 0.3$ m (i.e., 0.5H) needs to be increased. A new trial reinforcement length of $L_t = 0.36$ m (or 0.6H) is selected. Using the same procedure illustrated in Figure 5.47, the new eccentricity becomes

$$e\left(\text{for } L_t = 3.6 \text{ m}\right) = \frac{M_{centerline}}{R_V} = \frac{237}{359}$$
$$= 0.66\,(\text{m}) > \frac{L_t}{6} = \frac{3.6}{6} = 0.6\,\text{m}\,(\text{N.G.})$$

The tentative reinforcement length of $L_t = 0.36$ m (or 0.6H) still fails the eccentricity criterion of $e \leq L_t/6$, therefore L_t needs to be increased further.

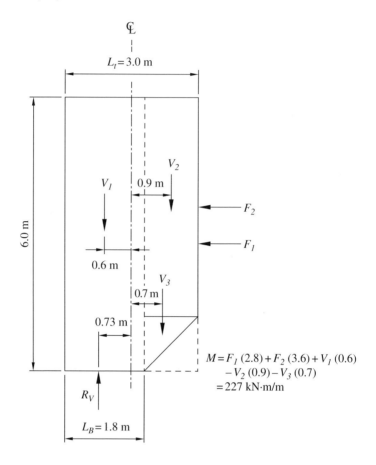

Resultants at base:

$R_V = 311\,\text{kN/m}$

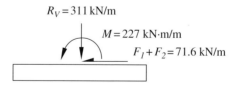

$M = 227\,\text{kN·m/m}$

$F_1 + F_2 = 71.6\,\text{kN/m}$

Figure 5.47 Resultant forces and moment on the base of reinforced soil mass, Design Example 3 of the GRS-NLB method

$L_t = 3.9$ m (or $0.65H$) is subsequently tried. Using the same procedure shown in Figure 5.47, the eccentricity for $L_t = 3.9$ m (or $0.65H$) is

$$e \text{ (for } L_t = 3.9\,\text{m)} = \frac{M_{centerline}}{R_V} = \frac{228}{381}$$

$$= 0.60\,(\text{m}) \le \frac{L_t}{6} = \frac{3.9}{6} = 0.65\,\text{m (OK)}$$

Therefore, reinforcement length $L_t = 3.9$ m (with $L_B = 0.18$ m) is adopted. Lateral sliding failure is re-calculated as follows to ensure $L_t = 3.9$ m is acceptable.

Resisting force against lateral sliding $= (V_1 + V_2 + V_3)\tan\delta$

$$= (381)\tan\left[35° \times (2/3)\right]$$

$$= 164 \text{ (kN/m)}$$

Driving force for sliding $= F_1 + F_2 = 37 + 13 = 50$ (kN/m)

Factor of safety against lateral sliding $FS_{sliding} = 164/50 = 3.3 > 1.5$ (OK)

Step 5: Check stability against foundation bearing failure

Foundation soil: $c_f = 9.5$ kPa and $\phi = 30°$; from Figure 5.38, for $\phi_f = 30°$, the bearing capacity factors are $N_c = 30$ and $N_\gamma = 18$

Vertical load $= R_v = V_1 + V_2 + V_3 = 381$ kN/m

Horizontal load $= F_1 + F_2 = 50$ kN/m

Angle of inclination for the resultant load with the vertical plane,

$$\lambda = \tan^{-1}(50/381) = 7.5° \ (< \phi = 35°)$$

hence, $i_\gamma = (1 - \lambda/\phi)^2 = (1 - 7.5°/35°)^2 = 0.62$

$$i_c = (1 - \lambda/90°)^2 = (1 - 7.5°/90°)^2 = 0.84$$

Average base pressure, $\sigma_v = R_v/(L_t - 2e) = 381/(3.9 - 2 \times 0.6) = 141.1$ kN/m^2

The safety factor against bearing capacity failure, $FS_{bearing}$, is (Eqns. (5-44) and (5-45))

$$FS_{bearing} = \frac{0.5\,\gamma_f\,(L_t - 2e)^2\,i_\gamma N_\gamma + c_f\,(L_t - 2e)\,i_c N_c}{\sigma_v + q}$$

$$= \frac{0.5(18.8)(3.9 - 2 \times 0.6)^2\,(0.62)(18) + (9.5)(3.9 - 2 \times 0.6)(0.84)(30)}{141.1 + 12}$$

$$= 9 \ (\geq 2.5, \text{ OK})$$

Step 6: Select vertical reinforcement spacing and block height

Uniform spacing of 0.2 m ($s_v = 0.2$ m) is selected. The height of each facing block is also to be 0.2 m.

Step 7: Determine required tensile stiffness and strength for reinforcement

The confining stress of the reinforced soil mass (σ_c) is:

$$\sigma_c = \gamma_b \cdot D \cdot \tan\delta_b = (12.6 \text{ kN/m}^3)(0.2 \text{ m})(0.4) = 1.0 \text{ (kPa)}$$

(for conservatism, use $\sigma_c = 0$ in design)

The maximum lateral stress (σ_h) is:

$$\sigma_h = (\gamma H + q) K_A = (\gamma H + q) K_A$$
$$= (18.0 \times 6.0 + 12)(0.27) = 32.4 \ (\text{kPa})$$

The required minimum reinforcement stiffness, $T_{@\varepsilon = 3.0\%}$, in the direction perpendicular to the wall face is:

$$T_{@\varepsilon = 3.0\%} = \left[\frac{\sigma_h - \sigma_c}{0.7 \left(\frac{s_v}{6 d_{max}} \right)} \right] s_v = \left[\frac{32.4 - 0}{0.7 \left(\frac{0.2}{6 \times 0.025} \right)} \right] (0.2) = 10.4 \ (\text{kN/m})$$

$PI = 2 \le 3$, thus the ductility and long-term factor, $F_{dl} = 3.5$. The required minimum value of ultimate reinforcement strength in the direction perpendicular to the wall face, T_{ult}, is:

$$T_{ult} = F_{dl} \cdot T_{@\varepsilon = 3\%} = (3.5)(10.4) = 36.4 \ (\text{kN/m})$$

Also, the ductility strain corresponding to T_{ult} $(\ge 36.4 \ \text{kN/m})$ of the selected reinforcement needs to be at least 5–7%.

Step 8: **Select geosynthetic reinforcement**
Select a geosynthetic product with tensile stiffness $T_{@\varepsilon = 3.0\%} \ge 10.4$ kN/m and tensile strength $T_{ult} \ge 36.4$ kN/m, both in the direction perpendicular to the wall face. A nonwoven polypropylene geotextile with MARV $= 38$ kN/m (at $\varepsilon = 10\%$), and $T_{@\varepsilon = 3.0\%} = 12$ kN/m is selected.

Step 9: **Estimate lateral movement of wall face**
Studies have suggested that for walls constructed with well-compacted free-draining granular backfill and over a firm foundation, there is not a significant difference between truncated-base walls and full-length walls in terms of lateral displacement of the wall face. Compared to a wall with full-length reinforcement, a truncated-base wall generally has slightly larger movement in the lower portion of the wall, and slightly smaller movement in the upper portion of the wall. The maximum deflection of a truncated-base wall is typically about 10% greater than that of a full-length reinforcement wall.

The lateral movement at the wall face, Δ_i, of a truncated-base wall at depth z_i is:

$$\Delta_i = 0.5 B_T \left[\frac{K_A (\gamma z_i + q) s_v - (\gamma_b D s_v)(\tan \delta)}{K_{reinf}} \right] (H - z_i) \times$$

$$\left[\tan \left(45° - \frac{\psi}{2} \right) + \tan (90° - \phi_{ds}) \right]$$

The reinforcement stiffness, K_{reinf} is equal to $T_{@\varepsilon=3.0\%}/(3\%) = 12\,kN/m/(3\%) =$ 400 kN/m. The values of the other parameters are: $K_A = 0.27$, $\gamma = 18.0\,kN/m^3$, $q = 12\,kN/m^2$, $s_v = 0.2\,m$, $\gamma_b = 12.6\,kN/m^3$, $D = 0.2\,m$, $\tan\delta = 0.4$, $H = 6.0\,m$, $\psi = 5°$, and $B_T = 1.0$ for $z_i < H/2$, $B_T = 1.1$ for $z_i \geq H/2$. The direct shear friction angle ϕ_{ds} can be estimated as:

$$\phi_{ds} = \tan^{-1}\left[\frac{\sin\phi_{ps}\cos\psi}{1-\sin\phi_{ps}\sin\psi}\right] = \tan^{-1}\left[\frac{(\sin 35°)(\cos 5°)}{1-(\sin 35°)(\sin 5°)}\right] = 31°$$

To account for face batter,

$$\Phi_{fb} = \left(\frac{K_{ab}}{K_{av}}\right)^{0.5} = 0.88$$

The term K_A in the equation for calculating Δ_i is to be replaced by $K_{ab} = (0.88)^2(0.27) = 0.21$.

The lateral movement profile is depicted in Figure 5.48. Note that the 10% correction of the calculated movement due to base truncation is made for the lower half of the wall, therefore the deflection profile contains a *kink* near the mid-height. A "reasoned" deflection profile near the kink is shown by a dotted line in Figure 5.48. The maximum lateral movement is found to be 27 mm, which is smaller than the specified performance limit of 60 mm. Note that a vertical wall without batter would suffice also.

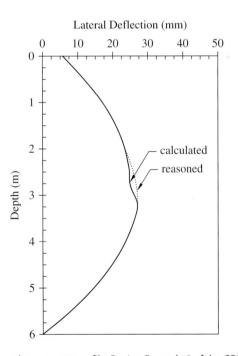

Figure 5.48 Calculated lateral movement profile, Design Example 3 of the GRS-NLB method (Note: The *kink* near mid-height is due to correction for base truncation)

Step 10: **Check stability against pullout failure**

The pullout resisting force (Pr_i) and driving force (Pd_i) for reinforcement *layer-i* are:

$$Pr_i = 2\gamma z_i \tan\delta \left[L_t - (H - z_i)\tan\left(45° - \frac{\phi}{2}\right) \right] \qquad \text{(for } z_i \le H - L_t + L_B)$$

$$Pr_i = 2\gamma z_i \tan\delta \left[H - z_i + L_B - (H - z_i)\tan\left(45° - \frac{\phi}{2}\right) \right] \qquad \text{(for all other } z_i)$$

$$Pd_i = \frac{1}{W} K_A (\gamma z_i + q) s_v$$

where $W = 0.7^{\left(\frac{s_v}{6 d_{max}}\right)} = 0.7^{\left(\frac{0.2}{(6 \times 0.025)}\right)} = 0.62$. Note that surcharge q appears only in the equation for driving force, Pd_i (see Figure 5.36(a)). The pullout resisting force and driving force for each layer are listed in Table 5.11.

Using the *rule of three*, i.e., the group pullout safety factor for *any three consecutive* reinforcement layers with its mid-layer located at depth z_i can be determined as:

$$Fs_{i\,(group)} = \frac{Pr_{i-1} + Pr_i + Pr_{i+1}}{Pd_{i-1} + Pd_i + Pd_{i+1}}$$

The group pullout safety factors for all reinforcement layers of the wall are also listed in Table 5.11. The group safety factor of *layer-i* is the safety factor with *layer-i* as the mid-layer of the group. It is seen that the group safety factors at all depths for $L_B = 1.8$ m and $L_t = 3.9$ m are greater than 1.5. The layout of reinforcement length is as follows: reinforcement length in the truncated zone is 1.8 m at the base, the length is to increase linearly to 3.9 m at 2.1 m from the base. The reinforcement in the non-truncated zone is a uniform length of 3.9 m.

Step 11: **Check stability against rotational slide-out failure**

A limiting-equilibrium based slope stability program (such as SLOPE-W, STABL WV, or ReSSA) can be used to analyze the stability against different modes of rotational failure (including overall slide-out failure and compound failure). The analysis can be carried out with the reinforced soil mass having the following soil–geosynthetic composite strength parameters (because the cohesion of the soil is quite small, $c = 0$ is assumed):

$$\phi R = \phi = 35°$$

$$c_R = \left[0.7^{\left(\frac{s_v}{6 d_{max}}\right)} \right] \frac{T_{ult}}{2 s_v} \sqrt{K_p}$$

$$= \left[0.7^{\left(\frac{0.2}{6 \times 0.025}\right)} \right] \left[\frac{38}{2(0.2)} \right] \sqrt{\frac{1}{0.27}} = 114 \,(\text{kPa})$$

If the smallest safety factor against any modes of rotational slide-out failure is less than 1.3, the reinforcement length may be increased to raise the safety factor. Alternatively, if the critical rotational failure plane is passing through the foundation

Table 5.11 Resisting and driving forces for pullout of each reinforcement layer and corresponding group pullout safety factor for $L_B = 1.8$ m and $L_t = 3.9$ m, Design Example 3

z_i (m)	Pr_i (kN/m)	Pd_i (kN/m)	$Fs_{(group)}$
0.2	2.7	2.1	2.5
0.4	6.1	2.4	2.6
0.6	10.1	2.7	3.8
0.8	14.8	3.0	5.0
1.0	20.1	3.3	6.1
1.2	26.1	3.7	7.2
1.4	32.7	4.0	8.3
1.6	40.0	4.3	9.4
1.8	47.9	4.6	10.5
2.0	56.5	4.9	11.5
2.2	65.7	5.2	12.6
2.4	75.5	5.5	13.7
2.6	86.0	5.8	14.7
2.8	97.1	6.2	15.8
3.0	108.9	6.5	16.9
3.2	121.4	6.8	17.9
3.4	134.5	7.1	19.0
3.6	148.2	7.4	20.0
3.8	162.6	7.7	20.8
4.0	171.4	8.0	21.0
4.2	173.7	8.4	20.8
4.4	175.4	8.7	20.2
4.6	176.5	9.0	19.6
4.8	177.1	9.3	19.0
5.0	177.0	9.6	18.4
5.2	176.3	9.9	17.8
5.4	175.1	10.2	17.1
5.6	173.2	10.5	16.4
5.8	170.8	10.9	15.7
6.0	167.7	11.2	15.4

Note: The group safety factor against pullout is the sum of the resisting pullout forces of any consecutive three layers divided by the sum of driving pullout forces of the same three layers. Note that there is no layer above the top layer (i.e., layer $i-1$ does not exist) or beneath the bottom layer (i.e., layer $i+1$ does not exist). The values of the non-existent layers (i.e., Pr_{i-1} and Pd_{i-1} for evaluation of the top layer, and Pr_{i+1} and Pd_{i+1} for evaluation of the bottom layer) are taken as zero.

soil, a viable ground improvement technique may be employed to improve the strength of the foundation soil and raise the safety factor.

Step 12:   **Determine facing requirements**

a) *Embedment*: Considering that there is no concern for erosion or other untoward events surrounding the planned location of wall base during the design life, the GRS wall is to be constructed with a nominal embedment of 0.1 m below the ground surface.

b) *Facing batter*: The facing of the GRS wall is to have a face batter of 1 (horizontal) to 10 (vertical). Note that a vertical wall without any batter will also suffice.

c) *Connection strength requirements*: All facing blocks are to be dry-stacked. Reinforcement is at every course of facing blocks and reinforcement spacing is 0.2 m, hence connection stability is ensured. The top two courses of facing blocks are to be bonded by filling the hollow cells with concrete mix. All other facing blocks are to be dry-stacked with their core filled with compacted fill.

Design Summary for Reinforcement:

Reinforcement lengths at base, L_B = 1.8 m, increasing linearly to 3.9 m at 2.1 m above the base, and a uniform length of 3.9 m for all layers above 2.1 m from the base

Reinforcement spacing (same for all depths), s_v = 0.2 m

Reinforcement type: woven polypropylene with T_{ult} = 38 kN/m (at ε = 10%) and $T_{@\varepsilon=3.0\%}$ = 12 kN/m

References

Abu-Hejleh, N., Wang, T., and Zornberg, J.G. (2000). *Performance of Geosynthetic-Reinforced Walls Supporting Bridge and Approaching Roadway Structures*. ASCE Geotechnical Special Publication No. 103, Advances in Transportation and Geoenvironmental Systems using Geosynthetics, pp. 218–243.

Adams, M.T., Lillis, C.P., Wu, J.T.H., and Ketchart, K. (2002). Vegas Mini Pier Experiment and Postulate of Zero Volume Change. *Proceedings, 7th International Conference on Geosynthetics, Nice, France*, pp. 389–394.

Adams, M.T., Schlatter, W., and Stabile, T. (2007). Geosynthetic-Reinforced Soil Integrated Abutments at the Bowman Road Bridge in Defiance County, Ohio. *Proceedings, Geo-Denver 2007, ASCE, Denver, Colorado.*

Adams, M.T., Nicks, J., Stabile, T., Wu, J.T.H., Schlatter, W., and Hartmann, J. (2011a). *Geosynthetic Reinforced Soil Integrated Bridge System Interim Implementation Guide.* Report No. FHWA-HRT-11-026, Federal Highway Administration, McLean, Virginia.

Adams, M.T., Nicks, J., Stabile, T., Wu, J.T.H., Schlatter, W., and Hartmann, J. (2011b). *Geosynthetic Reinforced Soil Integrated Bridge System Synthesis Report*. Report No. FHWA-HRT-11-027, Federal Highway Administration, McLean, Virginia.

Allen, T.M. and Bathurst, R.J. (1996). Combined Allowable Strength Reduction Factor for Geosynthetic Creep and Installation Damage. *Geosynthetics International*, 3(3), 407–439.

Allen, T. M. and Bathurst, R.J. (2001). *Application of the K_0-Stiffness Method to Reinforced Soil Wall Limit State Design*. Report No. WA-RD 528.1, Research Report, Washington State Department of Transportation.

Allen, T.M. and Bathurst, R.J. (2003). *Prediction of Reinforcement Loads in Reinforced Soil Walls*. Final Research Report, Washington State Department of Transportation and Federal Highway Administration, 290 pp.

Allen, T. M., Bathurst, R. J., Holtz, R. D., Lee, W. F., and Walters, D. L. (2004). New Method for Prediction of Loads in Steel Reinforced Soil Walls. *Journal of Geotechnical & Geoenvironmental Engineering, ASCE*, Eng., 10.1061/(ASCE)1090-0241(2004)130:11 (1109), pp. 1109–1120.

AASHTO (2002). *Standard Specifications for Highway Bridges*, 17th edition. American Association of State Highway and Transportation Officials, Washington, D.C.

AASHTO (2014). *LRFD Bridge Design Specifications*, 7th edition with 2016 Interims. American Association of State Highway and Transportation Officials, Washington, D.C.

Ballegeer, J. and Wu, J.T.H. (1993). *Intrinsic Load-Deformation Properties of Geotextiles*. Geosynthetic Soil Reinforcement Testing Procedures, ASTM STP 1190, American Society for Testing and Materials, pp. 16–31.

Barreire, W. and Wu, J.T.H. (2001). *Guidelines for Design and Construction of Reinforced Soil Foundations*. Research Report, Department of Civil Engineering, University of Colorado Denver, 109 pp.

Bastick, M.J. (1990). Reinforced Earth Narrow Walls and Abutments Correlation of Measured Performance with Design. In *Performance of Reinforced Soil Structures*, McGowan, Yeo, and Andrawes (eds.). British Geotechnical Society, Glasgow, pp. 59–63.

Bathurst, R.J., Vlachopoulos, N., Walters, D.L., Burgess, P.G., and Allen, T.M. (2006). The Influence of Facing Stiffness on the Performance of Two Geosynthetic Reinforced Retaining Walls. *Canadian Geotechnical Journal*, 43(12), 1225–1237.

Bathurst, R.J., Miyata, Y., Nernheim, A., and Allen, A.M. (2008). Refinement of K-stiffness Method for Geosynthetic Reinforced Soil Walls. *Geosynthetics International*, 15(4), 269–295.

Beauregard, M.S. (2016). *Facing of Geosynthetic Reinforced Soil Structures*. Doctoral dissertation, College of Engineering and Applied Science, University of Colorado Denver.

Berg, R.R. (2010). Fill Walls – Recent Advances and Future Trends. *Proceedings, Earth Retention Conference. ASCE, Bellevue, Washington*, pp. 1–19.

Berg, R.R. Christopher, B.R., and Samtani, N.C. (2009). *Design of Mechanically Stabilized Earth Walls and Reinforced Soil Slopes, Design & Construction Guidelines*. Report No. FHWA-NHI-00-043. Federal Highway Administration, 394 pp.

Bonaparte, R., Holtz, R.D., and Giroud, J.P. (1987). *Soil Reinforcement Design using Geotextiles and Geogrids*, Fluet (ed.). Geotextile Testing and the Design Engineer, ASTM STP 952, American Society for Testing and Materials, Philadelphia, Pennsylvania, pp. 69–116.

Brandl, H. (1998). Multi-anchored Soil Retaining Walls with Geosynthetic Loop Anchors. *Proceedings, 6th International Conference on Geosynthetics, Atlanta, Georgia, Volume 2*. Industrial Fabrics Association International (IFAI), pp. 581–586.

Broms, B.B. (1978). Design of fabric reinforced retaining structures. *Proceedings, Symposium on Earth Reinforcement*. ASCE, Pittsburgh, Pennsylvania, pp. 282. [Summarized in *Reinforcement of Earth Slopes and Embankments*, National Cooperative Highway Research Program (NCHRP) Report 290 (1987).]

BSI (1995). *Code of Practice for Strengthened/Reinforced Soils and Other Fills.* BS8006. Amended 1999. British Standards Institution, 161 pp.

Chew, S.H. and Mitchell, J.K. (1994). Deformation Evaluation Procedure for Reinforced Walls. *Proceedings, 5th International Conference on Geotextiles, Geomembranes and related Products, Singapore.*

Chew, S.H., Schmertmann, G.R., and Mitchell, J.K. (1991). Reinforced Soil Wall Deformations by Finite Element Method. In *Performance of Reinforced Soil Structures,* McGowan, Yeo, and Andrawes (eds.). British Geotechnical Society, Glasgow, pp. 35–40.

Chou, N.N.S. and Wu, J.T.H. (1993). *Investigating Performance of Geosynthetic-Reinforced Soil Walls.* Report No. CDOT-DTD-93-21, Colorado Department of Transportation, 197 pp.

Christopher, B.R., Gill, S.A., Giroud, J.P, Juran, I., Scholesser, F., Mitchell, J.K., and Dunnicliff J. (1990). *Reinforced Soil Structures, Volume 1: Design and Construction Guidelines.* Publication No. FHWA-RD-89-043, Federal Highway Administration, Washington, D.C.

Claybourn, A.F. (1990). *A Comparison of Methods for Geosynthetic-Reinforced Earth Walls.* M.S. Thesis, Department of Civil Engineering, University of Colorado Denver.

Claybourn, A.F. and Wu, J.T.H. (1992). Failure Loads of the Denver Walls by Current Design Methods. *International Symposium on Geosynthetic-Reinforced Soil Retaining Walls,* A.A. Balkema, Rotterdam, pp. 61–77.

Claybourn, A.F. and Wu, J.T.H. (1993). Geosynthetic-Reinforced Soil Wall Design. *Geotextiles and Geomembranes, Journal of International Geotextile Society,* 12(8), 707–724.

Collin, J.G. (1986). *Earth Wall Design.* Doctoral dissertation, Department of Civil Engineering, University of California, Berkeley.

Crouse, P. and Wu, J.T.H. (2003). Geosynthetic-Reinforced Soil (GRS) Walls. *Journal of Transportation Research Board,* 1849, 53–58.

Dallaire, G. (2001). Segmental Retaining Walls Come of Age. *Erosion Control, Journal of the International Erosion Control Association,* November–December.

Elias, V. (2000). *Long-Term Durability of Geosynthetics Based on Exhumed Samples from Construction Projects.* FHWA Report No. FHWA RD-00-157, 53 pp.

Elias, V., Christopher, B.R., and Berg, R.R. (2001). *Mechanically Stabilized Earth Walls and Reinforced Soil Slopes Design and Construction Guidelines.* NHI Course 132042, FHWA NHI-00-043, 394 pp.

Elton, D.J. and Patawaran, M.A.B. (2005). *Mechanically Stabilized Earth (MSE) Reinforcement Tensile Strength from Tests of Geotextile Reinforced Soil.* Technical Report, Alabama Highway Research Center, Auburn University.

Fukuoka, M., Imamura, Y., and Nishimura, J. (1986). Fabric Faced Retaining Wall with Multiple Anchors. *Geotextile and Geomembranes,* 4, 207–221.

GEO (2002). *Guide to Reinforced Fill Structure and Slope Design, Draft.* Geoguide 6. Civil Engineering Department, Geotechnical Engineering Office, Government of the Hong Kong Special Administrative Region, 240 pp.

Gerber, T.M. (2012). *Assessing the Long-Term Performance of Mechanically Stabilized Earth Walls.* NCHRP Synthesis 437, National Cooperative Highway Research Program, Transportation Research Board, Washington, D.C.

Giroud, J.P. (1989). *Geotextile engineering workshop – design examples.* Publication No. FHWA-HI-89-CO2, Federal Highway Administration, Washington, DC.

Greenwood, J.H. (2002). The Effect of Installation Damage on the Long-Term Strength Design Strength of a Reinforcing Geosynthetic. *Geosynthetics International,* 9(3), 247–258.

Hatami, K. and Bathurst, R.J. (2005). Development and Verification of a Numerical Model for the Analysis of Geosynthetic Reinforced Soil Segmental Walls. *Canadian Geotechnical Journal*, 42(4), 1066–1085.

Hatami, K. and Bathurst, R.J. (2006). Numerical Model for Reinforced Soil Segmental Walls under Surcharge Loading. *Journal of Geotechnical and Geoenvironmental Engineering, ASCE*, 132(6), 673–684.

Helwany, S., Reardon, G., and Wu, J.T.H. (1999). Effects of Backfill on the Performance of GRS Retaining Walls. *Geotextiles and Geomembranes*, 17(1), 1–16.

Helwany, S., Wu, J.T.H., and Meinholz, P. (2012). *Seismic Design of Geosynthetic-Reinforced Soil Bridge Abutments with Modular Block Facing.* NCHRP Web-Only Document 187, 251 pp.

Holtz, R.D. (2010). *Reinforced Soil Technology: from Experimental to the Familiar.* Terzaghi Lecture, GeoFlorida, Palm Beach.

Holtz, R.D., Kovacs, W.D., and Sheahan, T.C. (2011). *An Introduction to Geotechnical Engineering*, 2nd edition. Prentice Hall, New Jersey.

Hufenus, R., Ruegger, R., Flum, D., and Sterba, I.J. (2005). Strength Reduction due to Installation Damage of Reinforcing Geosynthetics. *Geotextiles and Geomembranes*, 23(5), 401–424.

Iwamoto, M.K., Ooi, P.S.K., Adams, M.T., and Nicks, J.E. (2015). Composite Properties from Instrumented Load Tests on Mini-Piers Reinforced with Geotextiles. *Geotechnical Testing Journal*, 38(4), 397–408.

Jewell, R.A. and Milligan, G.W. (1989). Deformation Calculation for Reinforced Soil Walls. *Proceedings, 12th International Conference on Soil Mechanics and Foundation Engineering, Volume 2*, pp. 1259–1262.

Ketchart, K. and Wu, J.T.H. (2001). *Performance Test for Geosynthetic-Reinforced Soil including Effects of Preloading.* Report No. FHWA-RD-01-018, Federal Highway Administration, 270 pp.

Ketchart, K. and Wu, J.T.H. (2002). A Modified Soil-Geosynthetic Interactive Performance Test for Evaluating Deformation Behavior of GRS Structures. *ASTM Geotechnical Testing Journal*, 25(4), 405–413.

Koerner, R.M. (2005). *Designing with Geosynthetics*, 5th edition. Prentice Hall, New Jersey, 816 pp.

Koerner, R.M. and Koerner, G.R. (2013). A Data Base, Statistics and Recommendations Regarding 171 Failed Geosynthetic reinforced mechanically stabilized earth (MSE) walls. *Geotextiles and Geomembranes*, 40(October), 20–27.

Lawson, C.R. and Yee, T.W. (2005). Reinforced soil retaining walls with constrained reinforced fill zones. In *Geo-Frontiers 2005*, ASCE Geotechnical Special Publication 140: Slopes and Retaining Structures under Seismic and Static Conditions.

Lee, K., Jones, C.J.F.P., Sullivan, W.R., and Trolinger, W. (1994). Failure and Deformation of Four Reinforced Soil Walls in Eastern Tennessee. *Géotechnique*, 4(3), 397–426.

Leshchinsky, D. and Perry, E.B. (1987). A Design Procedure for Geotextile-Reinforced Walls. *Proceedings, Geosynthetics '87 Conference, Volume 1*, New Orleans, pp. 95–107.

Lin, C.C., Hsieh, T.J., Tsao, W.H., and Wang, Y.H. (1997). Combining Multi-Nailings with Soil Reinforcement for Construction. In *Mechanically Stabilized Backfill*, Wu (ed.). A.A. Balkema, Rotterdam, pp. 255–257.

Ling, H.I. and Leshchinsky, D. (2003). Finite Element Parametric Study of the Behavior of Segmental Block Reinforced Soil Retaining Walls. *Geosynthetics International*, 10(3), 77–94.

Ling, H., Wu, J.T.H., and Tatsuoka, F. (1992). Short-Term Strength and Deformation Characteristics of Geotextiles under Typical Operational Conditions. *Geotextiles and Geomembranes*, 11(2), 185–219.

Liu, H. (2012). Long-term Lateral Displacement of Geosynthetic Reinforced Soil Segmental Retaining Walls. *Geotextiles and Geomembranes*, 32, 18–27.

Macklin, P.M. (1994). *A Comparison of Five Methods for Calculation Lateral Deformation of Geosynthetic-Reinforced Soil (GRS) walls*. M.S. Report, Department of Civil Engineering, University of Colorado Denver.

Meyerhof, G.G. (1963). Some Recent Research on the Bearing Capacity of Foundations. *Canadian Geotechnical Journal*, 1(1), 16–26.

Mitchell, J.K. and Villet, W.C.B. (1987). *Reinforcement of Earth Slopes and Embankments*. NCHRP Report 290, Transportation Research Board, Washington, D.C., 323 pp.

Morishima, H., Saruya, K., and Aizawa, F. (2005). *Damage to soils structures of railway and their reconstruction*. Special Issue on Lessons from the 2004 Niigata-ken Chu-Etsu Earthquake and Reconstruction, Foundation Engineering and Equipment (Kiso-ko), October, pp. 78–83 (in Japanese).

Morrison, K.F., Harrison, F.E., Collin, J.G., Dodds, A., and Arndt, B. (2006). *Shored Mechanical Stabilized Earth (SMSE) Wall Systems Design Guidelines*. Central Federal Lands Highway Division, FHWA Report No. FHWA-CFL/TD-06-001, 230 pp.

NAVFAC (1986). *Design Manual 7.02 Foundations and Earth Structures*. Bureau of Yards and Docks, U.S. Navy.

NCMA (2010). *Design Manual for Segmental Retaining Walls*, 3rd edition, Collin (ed.). National Concrete Masonry Association, 206 pp.

Perloff, W.H. (1975). Pressure Distribution and Settlement. Chapter 4 in *Foundation Engineering Handbook*, Winterkorn and Fang (eds.). Van Nostrand Reinhold, New York, pp. 148–196.

Perloff, W.H. and Baron, W. (1976). *Soil Mechanics Principles and Applications*. John Wiley & Sons, New York, 745 pp.

Pham, T.Q. (2009). *Investigating Composite Behavior of Geosynthetic-Reinforced Soil (GRS) Mass*. Doctoral Dissertation, Department of Civil Engineering, University of Colorado Denver.

Poulos, H.G. (2000). *Foundation Settlement Analysis–Practice versus Research*. The Eighth Spencer J. Buchanan Lecture, Texas A&M University.

Rowe, R.K. and Ho, S.K. (1993). A Review of the Behavior of Reinforced Soil Walls. In *Earth Reinforcement Practice, Volume 2*, Ochiai, Hayashi, and Otani (eds.). Proceedings, International Symposium on Earth Reinforcement Practice, Fukuoka, Kyushu, Japan, A.A. Balkema, Rotterdam, pp. 801–830.

Saghebfar, M., Abu-Farsakh, M., Ardah, A., Chen, Q., and Fernandez, B.A. (2017). Performance monitoring of Geosynthetic Reinforced Soil Integrated Bridge System (GRS–IBS) in Louisiana. *Geotextiles and Geomembranes*, 45, 34–47.

Schlosser, F. and Segrestin, P. (1979). Local Stability Analysis Method of Design of Reinforced Earth Structures. *Proceedings, International Conference on Soil Reinforcement, Volume 1, Paris, France*, pp. 157–162 (in French).

Schmertmann, G.R., Chouery-Curtis, V.E., Johnson, R.D., and Bonaparte, R. (1987). Design Charts for Geogrid-Reinforced Soil Slopes. *Proceedings, Geosynthetics '87 Conference, Volume 1, New Orleans*, pp. 108–120.

Segrestin, P. (1994). GRS Structures with Short Reinforcements and Rigid Facing – Discussion of Previous Papers Published by Prof. F. Tatsuoka, M. Tateyama and

O. Murata. In *Proceedings, Recent Case Histories of Permanent Geosynthetic-Reinforced Soil Retaining Walls*. Tatsuoka and Leshchinsky (eds.). A.A. Balkema, Rotterdam, pp. 323–342.

Smith, G.N. and Pole, E.L. (1980). *Elements of Foundation Design.* Granada Publishing, London, 222 pp.

Steward, J. E., Williamson, R., and Mohney, J. (1977, revised 1983). Earth Reinforcement. Chapter 5 in *Guidelines for Use of Fabrics in Construction and Maintenance of Low-Volume Roads.* U.S. Department of Agriculture, Forest Service, Portland, Oregon.

Tatsuoka, F. (1992). Roles of Facing Rigidity in Soil Reinforcing. In *Earth Reinforcement Practice*, Ochiai, Hayashi, and Otani (eds.). A.A. Balkema, Rotterdam, pp. 831–867.

Tatsuoka, F. (2008). Recent Practice and Research of Geosynthetic-Reinforced Earth Structures in Japan. *Journal of GeoEngineering*, 3(3), 77–100.

Tatsuoka, F., Murata, O., and Tateyama, M. (1992). Permanent Geosynthetic-Reinforced Soil Retaining Walls Used for Railway Embankments in Japan. In *Geosynthetic-Reinforced Soil Retaining Walls, Denver, Colorado, 8–9 August 1991*, Wu (ed.). A.A. Balkema, Rotterdam, pp. 101–130.

Tatsuoka, F., Tateyama, M., Uchimura, T., and Koseki, J. (1997). Geosynthetic-reinforced Soil Retaining Walls as Important Permanent Structures. In *Mechanically-Stabilized Backfill*, Wu (ed.). A.A. Balkema, Rotterdam, pp. 3–24.

Tatsuoka, F., Tateyama, M., Aoki, H. and Watanabe, K. (2005). Bridge abutment made of cement-mixed gravel backfill. In *Ground Improvement, Case Histories*, Elsevier Geo-Engineering Book Series, Volume 3, Indradratna and Chu (eds.). Elsevier, pp. 829–873.

Terzaghi, K., Peck, R.B., and Mesri, G. (1996). *Soil Mechanics in Engineering Practice*, 3rd edition. Wiley, New York.

Thomas, D. and Wu, J.T.H. (2000). *Analysis of Geosynthetic-reinforced Soil Walls with a Truncated Base.* Technical Report, Department of Civil Engineering, University of Colorado Denver, 191 pp.

TRB (1996). *Landslides.* Special Report 247, Transportation Research Board, National Research Council, Washington, D.C., 673 pp.

Valentine, R.J. (2013). An Assessment of the Factors that Contribute to the Poor Performance of Geosynthetic-Reinforced Earth Retaining Walls. In *Design and Practice of Geosynthetic-Reinforced Soil Structures*, Ling et al. (eds.). DEStech Publications, Bologna, pp. 318–327.

Vulova, C. (2000). *Effects of Geosynthetic Reinforcement Spacing on the Behavior of Mechanically Stabilized Earth Walls.* Ph.D. Dissertation, Department of Civil Engineering, University of Delaware.

Vulova, C. and Leshchinsky, D. (2003). *Effects of Geosynthetic Reinforcement Spacing on the Behavior of Mechanically Stabilized Earth Walls.* FHWA Report No. FHWA-RD-03-048, 226 pp.

Wu, J.T.H. (1991). Measuring Inherent Load-Extension Properties of Geotextiles for Design of Reinforced Structures. *ASTM Geotechnical Testing Journal*, 14(2), 157–165.

Wu, J.T.H. (1992). *Predicting Performance of the Denver Walls: General Report.* International Symposium on Geosynthetic-Reinforced Soil Retaining Walls. A.A. Balkema, Rotterdam, pp. 3–20.

Wu, J.T.H. (1994). *Design and Construction of Low Cost Retaining Walls: The Next Generation in Technology.* Publication No. CTI-UCD-1-94, Colorado Transportation Institute, Denver, Colorado, 152 pp.

Wu, J.T.H. (2001). *Revising the AASHTO Guidelines for Design and Construction of GRS Walls*. Report CDOT-DTD-R-2001-16, Colorado Department of Transportation, Denver, Colorado, 148 pp.

Wu, J.T.H. (2007). *Lateral Earth Pressure against the Facing of Segmental GRS Walls*. Geosynthetics in Reinforcement and Hydraulic Application, Geo-Denver 2007: New Peaks in Geotechnics, American Society of Civil Engineers, pp. 1–11.

Wu, J.T.H. (2012). *Design Protocol of GRS Walls for Non-Load-Bearing Applications*. Report to Turner-Fairbank Highway Research Center, Federal Highway Administration, McLean, Virginia.

Wu, J.T.H. (2018). A Generic Design Protocol for Geosynthetic Reinforced Soil Foundation. *Transportation Infrastructure Geotechnology*, 5(16), 1–15. DOI 10.1007/s40515-018-0061-2.

Wu, J.T.H. and Helwany, S. (1996). A Performance Test for Assessment of Long-Term Creep Behavior of Soil-Geosynthetic Composites. *Geosynthetics International*, 3(1), 107–124.

Wu, J.T.H. and Payeur, J.B. (2015). Connection Stability Analysis of Segmental Geosynthetic Reinforced Soil (GRS) Walls. *Transportation Infrastructure Geotechnology*, 2(1), 1–17. DOI 10.1007/s40515-014-0013-4.

Wu, J.T.H. and Pham. Q. (2010). An Analytical Model for Calculating Lateral Movement of a Geosynthetic-Reinforced Soil (GRS) Wall with Modular Block Facing. *International Journal of Geotechnical Engineering*, 4(4), 527–535.

Wu, J.T.H., Lee, K.Z.Z., Helwany, S.B., and Ketchart, K. (2006). *Design and Construction Guidelines for GRS Bridge Abutment with a Flexible Facing*. Report No. 556, National Cooperative Highway Research Program, Washington, D.C.

Wu, J.T.H., Ketchart, K., and Adams, M. (2008). Two Full-Scale Loading Experiments of Geosynthetic-Reinforced Soil (GRS) Abutment Wall. *International Journal of Geotechnical Engineering*, 2(4), 305–317.

Wu, J.T.H., Pham, T.Q., and Adams, M.T. (2010). *Composite Behavior of Geosynthetic-Reinforced Soil (GRS) Mass*. Research Report, Department of Civil Engineering, University of Colorado Denver, 277 pp.

Wu, J.T.H., Ma, C.Y., Pham, T.Q., and Adams, M.T. (2011). Required Minimum Reinforcement Stiffness and Strength in Geosynthetic-Reinforced Soil (GRS) Walls and Abutments. *International Journal of Geotechnical Engineering*, 5(4), 395–404.

Wu, J.T.H., Pham, T.Q., and Adams, M.T. (2013). *Composite Behavior of Geosynthetic Reinforced Soil Mass*. Publication No. FHWA-HRT 10-077, Federal Highway Administration, McLean, Virginia, 211 pp.

6

Construction of Geosynthetic Reinforced Soil (GRS) Walls

Geosynthetic reinforced soil (GRS) walls can be constructed anywhere a retaining wall is deemed warranted and practical. If the foundation soil is too weak to allow safe construction of a GRS wall, ground improvement may be needed before construction. Various ground improvement techniques have proven to be effective, and they have been presented in the literature, including recent books by Kirsch and Bell (2013) and Nicholson (2015). Typical applications of GRS walls include embankment walls, temporary or permanent widening or diversion embankment walls, earth walls on steep mountain slopes, slide stabilization walls, rock fall barrier walls, bridge abutment walls, small dams, sound barrier walls, and various urban earth wall projects where a sudden change in elevation is needed due to space constraints or aesthetics considerations.

This chapter addresses the construction of GRS walls, including construction procedures and general construction guides. Many types of GRS walls have been constructed (see Figure 3.23). This chapter, however, focuses on four common types of GRS walls: concrete block GRS walls, wrapped-face GRS walls, full-height precast panel facing GRS walls, and timber facing GRS walls.

6.1 Construction Procedure

This section describes typical construction procedures for four common types of GRS walls: concrete block GRS walls, wrapped-face GRS walls, full-height precast panel facing GRS walls, and timber facing GRS walls. These walls differ primarily in facing and connection between the facing and the reinforced soil mass.

Geosynthetic Reinforced Soil (GRS) Walls, First Edition. Jonathan T.H. Wu.
© 2019 John Wiley & Sons Ltd. Published 2019 by John Wiley & Sons Ltd.

6.1.1 Concrete Block GRS Walls

The typical construction sequence of concrete block GRS walls is illustrated in Figure 6.1 and can be described as follows:

Step 1: Level the wall site by removing vegetation, large rocks, stumps, etc.; excavate a shallow trench (about 450 mm or 18 in. wide and 75–150 mm or 3–6 in. deeper than the embedment depth of facing, as per construction plans of the wall project) along the planned location of the concrete block facing (Note: For recommended design embedment depth, see item *(a) Embedment*, Step 12, Section 5.7.1); then place and level a compacted *sand-gravel pad* of thickness 75–150 mm (3–6 in.) inside the trench.

Step 2: Position the first course of concrete facing blocks over the sand-gravel pad; check alignment and ensure that the blocks are level; (optional) place aggregates to fill the hollow cores of the blocks and the space between the blocks; and sweep away the debris on the top surface of the blocks.

Step 3: Place free-draining fill material behind the facing blocks all the way to the back of the wall, as per construction plans; compact the fill until it matches the top surface of the blocks.

Step 4: Lay down a layer of geosynthetic reinforcement to cover the compacted fill and the facing blocks (the front of the reinforcement should be slightly behind the front edge of the blocks and should extend to the design length as measured from the back of the blocks); pull the reinforcement taut.

Step 5: Position a subsequent course of concrete facing blocks over the previous course of blocks; backfill and compact the fill behind the concrete blocks until the compacted fill matches the top of the newly placed facing blocks.

Step 6: Repeat Steps 4 and 5 for subsequent courses of concrete blocks and compact the newly placed fill until the top surface of the fill reaches the design wall height.

Step 7: Add concrete mix in the cores of the uppermost 2–3 courses of blocks to help tie those blocks together. The geosynthetic reinforcement embedded between those blocks will need to be perforated with a knife or weed burner to allow the concrete mix to flow into the blocks underneath. For load-carrying applications (e.g., bridge abutment walls), the concrete mix will need to be reinforced by vertical rebar. The last course of blocks is usually capped according to the block manufacturer's recommendation. Weed burner may be used to remove visible excess reinforcement protruding from the wall face.

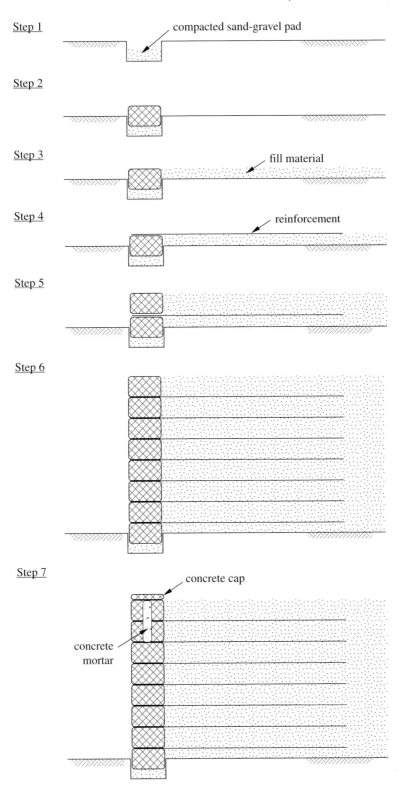

Figure 6.1 Typical construction sequence of concrete block GRS walls

Figure 6.2 shows photos taken during the construction of concrete block GRS walls. Below are a few notes about the construction of concrete block GRS walls:

- Before construction, measures for control of surface and subsurface water at and around the construction site (see Section 6.2.5) should be undertaken to prevent any potential drainage problems.

- Some wall builders use a concrete footer at the wall base to facilitate alignment of concrete facing blocks. A compacted sand-gravel pad, however, has been found to help in aligning the blocks and alleviating potential cracking of facing blocks better than a concrete footer.

- When a wall involves corners, it is important to compact the fill to the same degree as the rest of the reinforced fill. This may be achieved by employing heavy hand-operated tempers or other tools for fill compaction behind the corner. Use of a dry concrete mix in the cores of the corner blocks and reinforced with No. 4 vertical rebar has been proven effective in eliminating cracks or separation of blocks at corners.

- Although many designers routinely use a slight batter for facing blocks, vertical stacking works equally well for walls of low to medium height (i.e., heights not greater than 20 ft or 6.0 m). Facing batter can help reducing lateral movement and masking minor lateral movement; however, the reduction is generally not needed for walls of low to medium heights in non-load-bearing applications.

- Shorter reinforcement (1-m or 3-ft long), referred to as Colorado Transportation Institute tails, or simply *CTI tails*, may be used in every other reinforcement layer, as shown in Figure 6.3. When CTI tails are used, great caution needs to be exercised during construction to avoid confusion on sequencing of alternating reinforcement length. If there is any doubt that strict construction control may not be carried out, uniform full-length reinforcement is recommended, especially for smaller projects.

6.1.2 Wrapped-Face GRS Walls

There are two types of wrapped-face GRS walls, differing in the forming elements used during construction and their final appearance: (i) geotextile wrapped-face GRS walls, referred to as temporary form-work walls, and (ii) wire-mesh wrapped-face GRS walls, referred to as permanent form-work walls. For the former, the construction form is temporary L-shaped metal brackets in combination with wooden running boards (see Figure 6.4); for the latter, the construction form is continuous L-shaped wire-mesh (see Figure 6.5). Geotextile wrapped-face GRS walls were developed by the U.S. Forest Service in the 1970s. They are one of the least expensive types of earth retaining wall developed to date, and have been used in many parts of the world. Wire-mesh wrapped-face GRS walls, on the other hand, have only gained increasing popularity since the 2000s. They have a distinct advantage over geotextile wrapped-face GRS walls in that the form is left in place hence results in more rapid construction. The following describes the construction sequence of the two types of wrapped-face GRS wall.

(a)

(b)

(c)

Figure 6.2 Construction of concrete block GRS walls: (a) placement of facing blocks, (b) compaction of fill behind facing, and (c) near completion of a GRS abutment wall

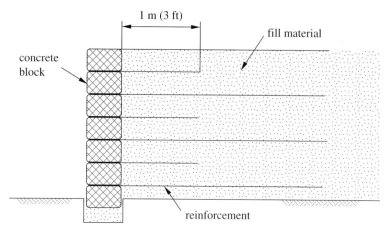

Figure 6.3 GRS wall with CTI tails (shorter reinforcement in place of full-length reinforcement in alternating layers)

(a)

(b)

Figure 6.4 Construction of a geotextile wrapped-face GRS wall (temporary form-work wall) with metal L-shaped brackets and wooden brace board: (a) top view and (b) front view (shown is a chain-link wall with wrapped-face geotextile)

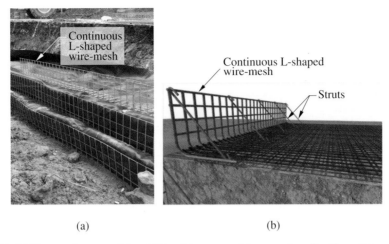

(a) (b)

Figure 6.5 (a) Wire-mesh wrapped-face GRS wall (permanent form-work wall) with continuous L-shaped wire-mesh brackets (courtesy of Bob Barrett) and (b) continuous L-shaped wire-mesh with struts (courtesy of Strata Systems, Inc.)

Geotextile Wrapped-Face GRS Walls (Temporary Form-Work Walls)

A schematic diagram of geotextile wrapped-face walls (temporary form-work walls) is seen in Figure 3.23, and a completed wall is shown in Figure 3.24(a). The typical construction sequence of geotextile wrapped-face GRS walls (temporary form-work walls) is illustrated in Figure 6.6, and can be described as follows:

Step 1: Level the wall site and install a series of L-shaped forms (of height slightly taller than design layer thickness) on the ground surface. The L-shaped form is composed of metal L-shaped brackets and continuous wooden brace boards along the wall face (see Figure 6.4).

Step 2: Lay down a sheet of geotextile over the soil surface and L-shaped forms by positioning the geotextile sheet in such a way that 1 m (3 ft) of the sheet is draped over the top of the form; backfill to about ½–¾ of the lift thickness, and compact the fill with conventional light-weight earth-moving equipment.

Step 3: Make a windrow 0.3–0.6 m (12–24 in.) from the wall face using a road grader or by hand; fold the loose end of the geotextile sheet over the L-shaped form into the windrow. When making windrows, care needs to be exercised not to damage the geotextile beneath or at the face.

Step 4: Backfill and compact the remaining lift thickness; remove the form from the layer below and reset it over the top of the soil layer. Typical layer thickness ranges from 0.2 to 0.45 m (8 to 18 in.), with 0.3 m (12 in.) being the most common.

Step 5: Repeat Steps 2 through 4 for subsequent layers until the design height is reached. For the final layer, the fold-back length should be extended to at least 1.8 m (6 ft) behind the wall face (preferably extend to the full length of the reinforcement layers) and covered with about 150 mm (6 in.) of fill.

Step 6:   If needed (e.g., for a more "permanent" wall), cover the exposed face of the wall with *shotcrete* (wet-mixed concrete or mortar conveyed through a hose and pneumatically projected at high velocity onto the wall surface), bituminous emulsions, or other asphalt products to prevent weakening of the geotextile due to UV exposure and vandalism (see Figure 6.7).

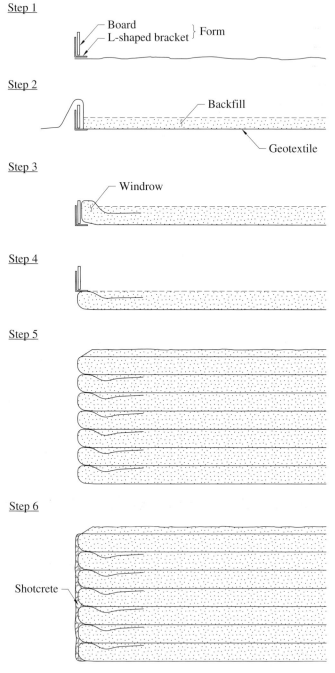

Figure 6.6 Typical construction sequence of geotextile wrapped-face GRS walls (temporary form-work walls)

(a)

(b)

Figure 6.7 Exposed face of a wrapped-face GRS wall covered with shotcrete: (a) application of shotcrete and (b) a completed wall face (rumor has it that the inspector remains trapped to this day) (courtesy of Bob Barrett)

Wire-Mesh Wrapped-Face GRS Walls (Permanent Form-Work Walls)

Two completed wire-mesh wrapped-face GRS walls (permanent form-work walls) are seen in Figures 3.24(b) and (c). The typical construction sequence of wire-mesh wrapped-face GRS walls (permanent form-work walls) is illustrated in Figure 6.8, and can be described as follows:

Step 1: Level the wall site and line up a series of continuous L-shaped wire-mesh on the ground surface along the planned location of the wall face.

Step 2: Lay down a sheet of geotextile over the soil surface and the L-shaped wire-mesh by positioning it in such a way that about 1 m (3 ft) of the geotextile sheet is draped over the top of the wire-mesh.

Step 3: Backfill and compact the fill with conventional light-weight earth-moving equipment; fold the loose end of the geotextile sheet over the wire-mesh and cover the soil surface.

Step 4: Install another series of L-shaped wire-mesh over the top surface of the previous soil layer at a prescribed setback (setback would be zero for a vertical wall).

Step 5: Repeat Steps 2 through 4 for subsequent layers until the design height is reached. For the final layer, the fold-back length should be extended to at least 1.8 m (6 ft) behind the back of the wall face (preferably extend to the full length of the reinforcement layers) and covered with about 150 mm (6 in.) of fill.

Step 6: If needed (e.g., for a more "permanent" wall), cover the exposed face of the wall with *shotcrete* (wet-mixed concrete or mortar conveyed through a hose and pneumatically projected at high velocity onto the wall surface), bituminous emulsions, or other asphalt products to prevent weakening of the geotextile due to UV exposure and vandalism (see Figure 6.7).

To increase the bending stiffness of the wire-mesh, many wall builders install *struts* to link across two ends of the L-shaped wire-mesh at a prescribed interval (say, every 16 in. or 0.4 m) in the longitudinal direction, see Figure 6.5(b), *during* backfill placement. In areas where climate permits, the setback area (if present) can be seeded to create a more natural-looking wall face, see Figure 3.26(b) as an example.

Wrapped-face walls of temporary and permanent form-work differ in the following aspects:

- The L-shaped continuous wire-mesh used in permanent form-work walls is left in place during construction, therefore construction time is substantially reduced. The left-in-place wire-mesh also increases slightly the rigidity of the wall face, especially when struts are employed, and thus reduces the lateral wall movement.

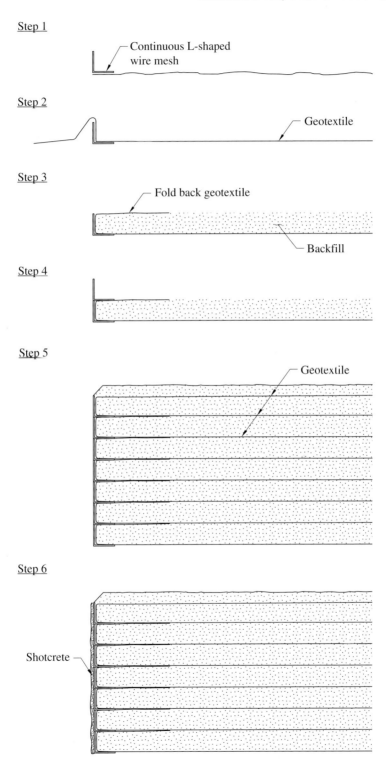

Figure 6.8 Typical construction sequence of wire-mesh wrapped-face GRS walls (permanent form-work walls)

- Permanent form-work walls can use geogrid as reinforcement when much stiffer reinforcement is needed. When geogrid reinforcement is used, the wall face should be vegetated or wrapped by shorter length geotextile sheets.

- The windrows in the construction of temporary form-work walls have been found unnecessary for construction of permanent form-work walls, thus the construction time is reduced appreciably.

6.1.3 Full-Height Precast Panel Facing GRS Walls

There are situations when a GRS wall with full-height concrete facing is desired as it gives the perception of a stronger and more permanent wall. A full-height concrete panel facing GRS wall, as shown in Figure 6.9, would serve that purpose. The wall can be constructed by two distinctly different methods: (i) construct a wrapped-face wall first, then install full-height precast facing panels over the wrapped-face wall or (ii) install full-height panels first, then construct a GRS soil mass behind the facing panels. In either method, the connection between the facing panels and the reinforced soil mass can be accomplished by face anchors (see Figure 6.9) installed in the reinforced soil mass during fill placement. The cast-in-place full-height facing (CIP-FHF) system described in Section 3.4.3 belongs to the first method, except that facing in the CIP-FHF system is cast-in-place. The wall system, referred to as an independent full-height facing (IFF) GRS wall, belongs to the second method. The IFF wall system has been described in some detail in Section 3.4.4.

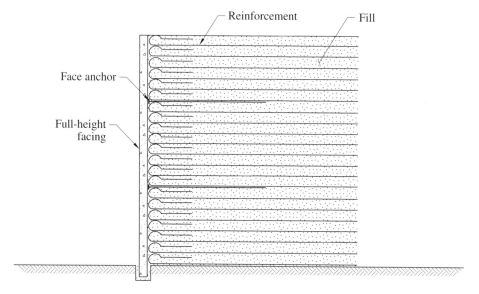

Figure 6.9 Schematic cross-sectional diagram of a full-height facing panel GRS wall

The typical construction sequence of the IFF GRS wall is shown in Figure 6.10, and can be described as follows:

Step 1: Level the wall site and excavate a shallow trench along the planned location of the wall face; the trench depth as suggested by the CTI is: depth (ft) $\cong$ [panel height (ft)/11] + 0.75.

Step 2: Position precast facing panels in the trench at a design offset, and use a *flow fill* (a mixture of concrete sand, cement, and water at a flowable consistency) and a temporary bracing system (in front of the facing panels, see Figure 6.11) to prop the panel in position.

Step 3: Backfill behind the wall facing, compact the fill in lifts to meet a specified minimum dry density, and lay down a reinforcement sheet on prescribed vertical spacing. For the fill within 1 m (3 ft) behind the facing panel, use only a hand-operated compactor for compaction.

Step 4: Install a row of face anchors (to connect the facing panel to the reinforced soil mass) at a prescribed interval along the longitudinal direction of the wall at a pre-selected elevation.

Step 5: Repeat Steps 3 and 4 by repeating the sequence of fill placement, fill compaction reinforcement placement, and face anchor installation as per construction plans until the design height is reached.

Step 6: Remove the temporary bracing; place about 1 m (3 ft) of surcharge fill over the finished grade for at least 60 hours; remove the surcharge and loosen the nuts slightly on all anchor bars.

Step 7: Back off the nuts between horizontally adjacent panels to align the facing panels as needed (see Figure 6.12).

As noted in Section 3.4.4, IFF full-height precast panel facing GRS walls have two distinct attributes: (i) the face anchors used to attach the facing panels to the reinforced soil mass need to allow the facing panel to move outward during fill placement and compaction, so that the reinforcing function of the reinforcement can be mobilized, and (ii) there needs to be a mechanism for adjusting the inclination of the facing panel to align the facing panels in the final phase of construction; this attribute is essential for the aesthetics of the wall. Note that the second attribute is also needed when employing the first construction method (i.e., constructing the GRS wall first, then installing the facing panels).

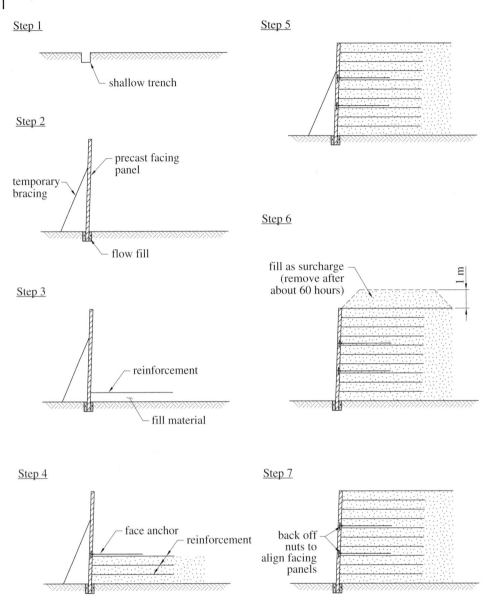

Figure 6.10 Typical construction sequence of IFF full-height precast panel GRS walls

(a)   (b)

Figure 6.11 Construction of an IFF full-height precast panel GRS wall: (a) facing panels situated in a shallow trench and (b) temporary bracing to prop up facing panels during construction (Ma and Wu, 2004)

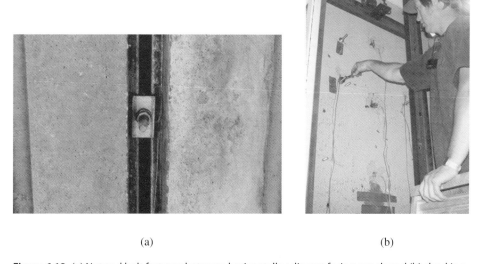

(a)   (b)

Figure 6.12 (a) Nut and bolt fastener between horizontally adjacent facing panels and (b) checking nut adjustment for alignment of facing panels in the laboratory (courtesy of Steve Foster)

6.1.4   Timber Facing GRS Walls

The typical construction sequence of timber facing GRS walls is shown in Figure 6.13, and can be described as follows:

Step 1: Level the wall site and install the first row of timber logs along the planned location of the wall face; lay down the first sheet of geosynthetic reinforcement layer with a minimum 0.3 m (12 in.) fold-back length (measured from the back of the timber facing).

Step 2: Secure the first layer of geosynthetic reinforcement to the back of the first row of timber logs by using forming elements (treated plywood, fiberglass, or plastic; 3.5 in. in height for 6 in. × 6 in. timber logs facing) and nails.

Step 3: Backfill to the top edge of the forming element; compact the fill; fold back the geotextile. Figure 6.14 shows reinforcement placement and fill compaction.

Step 4: Install the next two rows of timber logs (or a row of timber blocks and a row of timber logs, as shown in Figure 6.13); place the next layer of geosynthetic reinforcement; secure the reinforcement layer to the newly placed timber logs (or newly placed timber blocks and timber logs) by using forming elements and nails; backfill and compact the fill to the design lift thickness (typically 0.3 m or 12 in.). Shimming of timber logs to maintain verticality is usually permissible.

Step 5: Repeat Step 4 for all layers until the design height is reached. For the final layer, the fold-back tail length should extend at least 2.4 m (8 ft) behind the back of the timber facing. All exposed fabric is often coated with a latex paint matching the color of the timber if alternating rows of timber logs and timber blocks are used.

Figure 6.15 shows a timber facing GRS wall of which the facing comprises only timber logs; use of alternating rows of timber logs and timber blocks as illustrated in Figure 6.13 usually gives a more aesthetically pleasing wall face (see Figure 6.16).

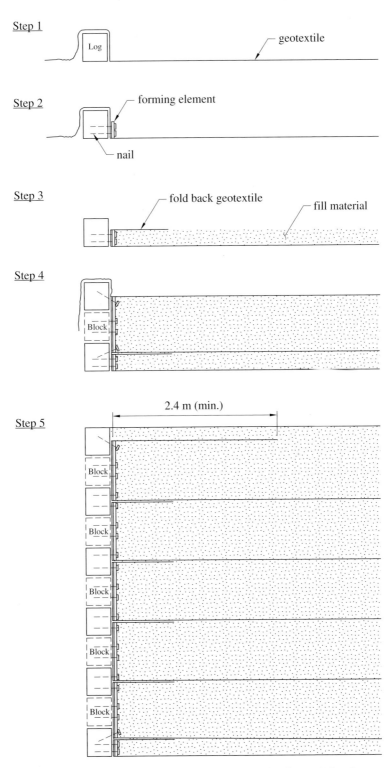

Figure 6.13 Typical construction sequence of timber facing GRS walls

(a)

(b)

Figure 6.14 Construction of a timber facing GRS wall: (a) placement of geosynthetic reinforcement and (b) fill compaction (courtesy of Bob Barrett)

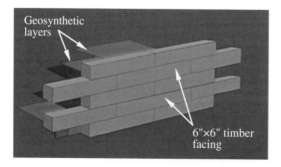

Figure 6.15 A timber facing GRS wall of which the facing comprises only timber logs (courtesy of Richard Andrew and Paul Macklin, in honor of John B. Gilmore)

Figure 6.16 A timber facing GRS wall with alternating rows of timber logs and timber blocks (courtesy of Bob Barrett)

6.2   General Construction Guidelines and Specifications

The earthwork construction guidelines and specifications for GRS walls are essentially the same as those for conventional earth retaining structures. Additional guidelines and specifications related to facing and geosynthetic reinforcement, however, are unique for GRS and GMSE walls.

The construction guidelines and specifications given in this section are syntheses of those stipulated by the American Association of State Highway and Transportation Officials, AASHTO (2002), the National Concrete Masonry Association, NCMA (2002), the Federal Highway Administration, FHWA (Elias et al., 2001), the Colorado Transportation Institute, CTI (Wu, 1994), the Swiss Association of Geotextile Professionals, SAGP (1981), and the Japan Railways, JR (1998), as well as observations and experiences gained from construction of GRS walls and abutments by the author and his wall builder friends. Construction guidelines and specifications are usually project specific. The guidelines and specifications given here are for general reference only.

The construction guidelines and specifications presented in this section are grouped into six subsections: (i) site and foundation preparation, (ii) geosynthetic reinforcement and reinforcement placement, (iii) fill material and fill placement, (iv) facing, (v) drainage, and (vi) construction sequence. It is important to point out that backfill material, backfill placement, and drainage have often been identified as the three most important factors affecting the performance of GRS walls.

6.2.1   Site and Foundation Preparation

- Before construction, the site of a GRS wall should usually be graded to provide a fairly smooth and level surface. When feasible, the ground surface should be cleared of shrubs, trees, large rocks, stumps, etc. Depressions should be filled, soft spots should be dug out and replaced with the fill material to be used for the wall, and the site may need to be proof rolled.

- If seepage from beneath the base of the wall is of concern, a granular soil bed of nominal thickness (about 100–150 mm or 4–6 in.) may be formed at the base of the wall to alleviate potential seepage-related problems.

- If the foundation soil contains a fair amount of frost-susceptible soils (e.g., silt, silty tills, soils with clay content of 15–25%), excavate the foundation soil to at least the depth of extreme frost penetration (e.g., Figure 2.35) and replace it with a non-frost-susceptible soil.

- If the foundation soil is considered marginally competent, improvement of the foundation soil will likely be needed. A number of methods, such as deep vibration (for granular deposits) or drainage combined with surcharge (for cohesive deposits), may be employed for ground improvement. If improvement of the load-bearing capability of the foundation soil is needed, a *reinforced soil foundation* may be formed in the top 1 m (3 ft) of the foundation soil by excavating the existing foundation soil and replacing it with well-compacted granular fill reinforced with 3–4 equally spaced layers of a geosynthetic reinforcement of wide-width tensile strength ≥ 70 kN/m or 4,800 lb/ft.

- If excavation is needed as part of the wall construction, it should be carried out to the lines and grades shown on the project grading plans. Overexcavation is generally not needed.

- In a stream environment, proper measures should be undertaken to protect the GRS wall from noticeable scour by water and abrasion by ripraps (e.g., Suaznabar et al., 2017).

6.2.2 Geosynthetic Reinforcement and Reinforcement Placement

- Geosynthetic products used as reinforcement for GRS walls are typically high-tenacity woven geotextiles or geogrids specifically manufactured for soil reinforcement applications. Most guidelines stipulate that badly damaged or improperly handled geosynthetics should be rejected. Most geosynthetics tend to be weakened when exposed to direct sunlight or extreme temperatures for an extended period of time.

- Geosynthetic reinforcement is preferably installed under slight tension to mobilize more readily the reinforcing function. Nominal tension, although not mandatory in most guidelines, may be applied to the reinforcement with the aid of staples, stakes or hand tensioning until after the reinforcement is covered by about 150 mm (6 in.) of fill.

- In the direction perpendicular to the wall face, geosynthetic reinforcement is preferably of one continuous sheet. Overlapping of reinforcement along that direction should be avoided when feasible. Otherwise, adjacent segments of geosynthetic reinforcement should have sufficient overlap as specified in the construction plans. The "stronger" direction of a geosynthetic product (often the machine direction) should be oriented perpendicular to the wall face. It would be wise to select a geosynthetic product of which the stronger direction is its cross-machine direction, and a choose roll width equal to the design reinforcement length plus the depth of the facing. In which case, the geosynthetic reinforcement can be installed by simply unrolling the reinforcement parallel to the wall face, hence expediting reinforcement placement.

- Tracked construction equipment is generally not allowed to operate directly over uncovered geosynthetic reinforcement. A minimum fill of 150 mm (6 in.) in thickness is commonly required before allowing operation of tracked vehicles over geosynthetic reinforcement. Turning of tracked vehicles should be kept to a minimum to prevent displacing the fill or damaging/displacing the geosynthetic reinforcement. Rubber-tired equipment is usually allowed to pass over geosynthetic reinforcement, but only at a slow speed (less than 17 km/hr, or 10 miles/hr). Sudden braking and sharp turning should be avoided.

- For construction of concrete block GRS walls that rely on frictional resistance to gain facing stability, the front edge of the geosynthetic reinforcement layer should be nearly matching the front edge of the facing blocks to achieve the largest contact surface area.

- For GRS bridge abutments with relatively rigid facing, the geosynthetic reinforcement behind the facing is preferably wrapped. Wrapped geosynthetic reinforcement will help minimize fill sloughing and maintain the integrity of the fill material (preventing fill from falling into the *gaps* formed between the back of the abutment wall and the reinforced fill resulting from seasonal temperature change, see Figure 3.20). Wrapped return of at least 0.45 m (18 in.) and anchored at least 0.1 m (4 in.) into the fill material is recommended. Wrapped return of the top geosynthetic layer should extend as far as practical, 1.8 m (6 ft) minimum, for load-bearing bridge abutment walls.

- For GRS bridge abutment walls, if the lateral earth pressure is expected to be very large, a compressible/collapsible layer (e.g., low to medium density expanded polystyrene) of approximately 50 mm (2 in.) in thickness may be installed over the backface of the abutment wall to relieve lateral earth pressure and minimize cracking of the wall (Monley and Wu, 1993).

- For concrete block GRS walls, installing reinforcement on every facing block to keep reinforcement spacing to a small value is generally recommended. Otherwise, secondary reinforcement (*tails*) (see Figure 6.3) alternated between full-length reinforcement layers may be incorporated if strict adherence to the sequence of reinforcement placement can be assured during construction. The length of tails is typically 1 m (3 ft) behind the backface of facing blocks. The cross-section of a typical concrete block GRS wall with tails in the upper portion and a truncated base in the lower portion is shown in Figure 6.17. Note that the cross-section assumes that the existing earth behind the GRS wall is sufficiently stiff to maintain a stable near-vertical face; otherwise, soil nailing with a top-down construction procedure may be employed. The cross-section also assumes that the fill material used in the GRS wall is free-draining (see Section 6.2.3). If the fill material is considered only marginally free-draining, geocomposite drainage strips with a high in-plane permeability may be installed along the surface of the excavation to help with drainage of subsurface water (see Section 6.2.5).

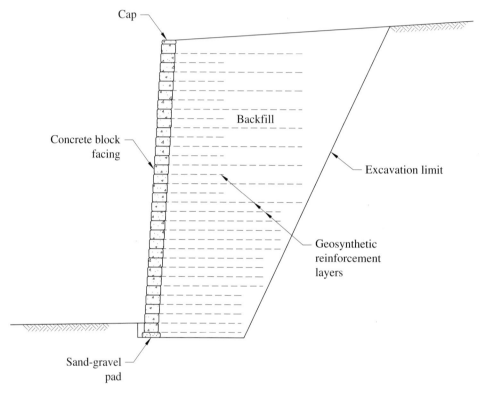

Figure 6.17 Typical cross-section of a concrete block GRS wall with alternating tails (shorter reinforcement layers) in the upper portion and a truncated base in the lower portion

6.2.3 Fill Material and Fill Placement

- Typical criteria of fill material in the construction of GRS walls are: 100% passing 4 in. (100 mm) sieve, 0–60% passing No. 40 (0.425 mm) sieve, and 0–15% passing No. 200 (0.075 mm) sieve (i.e., fines limit = 15%), with the plasticity index (*PI*) of the soil passing No. 40 sieve not exceeding 6. Fill material that meets these criteria is generally considered *free-draining*. Note that some guidelines have suggested a much higher fines limit (e.g., 35% by NCMA, 2002).

- For wrapped-face GRS walls that are not permanent in nature, deformation of the wall face is generally not a critical matter; fill materials of fines content of 30–50% have shown adequate performance, especially when the *plasticity index* does not exceed 8.

- The fill material should have a minimum angle of internal friction of 34°, with little or no cohesion, as determined by the standard direct shear test on the portion finer than No. 10 (2 mm) sieve, compacted to 95% of AASHTO T-99, Methods C or D, and at optimum moisture content. Some guidelines (e.g., AASHTO, 2002, 2014) do not recommend using a friction angle of more than 40° in design due to a controversial concern that reinforcement loads may be underestimated. Usually no testing is required for soils of which 80% or more of soil grains (by weight) are 19 mm (3/4 in.) or greater.

- The fill material shall be substantially free of shale or other poor-durability particles (e.g., shale, mica, gypsum) or organic material (usually should not be more than 1%). For permanent walls, the backfill should have a pH value between 3 and 9 for polyester reinforcement, and greater than 3 for polyolefin (polypropylene and HDPE) reinforcement. The pH limits may be relaxed to be between 3 and 11 for temporary walls.

- Strict attention should be given to ensure good compaction of fill in the construction of GRS walls, especially in areas near the wall face. Good compaction is especially important behind corners when they are present in a wall. Placement of fill material in the reinforced zone must meet the conditions specified in the construction plans. When it comes to improving performance of GRS, there is no substitute for higher soil stiffness (Wu et al., 2018). Provided the fill is sufficiently strong, all other factors (such as reinforcement stiffness, reinforcement spacing, and soil strength parameters) would have small influence on the stress–deformation behavior of GRS. Soil stiffness can be increased by proper selection of fill material type and better fill compaction.

- The fill material in the reinforced zone should generally be compacted within 2% dry of the optimum. If the reinforced fill is *free draining* (which is highly recommended for any type of GRS walls), the placement water content of the fill may be relaxed to within ±3% of the optimum. If the fill material contains an appreciable amount of fines (say, up to 35%), which should only be used if there are strong economic reasons for adopting the fill material and if a proven effective drainage system is installed, the placement moisture content of the backfill should generally be kept between the optimum and 4% wet-of-optimum.

- Compaction to a minimum dry density of 95% of the AASHTO T-99 is recommended for non-load-bearing GRS walls and 100% for load-bearing walls. A procedural specification is commonly used when a high percentage of coarse material (say, greater than 30% retained on ¾ in. or 19 mm sieve) prevents the use of the AASHTO T-99 or T-180 test methods. For procedural specification, 3–5 passes with conventional vibratory roller compaction equipment is generally adequate. Actual requirements on number of passes should be determined based on field trials.

- For compaction of a uniform fine to medium sandy fill material (say, with more than 60% passing No. 40 sieve), use of a smooth-drum static roller or light-weight walk-behind vibratory roller should suffice. Use of larger vibratory compaction equipment for this type of fill material tends to have difficulty in achieving proper wall alignment, especially for light-weight concrete block GRS walls.

- The fill material should be unloaded, spread, and compacted in a manner that will help minimize development of wrinkles or displacement of geosynthetic reinforcement. Placement of reinforced fill near the front of a wall should generally not lag behind the remainder of the wall by more than one compaction lift.

- Generally speaking, only hand-operated light-weight tampers or plates are allowed within 0.15 m (6 in.) of the back of a wall. For concrete block GRS walls, a unique tamping procedure suggested by Bob Barrett may be adopted to compact the fill within 15 mm (6 in.) behind the facing blocks. The procedure entails the wall builder placing one foot over the top of the block (of which the fill behind is being compacted) while using the other foot to forcefully tamp the fill.

- Selection of compaction machine is usually left to the contractor; sheepsfoot rollers, tamping foot rollers, or grid rollers are usually not to be used for compaction of fill material within the limits of the reinforced soil zone.

- When product specifications are adopted for fill compaction, field density and moisture testing of the reinforced fill should be performed on a regular basis during construction. A minimum frequency of one test per 1.5 m (5 ft) of wall height for every 30 m (100 ft) length of wall is common.

- At the end of each construction day, the last lift of backfill should be sloped away from the wall face to direct surface runoff away from wall face. Encroachment of surface runoff from adjacent areas toward the wall site should be prevented.

6.2.4   Facing

Dry-Stacked Concrete Block Facing

- As a general rule, the masonry concrete blocks used in construction of reinforced soil walls are required to meet a minimum compressive strength of 14 MPa (2,000 psi) and a water absorption limit of 5%. Concrete blocks used in freeze-thaw-prone areas are required to be tested for freeze-thaw resistance and survive 300 freeze-thaw cycles without failure, as per ASTM C666. Properly shaped concrete caps (to facilitate drainage) should be placed over the top course of facing blocks to alleviate durability issues associated with freeze-thaw. Facing blocks that may be exposed to spraying of pavement deicing agents are required to be sealed after construction with water-resistant coating or be manufactured with coating or additives to increase freeze-thaw resistance. Also, concrete masonry units (CMU) facing blocks or related products should meet the requirements of ASTM C90 and C140 for inspection, sampling, and testing. All facing blocks should be sound and free of cracks or other defects that may interfere with the placement of facing blocks or significantly impair the strength or permanence of a wall.

- Leveling of the first course of concrete facing blocks is especially important for proper alignment of wall face. Use of a string line set over pins from one end of the wall to the other will help with leveling the blocks.

- Most wall builders opt for constructing a concrete leveling pad under the first course of concrete blocks when the foundation soil is relatively incompressible and not susceptible to significant shrinkage/swell due to change in moisture. This measure is undertaken because a properly poured and leveled concrete pad can help speed up construction, ease the leveling process, and facilitate construction of a straighter wall. Other wall builders, however, have opted for installing a pad of compacted sand-gravel (of approximately 75–150 mm or 3–6 in. thick and 450 mm or 18 in. wide) in place of a concrete pad. The sand-gravel pad, being more flexible, tends to minimize cracking of the facing blocks.

- Facing blocks should be placed and supported as needed to ensure that their final position is vertical or battered as shown in the construction plans or in approved working drawings with a tolerance acceptable to the engineer.

- When light-weight concrete blocks are used, it is recommended that the cores of the top 2 to 3 courses of facing be filled with concrete mix and (for bridge abutments) reinforced vertically with rebar. This measure is recommended because frictional resistance between these blocks alone may not offer sufficient frictional resistance to ensure connection stability due to low normal loads in these blocks.

- When a curved wall face is involved, blocks on the curve should be laid first as their alignment is more critical and less forgiving. Tight curves often require cutting blocks to fit or breaking off the block tail. A diamond-tipped blade saw is convenient for cutting. The blocks should be laid from one end of the wall all the way to the other end to avoid laborious block cutting and fitting in the middle.

- If a wall involves sharp corners, as seen in Figure 6.18, a measure suggested by Bob Barrett is to fill the cores of corner blocks with concrete, along with No. 4 vertical rebar. It is very important that backfill in the area around the corner be compacted well to prevent opening up or cracking of blocks near the corner.

Figure 6.18 A tight corner of a concrete block GRS wall

- After wall construction is complete, weed burner is commonly used to remove any visible excess geosynthetic reinforcement protruding from the wall face to improve aesthetics.

- Split-face concrete blocks (blocks with a concave textured surface), manufactured by using a larger mold that holds two blocks face to face and broken apart at a precise time in the curing process, are often used as facing elements to help mask minor imperfections and slight lateral movement of the facing.

- The overall tolerance at the wall face of both vertical and battered walls should generally not exceed ± 30 mm (1.25 in.) over a 3 m (10 ft) distance (i.e., about 1%), and 75 mm (3 in.) total lateral movement at a maximum.

Wrapped-Face Geotextile Facing

- Compaction of fill material should be carried out with equipment that is less likely to damage the geotextile; generally, no machine compaction is allowed within 0.3 m (1 ft) of the wrapped face. Compaction lift thickness typically ranges from 0.2 m to 0.45 m (8 in. to 18 in.), with 0.3 m (12 in.) being the most common.

- If face deformation is to be kept to a minimum, reinforcement spacing of 0.15 m (6 in.) is recommended; this spacing has also been found easy to work with. Face alignment and fill compaction can be significantly improved with the use of external temporary forms. The use of scaffolds is usually necessary in front of the wall face for construction of walls taller than 1.8 m (6 ft).

- When making a windrow during construction (see Section 6.1.2), care should be exercised to avoid digging into the geotextile underneath or the wrapped face.

- Before applying shotcrete coating to the face of a vertical or near-vertical wall, a wire-mesh anchoring on the face is often needed to help the coating adhere to the wall face.

Precast Full-Height Panel Facing

- Precast panel facing GRS walls comprise two major components: facing panels and reinforced soil mass. The two components are typically connected by metal anchors which are embedded in the reinforced soil mass during fill placement. Facing panels can be installed either before or after construction of the reinforced soil mass. For the latter, the reinforced soil mass usually needs to be wrapped at the face, but this is not necessary for the former. The lateral earth pressure exerted by the reinforced soil mass on the facing is generally much smaller in the latter than in the former, if the latter does not involve having to "force" the panels into the reinforced soil mass.

- A wide variety of shapes, thicknesses, colors, textures, and finishes of facing panels can be manufactured to accommodate the designer's requirements. The exterior surface of precast concrete can vary from a highly ornamental exposed aggregate finish to a form face finish similar to cast-in-place.

- Precast-concrete facing panels by themselves must be able to resist lateral loads directly imparted on them, as well as vertical loads resulting from their self-weight. Other loads resulting from erection, impact, construction, and transportation of or on the facing panels should also be taken into account in design. It is important to evaluate the design, detailing, and erection of precast panels to avoid imposing unwanted loads on the panels.

- As in the IFF GRS wall described in Section 3.4.4, the base of the facing panel may be situated inside a narrow trench at a small batter with a temporary bracing system during construction. Flowable fill, a controlled low-strength material, may be employed in the trench to help prop facing panels during construction. Flowable fill is a mixture of cement, water, aggregate, and sometimes fly ash in a runny weak concrete (slurry) consistency.

- There needs to be a mechanism in a panel facing GRS wall system to adjust inclination of the facing panel to align all panels before completion of the wall construction.

Timber Facing

- Timber used in timber facing GRS walls should be treated with copper chromate or an approved preservative to an acceptable level. The bottom row of timber needs to be treated for direct burial. The color of the timber may be green or brown, but usually not mixed. All exposed fabric is often coated with a latex paint matching the color of the timbers.

- Typical dimensions of timber logs used in construction are 150 mm × 200 mm (6 in. × 8 in.) or 150 mm × 150 mm (6 in. × 6 in.). Shimming of timber to help maintain verticality is usually permitted.

- Forming elements in the back of timber logs may be plywood (with a minimum nominal thickness of 25 mm or 1 in., treated with preservative to an acceptable level), fiberglass, plastic, or other approved material.

- Typical reinforcement used for timber facing GRS walls is nonwoven geotextile, although other types of geosynthetics satisfying design criteria may also be used.

- Nails used in construction should be 16d galvanized ring shank nails and should be driven onto the top and bottom of the timbers at no greater than 0.3–0.6 m (1–2 ft) intervals along the longitudinal direction of the wall.

- To improve connection strength on the top lifts, geotextile reinforcement may be wrapped around the timber then covered/protected with wooden panels, as suggested by Keller and Devin (2003).

- Compaction of backfill should follow project specifications; generally no compaction is allowed within 0.3–0.6 m (1–2 ft) of wall face.

Natural Rock Facing

- The height and slope angles delineated in the construction plans should not be exceeded without clear evidence that higher or steeper features will be stable.

- Rocks should be placed by skilled operators and should be placed in fairly uniform lifts.

- Care must be exercised when placing infill in the voids between rocks. The infilling needs to be as complete as possible.

6.2.5 Drainage

- Drainage design for surface and subsurface water, including seasonal change of drainage conditions, must be a component of a design. Inadequate drainage has been identified as one of the leading causes for poor performance (especially excessive deformation) of reinforced soil walls (Peters, 2017).

- Percolation of surface runoff into the backfill that is less than free draining during the service life of a GRS wall can lead to significant wetting-induced settlement and reduced shear strength of the backfill. To reduce infiltration of surface runoff, the top surface of a GRS wall should be graded to direct runoff away from the wall. Interceptor drains on the back slope may also be used. Periodic maintenance may be needed to minimize surface runoff infiltration.

- For walls constructed below the free water level, dewatering may be needed to provide a dry working platform during construction. Many dewatering techniques (e.g., well points, horizontal drains, etc.) are available for this purpose, with the simplest technique being to construct perimeter trenches and connect them to sumps. In some cases, an impermeable barrier or other measures to minimize inflow of ground water into the wall site may be more effective than dewatering. Selection of the technique for ground water control during construction is usually left to the contractor.

- If the reinforced fill is not considered sufficiently free-draining (coefficient of permeability $\leq 10^{-4}$ cm/sec or 0.15 in./hour), preventing ground water from entering the reinforced soil zone is usually more efficient than allowing water to come into the reinforced zone then attempting to drain it out. Installation of geocomposite drainage strips along the excavation limit may be employed (see Figure 6.19). Subsurface drainage within the reinforced soil zone and behind the wall face is generally not needed. However, when the backfill contains an appreciable amount of fines (say, more than 35% of fines), measures should be taken to minimize wetting/drying of the fill.

6.2.6 Construction Sequence

- For GRS bridge abutments that involve an upper wall (with the lower wall being a load-carrying wall), it is better to construct the upper wall and compact the fill before placing the bridge girder on the lower wall. This sequence will produce a more favorable stress condition in the load-bearing wall, increase load-carrying capacity, and reduce settlement of the bridge abutment (Wu et al., 2006).

Figure 6.19 Geocomposite drainage strips installed at prescribed intervals over the face of the excavation limit

- Similarly, when constructing a GRS wall with rigid facing (e.g., high rigidity panel facing or cast-in-place concrete facing), it is better to construct the reinforced fill (commonly with a wrapped face) before constructing the rigid wall face. This will generally result in much reduced lateral earth pressure on the wall face and significantly smaller post-construction settlement of the wall.

References

AASHTO (2014). *LRFD Bridge Design Specifications*, 7th edition with 2016 Interims. American Association of State Highway and Transportation Officials, Washington, D.C.

American Association of State Highway and Transportation Officials (AASHTO) (2002). *Standard Specifications for Highway Bridges*, 17th edition. HB-17, American Association of State Highway and Transportation Officials, Washington, D.C.

Elias, V., Christopher, B.R., and Berg, R.R. (2001). *Mechanically Stabilized Earth Walls and Reinforced Soil Slopes, Design & Construction Guidelines*. Report No. FHWA-NHI-00-043, Federal Highway Administration, 394 pp.

Japan Railway (JR) Technical Research Institute (1998). *Manual on Design and Construction of Geosynthetic-Reinforced Soil Retaining Wall*. 118 pp.

Keller, G.R. and Devin, S.C. (2003). Geosynthetic-Reinforced Soil Bridge Abutments. *Transportation Research Record 1819*, Volume 2, 362–368.

Kirsch, K. and Bell, A. (eds.) (2013). *Ground Improvement*. CRC Press, 511 pp.

Ma, C. and Wu, J.T.H. (2004). Performance of an Independent Full-Height Facing Reinforced Soil Wall. *Journal of Performance of Constructed Facilities, ASCE*, Vol. 18(3), 165–172.

Monley, G.J. and Wu, J.T.H. (1993). Tensile Reinforcement Effects on Bridge-Approach Settlement. *Journal of Geotechnical Engineering, ASCE*, Vol. 119(4), 749–762.

National Concrete Masonry Association (NCMA) (2002). *Design Manual For Segmental Retaining Walls*, 2nd edition. J.G. Collin (ed.), 289 p.

Nicholson, P.G. (2015). *Soil Improvement and Ground Modification Methods*. Elsevier, 455 pp.

Peters, B.M. (2017). *Design Considerations to Improve Performance of Geosynthetic Reinforced Mechanically Stabilized Earth Walls in Transportation Projects*. M.S. Thesis, Department of Civil Engineering, University of Colorado Denver.

Swiss Association of Geotextile Professionals (SAGP) (1981). *The Geotextile Handbook*, translation by US Bureau of Reclamation. Association of Geotextile Professionals, Denver, Colorado, 411 pp.

Suaznabar, O., Huang, C., Xie, Z., Shen, J., Kerenyi, K., Bergendahl, B., and Kilgore, R. (2017). *Hydraulic Performance of Shallow Foundations for the Support of Vertical-Wall Bridge Abutments*. FHWA-HRT-17-013, Federal Highway Administration, Washington, D.C., 123 pp.

Wu, J.T.H. (1994). *Design and Construction of Low Cost Retaining Walls: The Next Generation in Technology*. Publication No. CTI-UCD-1-94, Colorado Transportation Institute, Denver, Colorado, 152 pp.

Wu, J.T.H., Lee, K.Z.Z., Helwany, S.B., and Ketchart, K. (2006). *Design and Construction Guidelines for GRS Bridge Abutment with a Flexible Facing*. Report No. 556, National Cooperative Highway Research Program, Washington, D.C.

Wu, J.T.H., Tung, C-Y, Adams, M.T., and Nicks, J.E. (2018). Analysis of Stress–Deformation Behavior of Soil–Geosynthetic Composites in Plane Strain Condition. *Transportation Infrastructure Geotechnology*, Vol. 5(3), 210–230.

Index

Geosynthetic Reinforced Soil (GRS) Walls, First Edition. Jonathan T.H. Wu.
© 2019 John Wiley & Sons Ltd. Published 2019 by John Wiley & Sons Ltd.